Cross-Cultural Comparisons on Surrogacy and Egg Donation

Sayani Mitra · Silke Schicktanz
Tulsi Patel
Editors

Cross-Cultural Comparisons on Surrogacy and Egg Donation

Interdisciplinary Perspectives from India, Germany and Israel

palgrave
macmillan

Editors
Sayani Mitra
Department of Medical Ethics and History of Medicine
University Medical Center Göttingen
Göttingen, Germany

Tulsi Patel
Department of Sociology
University of Delhi
New Delhi, India

Silke Schicktanz
Department of Medical Ethics and History of Medicine
University Medical Center Göttingen
Göttingen, Germany

ISBN 978-3-030-08760-9 ISBN 978-3-319-78670-4 (eBook)
https://doi.org/10.1007/978-3-319-78670-4

Printed on acid-free paper

This Palgrave Macmillan imprint is published by the registered company Springer International Publishing AG part of Springer Nature
The registered company address is: Gewerbestrasse 11, 6330 Cham, Switzerland

Acknowledgements

The editors would like to thank all scholars and peer reviewers who have carefully read and constructively commented on the chapters. Each chapter was peer-reviewed by two reviewers without whose input we would not have been able to put together such an interdisciplinary collection of work. They helped us strive for a balanced and topical composition of each chapter. Furthermore, we would like to thank all contributors and the various helping and critical friends who made this volume possible in a relatively short time. We would especially like to acknowledge and thank Marthe Irene Eisner for her patient and intensive work with the layout, editing and correspondence with authors; Pia Liebetrau for her meticulous language editing of some of the chapters at the very final stage of the book; and Sunita Reddy, Yael Hashiloni-Dolev and two anonymous reviewers for comments on the early version of the volume proposal. We are also grateful for the funding we received from the DAAD/UGC PPP program (German Academic Exchange Service in collaboration with the University Grants Commission, India) during 2014–2016 to prepare this project through joint research visits and research workshops.

Contents

Notes on Contributors

Katharina Beier is a postdoctoral researcher at the University Medical Center Göttingen, Department of Medical Ethics and History of Medicine. Her work focuses on research ethics, particularly in the field of biobanking, the ethics of reproduction and related concepts of (reproductive) autonomy, responsibility, and trust.

Sabrina Dücker is a Ph.D. researcher. She obtained her law degree from Humboldt-Universität Berlin, where she also worked as a research assistant at the faculty of law. At present, she is about to finish her Ph.D. research with a thesis on Preimplantation Genetic Diagnostics in Germany and England.

Tatjana Hörnle is professor of criminal law, criminal procedure, comparative criminal law and legal philosophy at the Humboldt-Universität Berlin. Her research interests are theories of criminalization and multicultural issues in criminal law.

Vasanti Jadva is currently a Senior Research Associate and an Affiliated Lecturer at the Department of Psychology, University of Cambridge and a member of the National Gamete Donation Trust's advisory council. Her research examines the psychological well-being of parents and children in families created by IVF, egg donation, sperm donation and surrogacy. She is currently working on a number of different studies including a longitudinal study of families created using egg and sperm donation and surrogacy.

Anika König is a postdoctoral researcher at University of Lübeck. She is a cultural anthropologist whose work focuses on medicine and violence. She received her Ph.D. from the Australian National University with a thesis on ethnic violence in Indonesia. Her current research deals with transnational gestational surrogacy commissioned by parents from the German-speaking region.

Prabha Kotiswaran is Reader in Law and Social Justice at the Dickson Poon School of Law, King's College London. Her research interests include criminal law, transnational criminal law, sociology of law, legal ethnography, postcolonial theory and feminist legal theory.

Malene Tanderup Kristensen has been engaged in reproductive health through interdisciplinary research during the last 5 years. She works as a physician and has published extensively on her research on surrogacy in India. Her research areas are surrogacy, reproductive ethics and transnational reproduction.

Nishtha Lamba has recently finished her Ph.D. from the Center for Family Studies, University of Cambridge. Her thesis focuses on studying the experiences, motivations and psychological well-being of Indian surrogate mothers catering to international intending parents. In addition to her thesis, she is working on a project focusing on the experiences of egg donors in India.

Vrinda Marwah is a doctoral candidate in the Department of Sociology at the University of Texas at Austin. Her master thesis focused on hijras in India, and examined debates around sexual subjectivity, identity, and terminology in the context of HIV/AIDS, queer mobilisation and legal reform. Her primary research interests are in reproductive health and women's labour in contemporary India.

Sayani Mitra is a postdoctoral researcher at the Department of Medical Ethics and History of Medicine, Göttingen. She has recently completed her Ph.D. in Social Sciences from the University of Göttingen. Her thesis examined aspects of risks and reproductive disruptions during commercial surrogacy in India. Her research interests lies in the fields of assisted reproductive technologies, political economy, comparative health policies, gender and medicine, to state a few.

Sarojini Nadimpally has been working on women's health and rights for over 20 years and is also one of the founder members of Sama Resource Group for Women and Health in Delhi, India. She has over 10 years of experience of engaging with research and policy on assisted reproductive technologies (ARTs) and surrogacy. She has coordinated two national level studies on ARTs including surrogacy and their implications on women.

Michal Nahman is an anthropologist and Senior Lecturer at the University of the West of England. She is also the author of *Extractions: An Ethnography of Reproductive Tourism*. Her work is at the intersections of political and medical anthropology. She is currently researching cross-border egg donation in Europe.

Birgitte Bruun Nielsen is a consultant and specialist in Obstetrics and Gynaecology at Copenhagen University Hospital. She is an Associate Professor. She has been engaged in research in reproductive health in developing countries for 20 years. Her research interest focuses on maternal health, delivery care, sex selection and cross-border reproductive care.

Tulsi Patel is Professor of Sociology and teaches at the Department of Sociology, Delhi School of Economics, University of Delhi. Her research interests include gender and society, anthropology of reproduction and childbirth, medical sociology, kinship and family, and old age.

Sunita Reddy is an Associate Professor at the Centre of Social Medicine and Community Health, School of Social Sciences, Jawaharlal Nehru University, New Delhi, India. She is an anthropologist, specialised in medical anthropology, researching on medical tourism, surrogacy and disaster issues.

Nitzan Rimon-Zarfaty is a research fellow at the University Medical Center Göttingen, Department of Medical Ethics and History. Her work focuses on cross-cultural comparative bioethics, bioethics of reproductive medicine and related concepts of personhood, responsibility and time perceptions.

Judit Sándor is a full professor at the Faculty of Political Science, Legal Studies and Gender Studies of the Central European University (CEU), Budapest. Since September 2005 she is a founding director of the Centre for Ethics and Law in Biomedicine (CELAB) at the Central European University.

Silke Schicktanz is full professor of Cultural and Ethical studies of Biomedicine at the University Medical Center of Göttingen. She studies and teaches in the interdisciplinary field of bioethics and cultural studies of medicine. Her research focuses on cross-cultural bioethics, lay-expert-interaction, and concepts of responsibility in various fields of modern medicine (dementia, transplantation, genetics etc.).

Carmel Shalev is a retired public interest and academic lawyer. Throughout her career she specialized in health rights and bioethics, and in medically assisted reproduction in particular. Her book *Birth Power* (Yale University Press, 1989) was the first legal treatise on the subject of surrogacy. She now co-leads Wisdom of Aging—a movement for aware and engaged aging.

Elly Teman is a senior lecturer of cultural anthropology in the Department of Behavioral Sciences at Ruppin Academic Centre, Israel. She is the author of the celebrated monograph *Birthing a Mother: The Surrogate Body and the Pregnant Self* (2010). Her research focuses on gestational surrogacy arrangements and on the personal experiences of surrogates and intended parents. Her current research is a comparative study of Ultra-Orthodox Jewish women from Israel and the US and their experiences with pregnancy and prenatal diagnosis.

Julia Teschlade is a doctoral candidate in the International Research Training Group "Human Rights under Pressure" at Free University Berlin and Hebrew University Jerusalem. Her doctoral research on

"Human Rights and Reproduction—Gay Parenthood and Surrogacy in Israel and Germany" is a comparative analysis of male gay couples from Israel and Germany, who commissioned gestational surrogacy in the US.

Gabriele Werner-Felmayer is an Associate Professor of Medical Biochemistry at the Division of Biological Chemistry, Medical University of Innsbruck, and runs the bioethics network Ethucation affiliated to the International Network of the UNESCO Chair in Bioethics (Haifa). In her research, she explores concepts of identity, hope and promise in biomedicine, focusing on genomics, and reproductive as well as regenerative medicine.

1

Introduction: Why Compare the Practice and Norms of Surrogacy and Egg Donation? A Brief Overview of a Comparative and Interdisciplinary Journey

Sayani Mitra, Silke Schicktanz and Tulsi Patel

Introduction

Assisted reproductive technologies (ART) are challenging the traditional perceptions of 'natural' kin-ties with increasingly dynamic processes of 'kinning' by moving reproduction from the domain of 'nature' to the domain of 'science'. Since the development and widespread usage of ART, starting with in vitro fertilisation (IVF) in the late 1970s, the separation of reproduction from sexuality has led to a new form of 'medicalisation of reproduction' and initiated the first wave of discussion on the 'natural' vs. 'artificial' realm of reproduction. ART were initially

S. Mitra (✉) · S. Schicktanz
Medical Ethics and History of Medicine, University Medical Center Göttingen, Göttingen, Germany
e-mail: sayani.mitra@medizin.uni-goettingen.de

S. Schicktanz
e-mail: silke.schicktanz@medizin.uni-goettingen.de

T. Patel
Department of Sociology, University of Delhi, New Delhi, India

S. Mitra et al. (eds.), *Cross-Cultural Comparisons on Surrogacy and Egg Donation*,
https://doi.org/10.1007/978-3-319-78670-4_1

developed to deal with the problem of infertility. But the involvement of third parties through the usage of gametes (sperms or eggs) or even wombs, as in the case of surrogacy, has led not only to new notions of motherhood, fatherhood, family and kinship, but have also given rise to markets around reproduction. The first cases of commercial surrogacy through IVF, back in the mid-1980s initiated an international debate. Various national laws began to either ban or permit surrogacy (McEwen 1999). Subsequently, legal and ethical debates evolved around the question of the extent to which a modern democratic state is to regulate the citizens' right to reproduce by means of ART. The increasing phenomenon of commercialisation across the borders of a nation state through the selling and buying of gametes as well as the renting of wombs, labelled as 'reproductive tourism', furthermore triggered a worldwide debate on whether countries should allow or restrict access to such transnational fertility markets. Countries with permissive regulations allow commercial surrogacy and/or egg donation, while restrictive regimes forbid surrogacy and/or egg donation or allow the so-called altruistic models of practice. These legal debates on market regulation are closely interlinked with the ongoing debates of cultural and moral values, gender, kinship, class/caste/ethnicity issues and professional medical ethics related to reproductive care and women's health in addition to eugenics. In an increasingly globalised world, connected through biomedicine and media, these ethico-legal stances are however not fixed, but are continuously negotiated.

With the expansion of technological research, its use and popularity, ART have opened up a globalised market in which the demand for eggs and their accessibility is skewed in favour of those economically well off. During the last decade, India was the leading nation to provide cross-border commercialised surrogate and gamete selling services because of its rather low prices (in comparison to western countries) and excellent internationalised health services. The country underwent a legal transformation in 2015, restricting cross-border surrogacy and allowing it only for intended parents (IPs) of Indian origin. Contrastingly, some industrialised countries like the UK and Canada allow only altruistic, non-commercial transfer of egg and sperm for the purpose of fertility treatment or research and maintain databases of all egg and sperm donors through strict licensing of fertility centers. Again, countries like Germany, Austria and Italy allow neither egg-sharing nor commercial procurement of eggs. Countries and states such as Germany, France, Belgium, Italy, Switzerland, Austria, Norway, Sweden, Iceland, China, Japan, Quebec (in Canada), Arizona, Michigan, Indiana and North Dakota (in the US) prohibit the practice of both commercial and altruistic surrogacy, whereas countries such as Australia, Canada (except for Quebec), the UK, the Netherlands, Denmark and Hungary allow the practice only with altruistic approaches. Israel, with a strong Jewish *halakhic* (rabbinic) tradition, has reached a middle-ground by limiting surrogacy under strict rules for those with the same religion. Russia, Ukraine, Georgia, Laos and Kenya along with selected states of the US have become the new destination for commercial transnational surrogacy, while Spain, Czeck Republic, South Africa and California (in the US) are known as the popular centres for egg donation services.

A Need for Comparison

Studies from bioethics, social and medical anthropology, as well as from a human rights perspective (e.g. Inhorn and Patrizio 2012; Crozier 2010; Shenfield et al. 2010) have identified major, unresolved problems in cross-border ART. These problems refer to unfair exploitation of surrogate mothers, lacking professional medical ethics, protection of children's rights and various legal loopholes in cases of infringement of contracts between different unequal parties. These practices are often considered context-specific phenomena, and yet they are simultaneously related to the globalisation of ART. However, the concept of globalisation is not sufficient for analysing and understanding the complex interlinkages concerning how this technology spreads globally and locally adapts to fit into a specific cultural context. Therefore, we rely on conceptional approaches such as 'technoscape' (Appadurai 1996), which entails the global configuration of technologies including their economic and legal regimes in their connectivity as well as in their disjunctions. Another helpful concept is 'glocalisation' (Alexander 2003; Roudometof 2016), here understood as the process of negotiation, refraction and mimicry between globalisation and localisation. Such conceptualisations are useful for analysing how the global and the local

ART practices meet and how social, moral and religious conditions are negotiated within the global market of ART to create unique conditions for its adoption at various locations.

As this volume illustrates, the practices of surrogacy and gamete donation differ worldwide with regard to their ethico-legal frameworks. Economic parameters of ART, i.e. whether the use of ART is covered by public health insurance or is paid out of pocket, are also an obvious driving force for its global spread. While some public health systems cover all costs for IVF for every woman, others cover only a limited number of treatments only for heterosexual couples. The dramatic differences in costs for surrogacy and egg donation across the globe also explain the cross-national moves of IPs or even health care professionals in this field.

However, until now, no systematic comparison of such regimes and contextualised problems has been done. It has been rarely examined how the ethical, legal and sociocultural boundaries are negotiated within the different restrictive vs. permissive regimes. The making and unmaking of such ethico-legal regimes as 'macro-politics' needs to be examined along the 'micro-politics' of gender, class and ethnicity issues related to ART. Therefore we need to ask: Who is considered vulnerable or protectable on the grounds of different understandings of vulnerability? Who is granted what kind of reproductive rights or choices? Which understandings of reproductive needs, kinship or fertility underlie which type of regime? How is each current regime debated as consistent or incoherent? How do national, ethical and legal frameworks refer to cross-border or foreign reproductive care? What are the resulting paradoxes? This volume attempts to answer some of these questions. By choosing to discuss two of the widely used forms of ART practices—surrogacy and egg donation, it aims at filling this striking gap by comparing the ethico-legal and sociocultural debates in three different countries—India, Germany and Israel.

The selection of these three countries has been carefully made. India for a long time had a rather permissive, market-oriented regime, which became restrictive and underwent a dramatic change with the banning of commercial surrogacy for foreigners in 2015. At present, cross-border commercial surrogacy is only available for non-resident married (heterosexual) Indians and persons of Indian origin. Further, if the Draft Surrogacy Bill 2016 is legislated, surrogacy in India would only be permitted in its altruistic form. But questions of regulation and rights of surrogates still remain unresolved in India. Also the rights of egg donors and other actors who are part of other forms of ART such as IVF with egg or sperm donation, continue to go unregulated. Germany, in contrast, has had a very restrictive approach to ART since the 1990s. The German law permits neither commercial nor altruistic egg donation or surrogacy. However, commercial sperm donation is allowed in Germany and is practised in a legal grey-zone (Klotz 2013). But such a restrictive attitude towards the use of ART does not prevent German citizens from crossing borders to avail fertility services overseas. In fact, German heterosexual as well as homosexual couples and single parents constitute a large proportion of the so-called cross-border reproductive travellers. Israel, in contrast, presents a rather unique case of a state-supported and regulated regime of surrogacy and egg donation, in comparison with the two extreme forms of the practice in India and Germany. The Israeli context is often seen as very permissive and supportive of reproductive technologies. It was the first country to regulate ART from 1996 onwards by allowing gamete donation and surrogacy but only under strict regulation by a state committee assessing every single case. Yet the practice in Israel is marked by strong ideologies of heteronormativity and a Jewish cultural and religious presumption of kinship, motherhood and citizenship.

Strikingly, over the last two decades, these three countries have either undergone shifts in their policies on surrogacy and egg donation or were challenged within the respective local political and ethico-legal framework, as will be discussed in this volume. These three countries present 'prototypes' of diverse forms of governance and sociopolitical attitudes towards ART, although they face very similar challenges concerning governance and the ethical issues raised. Moreover, all three countries can be seen as interconnected by ground-breaking cases of cross-border surrogacy, which have led to intensified public and policy debates. Using these three countries as striking examples of diversity of practices, this volume traces back the development of these respective regimes, compares the transition processes and examines what can be learnt from comparing the different politics, practices and norms within and between them. Moreover, it shows how local, national developments cannot be isolated from global, cross-border events and vice versa. Therefore, the book strives for a strong systematic comparative approach in order to provide a broader picture of bioethics and sociocultural aspects of surrogacy and egg donation in current times. For this purpose, the book brings together designated scholars from bioethics, medical sociology and anthropology, cultural anthropology, psychology, legal and policy studies. All the scholars are committed to interdisciplinary and international exchanges and contributed a particular perspective to this collective venture of comparison. Each part of this book is structured to include legal, ethical, medical, psychological, ethnographic or social–theoretical work from different angles towards creating a comparative picture.

Contents of the Volume

Part I presents *comparative views on surrogacy and egg donation from a wider transnational perspective.*

This first part starts with Gabriele Werner–Felmayer's overview (Chapter 2) on how globalisation and market orientation are challenging reproductive medicine. According to her analysis, current reproductive health care is at risk of overusing ART in response to a perceived increase of infertility and numerous market opportunities. Her chapter unfolds some of the ramifications of the ART landscape highlighting 'side' effects on the health of women and their children born to such arrangements. Judit Sándor's overview of legal and ethical issues on transnational surrogacy (Chapter 3) discusses the legal diversities and controversies that occurred in the domestic recognition of family ties after transnational surrogacy. She demonstrates these controversies by discussing relevant legal cases and argues for the need to develop international standards to oversee surrogacy arrangements. By comparing Israeli and Indian ethnographic studies on surrogacy, Elly Teman discusses in Chapter 4 how ethnographic comparative analysis can help to formulate empirically based criteria towards regulating surrogacy. She demonstrates how a restrictive regulation of surrogacy might create the grounds for a more ethical practice. Her chapter is followed by Sayani Mitra's analysis of cross-border 'reproflows' (Chapter 5). She shows how national legislations along with the actors' social and cultural attitudes create distinct forms of consumerism, choice, service models and labour relations during cross-border surrogacy and egg donation. She analyses how these diverse patterns and flows within the global 'reproscapes' create further stratifications and inequalities. The last chapter in this cross-cultural section is an ethical-comparative analytical chapter by Silke Schicktanz (Chapter 6). She discusses ethical concerns such as exploitation, the best interest of the child and the inconsistencies produced by a too biology-oriented understanding of parenthood. By comparing the norms and the moral practices, she provides helpful insights on how to get beyond cultural relativism and identify ethical issues that should be addressed on a larger scale.

Part II presents *the case of surrogacy in India as a system undergoing legal transition.*

This part of the book begins with Prabha Kotiswaran's legal analysis of the Indian governance of the surrogacy market (Chapter 7). Kotiswaran maps the legal and discursive shifts through a sociolegal understanding of several legislative proposals to reflect on the continued lack of settlement of legal norms in the surrogacy sector. It is followed by a critical analysis of ART bills and the surrogacy bill 2016 by Sunita Reddy, Tulsi Patel, Malene Tanderup Kristensen, Birgitte Bruun Nielsen (Chapter 8). The authors discuss how various stakeholders are advocating for a reversal of the ban on commercial surrogacy and the impacts of socio-economic issues on the governance debate. Chapter 9 by Nishtha Lamba and Vasanti Jadva adds a psychological dimension to these social and legal views by discussing the significance of whether the surrogate sees or meets the newborn(s) and IP(s), social stigma, the availability of social support from family and other surrogates, and the role of financial compensation on the surrogates' psychological well-being. The section is completed by a perspective from the Indian women's health activists provided by Vrinda Marwah and Sarojini Nadimpally (Chapter 10). They embed the political discourse of surrogacy in broader framework of commercialisation, sex work and the labour market and reflect thematically on interviews with activists and experts from key social movements in India. They aim to bring new perspectives and movements into conversation and collaboration so as to respond to the challenges posed by surrogacy.

Part III focuses on *a restrictive system: the case of Germany and discusses the rather restrictive legal, social and moral take of Germany on surrogacy and egg donation.*

This part of the book starts again with a legal overview and critical assessment of the current legal situation (Chapter 11) provided by Sabrina Dücker and Tatjana Hörnle. They describe the German legal prohibitions against egg donation and surrogacy and the concrete implications for physicians and other persons if they provide information about transnational surrogacy. Their chapter is backed by a detailed ethical analysis of arguments against surrogacy in German debates (Chapter 12) by Katharina Beier. She examines the soundness of objections by highlighting their underlying premises and confronting them with insights from international analyses. The ethico-legal part is then complemented by two ethnographic studies. Anika König (Chapter 13) interviewed German IPs who chose to commission surrogacy abroad. It deals with their experiences in a legally restrictive national context and their strategies to circumvent this, and with their ways of establishing kinship and parenthood. In her study with gay couples on egg donation, Julia Teschlade (Chapter 14) discusses the motives how these couples engage with and reproduce normative family ideals to avoid discriminatory judgements about their non-traditional family.

Part IV discusses *the case of Israel as a state-supported system.*

The legal scholar Carmel Shalev (Chapter 15) embeds the current developments in a larger picture of how ART have developed since the 1980s and especially how Israel has embraced ART and surrogacy legally and culturally but often in a particular way. The legally permissive situation in Israel however does not prevent it from repro-migration as Michal Nahman (Chapter 16) shows in her ethnographic study. She provides a critical analysis of how the politics of race and borders are also mirrored in the practice and moral opinions of users in order to help us reflect about its present day manifestations. The religious particularities in Israeli legislation for surrogacy and egg-donation legislation are discussed finally by Nitzan Rimon–Zarfaty (Chapter 17). She analyses the connection between the legislation's restrictions and rabbinic concerns regarding illegitimacy, incest, religious identity and family integrity and shows how the legal restrictions further represent mechanisms of sociopolitical power relations, highlighting categories of religiosity, religious affiliation and nationality.

Conclusion

The perspective presented by these three different country-specific parts of the book as well as the comparative part brings out the merit of having a comparative or simultaneous look at different forms and practices of surrogacy and egg donation across the world. Due to its global interconnectivity, we would like to suggest that future research agendas in the field of ART can gain enormously by taking up a cross-cultural or comparative perspective. Such projects could gain by developing parameters to carry out not just a socio-legal analysis, as is usually done, but also develop ethical–moral as well as ethnographic comparisons. The methodological and logistic rigour that such comparisons demand can also potentially pioneer new methodologies for ethical, legal and cultural studies in the field of ART.

Some final thoughts: Some readers might prefer to receive very concrete recommendations or straightforward ethical guidance regarding surrogacy or gamete donation. We would like to point out that our comparative chapters in Part I provide a long and complex list of points that can be considered for such future ethico-legal debates. However, we would like refrain from any simplistic policy advice on such complex interconnectivities, as the book reveals. There is a need for further comparative research before we advise the global public or political debates about potential solutions because it is obvious that we need translocal solutions and transnational guidance. We suggest understanding this volume as very concrete starting point for future debates.

Bibliography

Alexander, J. C. (2003). *The meanings of social life: A cultural sociology*. Oxford: Oxford University Press.

Appadurai, A. (1996). *Modernity at large. Cultural dimensions of globalization*. Minneapolis: University of Minnesota Press.

Crozier, G. K. D. (2010). Protecting cross-border providers of ova and surrogacy services? *Global Social Policy, 10*(3), 299–303.

Inhorn, M. C., & Patrizio, P. (2012). The global landscape of cross-border reproductive care: Twenty key findings for the new millennium. *Current Opinion in Obstetrics and Gynecology, 24*(3), 158–163.

Klotz, M. (2013). Genetic knowledge and family identity: Managing gamete donation in Britain and Germany. *Sociology, 47*(5), 939–956.

McEwen, A. G. (1999). So you're having another women's baby: Economics and exploitation in gestational surrogacy. *Vanderbilt Journal of Transnational Law, 32*(1), 271–304.

Roudometof, V. (2016). *Glocalization: A critical introduction*. London and New York: Routledge.

Shenfield, F., De Mouzon, J., Pennings, G., Ferraretti, A. P., Nyboe Andersen, A., De Wert, G., et al. (2010). Cross border reproductive care in six European countries. *Human Reproduction, 25*(6), 1361–1368.

Part I

Comparative Views

2

Globalisation and Market Orientation: A Challenge Within Reproductive Medicine

Gabriele Werner-Felmayer

Introduction

In vitro fertilisation (IVF) and related assisted reproductive technologies (ART)[1] have developed into a thriving field of innovation and a rapidly growing segment of the health global market. An important driver for this development is the efficient and fast translation of research outcomes from the IVF laboratory to clinical care. IVF can thus be perceived as 'the perfect example of translational research' (DeCherney and Barnett 2016, p. 1634). Other than in the global market for pharmaceuticals or medical devices, this field is also a 'bio-economy' based on women who provide oocytes or carry pregnancies for others (and undergo invasive treatments in this process) and men who provide sperm. The challenges for and within life sciences and medicine are thus myriad. They relate to several layers of complexity pertaining to invasive procedures that manipulate gametes, embryos and women's bodies and

G. Werner-Felmayer (✉)
Medical University of Innsbruck, Innsbruck, Austria
e-mail: gabriele.werner-felmayer@i-med.ac.at

S. Mitra et al. (eds.), *Cross-Cultural Comparisons on Surrogacy and Egg Donation*,
https://doi.org/10.1007/978-3-319-78670-4_2

the disruption of social and cultural norms by some of the procedures. Globalisation and market orientation in ART also can seriously corrupt good medical and scientific practices since they make use of legal and economic asymmetries, often operate with unreliable, i.e. overly promissory advertisements and information while down-playing risks and possible harms, and depend on cross-border arrangements that sometimes fulfil several criteria of human trafficking (Shalev, Chapter 15). In these settings, women involved as donors for oocytes or as surrogates are usually deprived of the standard of care provided to patients. This latter phenomenon is not restricted to women from developing countries but has also been documented for, e.g. the US (Riben 2016) and Canada (Gruben 2013) in the contexts of surrogacy and oocyte donation.

Market orientation is also changing reproductive health care within high-income countries leading to a risk of overusing ART and to establishing a practice of using risky and invasive procedures for growing target groups without even understanding side effects and long-term outcome (Evers 2016; te Velde et al. 2017). This is of particular concern given the fact that there is still a significant lack of evidence-based and standardised practice in the IVF lab (Sunde et al. 2016). Please note that all this serious critique (see also below) comes from medical experts in the field who—like Hans Evers or Arne Sunde—have or had leading positions within the European Society for Human Reproduction and Endocrinology (ESHRE).

In this chapter, I attempt to find some tracks in the jungle that we have created by entangling ourselves in technical, medico-legal and global business approaches to the intimacy of 'making babies' and—such is the current vision—of 'making' them healthier as ever. I will highlight the following aspects: infertility as global health issue; the current ART landscape and new target groups; developing global networks of reproduction; marketing the wish for a child; the particular significance of oocytes for ART and research; and some threats of increasing commercialisation of ART for good clinical and scientific practice. I explore these aspects through the biomedical scientist's lens interested in bioethical issues of new technologies and emphasise the deficit of information and systematic research regarding risks and harms of ARTs, the rapid pace of controversial research with regard to bioengineering embryos and gametes, and the increasing practice of engaging humans as resource for fulfilling the reproductive goals of others.

A Wider View on Infertility

Infertility[2] is seen nowadays as a global health issue. Universal access to reproductive health was expressed in 2000 as United Nations Millennium Development Goal 5B and is also on the agenda of the Sustainable Development Goals from 2015. Being crucially connected to maternal mortality, reproductive health comprises numerous highly complex issues such as family planning services, access to contraception and safe abortion, the improvement of health system capacity, including coverage of routine reproductive health care and of more advanced as well as emergency obstetric care, as was outlined in a recent report on the global burden of maternal mortality (GBD 2015 Maternal Mortality Collaborators 2016). According to a comprehensive review of available data on global infertility prevalence (Mascarenhas et al. 2012), 48.5 million couples worldwide were unable to have a child after five years. The study detailed that worldwide 1.9% of women aged 20–44 years (~19.2 million couples) were unable to have their first live birth (primary infertility), and 10.5% of women (~29.3 million couples) who had already one child were unable to have another one (secondary infertility). A different distribution of primary and secondary infertility was observed with regard to age, as primary infertility was more prevalent in women aged 20–24, whereas the latter was higher in women older than 24 years. In South Asia, sub-Saharan Africa, North Africa/Middle East, Central/Eastern Europe and Central Asia infertility prevalence was higher than in other regions. For example, secondary infertility prevalence ranged from 7.2% of women (aged 20–44 years) in the high-income regions as well as in the North Africa/Middle East region to 18.0% in the Central/Eastern Europe and Central Asia regions (Mascarenhas et al. 2012). It should be noted that the definition of infertility is crucial as infertility prevalence 'measured using a shorter exposure period would have a similar geographic and temporal pattern, but would be approximately twice as high as our estimates' (Mascarenhas et al. 2012, p. 9). Therefore, higher infertility prevalence might be reported by other studies, e.g. around 9% of reproductive-aged couples worldwide, reaching up to 30% in some regions of the world (Inhorn and Patrizio 2015) or 80 million couples (Chambers et al. 2013). There are multiple reasons for infertility (Mascarenhas et al. 2012) but infectious diseases, particularly those affecting the urogenital tract such as sexually transmitted infections, tropical diseases and genital tuberculosis, play a major role for both women and men (Pellati et al. 2008). Thus, prevention of infectious disease as well as safe abortion is important contribution to tackle the global prevalence of infertility (Mascarenhas et al. 2012).

Importantly, the study by Mascarenhas et al. (2012) indicates that 'global levels of primary and secondary infertility hardly changed between 1990 and 2010' (Mascarenhas et al. 2012, p. 12). From a global and particularly from a women's health perspective, infertility is 'a critical but much neglected aspect of reproductive health' (Mascarenhas et al. 2012, p. 12) reflecting mostly the developmental status of a region, global injustice and the health gap. In contrast, fertility decline in developed countries is mostly connected to complex sociocultural rather than medical reasons (te Velde et al. 2017). Despite the fact that biological fertility in Europe is stable since the 1950s there is a perceived increase of infertility as is reflected not only in media headlines and ARTs marketing but also by an ever-growing demand for IVF (te Velde et al. 2017; see also next section).

In light of all this, we should keep in mind various issues: (i) Improving overall health care and women's health status would help to decrease involuntary infertility in regions where—on the other hand—public health policy tries to reduce population growth and the birth rate of the population is much higher than in high-income regions. Sadly and for good reasons, WHO frames pregnancy as a risk (due to maternal mortality, see above) and infertility as a disability as an 'estimated 34 million women, predominantly from developing countries, have infertility which resulted from maternal sepsis and unsafe abortion (long-term maternal morbidity resulting in a disability). Infertility in women was ranked the 5th highest serious global disability [World Report on Disability] (among populations under the age of 60)' (WHO 2017). (ii) Many regions of the world are underserved with regard to infertility treatment and this is particularly problematic in regions where childlessness is a stigma and a socio-economic disadvantage (Inhorn and Patrizio 2015). (iii) In high-income societies, declining birth rates can be explained by numerous sociocultural and socio-economic rather than by unresolved medical issues or an infertility epidemic (te Velde et al. 2017; Schicktanz, Chapter 6).

Infertility Treatment by ARTs: From Care to Service

Louise Brown's birth in 1978 after IVF is generally perceived as the starting signal for establishing the field of reproductive medicine. Since then, medically assisted reproduction has become a healthcare practice which has helped many people to start or enlarge their family. In addition, untangling the mysteries of reproduction is a fascinating scientific field. For example, research on human reproduction helped to better understand molecular mechanisms of conception, early embryonic development, maturation of oocytes and the complex interplay of hormones. Many mechanisms have only been elucidated after 1978, when a deeper understanding of early steps in embryogenesis got facilitated by genomics and the insights from stem cell research and other fields of biomedicine. Initially intended to help women with sterility due to, e.g. occluded fallopian tubes, treatment indications and the spectrum of ARTs have been significantly expanded by now. Technologies allowing storing gametes (see later) and selecting embryos for certain traits or maybe in the near future even 'correcting' genes in gametes and embryos have significantly impacted the field. They opened the possibility to turn oocyte donation into a standard practice of medically assisted reproduction which is of particular significance for managing age-related subfertility (Argyle et al. 2016) and to establish IVF as a possibility to avoid the transmission of certain genetic diseases. Moreover, ARTs are increasingly used to overcome biological limitations of procreation, such as same sex or single parenthood, or to 'preserve' fertility to a later stage in life (see below).

According to current estimates by ESHRE, more than 7 million babies have been born with the aid of ARTs so far. Most ART treatment cycles[3] are performed in Europe (~800,000 in 2014, the latest year for which data are available), followed by Japan (368,000 in 2013), the US (150,000) and Australia/New Zealand (65,000) (ESHRE 2018). The number of cycles performed in different countries correlates, among other reasons, to funding policies and actual costs for the patient (Chambers et al. 2013). In Belgium, Czech Republic, Denmark, Estonia and Slovenia, more than 4.0% of all babies born are nowadays conceived by ARTs. All in all, around 2.4 million ART cycles are performed each year worldwide and estimated 500,000 babies are born (ESHRE 2018).

Despite the marked increase of ARTs use, it seems that still less than '20% of the demand for ART treatment [is] being met' (Dyer et al. 2016, p. 1606). Several factors need to be considered when explaining the increasing demand for IVF in Europe and other high-income regions (see also above): the widespread availability of birth control and hence the ability for planning childbearing; the impact of changing definitions of infertility; the false but recurring alarm of a sperm crisis; the commercialisation of infertility care (te Velde et al. 2017). For example, with the '1-year infertility-is-a-disease definition' which in fact 'disregards the overwhelming evidence about the ability of couples to conceive naturally after one year of non-conception [...] many more couples seem to require treatment nowadays than before' (te Velde et al. 2017, pp. 205–206). The infertility definition was only changed in 2008, when defining infertility by a two year span of not-conceiving naturally was shortened to one year '[w]ithout any explanation', the prognostic interpretation of infertility changed into a diagnostic interpretation, and infertility was 'redefined as a disease' (te Velde et al. 2017, pp. 205–206). The trend of increased ARTs use is also substantially triggered by increasing commercialisation which relies on various mechanisms to promote the field such as offering new often unproven technologies (see also later in this chapter), incentive schemes for successful doctors, and funding reports and conferences 'that highlight the potential of IVF to combat fertility decline and population ageing, thus exerting pressure on governments, the media and public opinion to expand IVF provisions' (te Velde et al. 2017, p. 206).

Global Networks of Reproduction

Among those living in high-income regions (including expatriates from lower-income regions) and the affluent social class of lower-income societies, seeking medically assisted reproduction across borders has become a growing practice over the past decades. There are several motivations for going across borders in the context of medically assisted reproduction, e.g. access to fertility treatment, the spectrum of ARTs provided, circumventing legal hurdles or searching good value for money (Inhorn and Patrizio 2015; König, Chapter 13; Teschlade, Chapter 14; Nahman, Chapter 16). This activity is often named cross-border reproductive care (CBRC) in order to refrain from more biased notations like 'fertility tourism'. However, the term CBRC insinuates that this cross-border activity always involves 'patients', that the reason is always 'infertility' in the medical sense and that hence the practice is always 'care'. This is, however, not the case (for discussion see Inhorn and Patrizio 2015; Shalev, Chapter 15). The debate over terminology indicates the difficulty to frame this activity which is 'a growing industry, with new global hubs, new intermediaries, new media, and new spaces of interaction' (Inhorn and Patrizio 2015, p. 158) characterised by a lack of oversight (Shalev, Chapter 15) and creating new versions of 'embodied labour' (Parry 2015, p. 33).[4] Typically, the considerable suffering involved in this kind of labour, for example when experiencing missed or failed conceptions in commercial surrogacy arrangements, appears as 'non-events within the discourse of the surrogacy industry' (Mitra and Schicktanz 2016, p. 1).

Some destinations have become global 'reprohubs' (Inhorn 2016, p. 6), e.g. India for surrogacy (Sarojini and Venkatachalam 2016) or Dubai for ICSI for infertile Muslim men (Inhorn 2016). Another market opportunity is preimplantation sex selection (mostly male) for non-medical reasons, e.g. traded between Australia (where it is illegal) and Thailand (Whittaker 2011). Also US fertility centres lucratively engage in this practice for national and international clients (Bumgarner 2007). Another case is Pakistan where ARTs are obtrusively marketed, offering non-medical sex selection to wealthy elites as well as to foreigners/expatriates without regulation or public debate on reproductive health issues (Irshad and Werner-Felmayer 2016).

Within Europe too, fertility treatment across borders for oocyte donation, preimplantation genetic diagnosis (PGD) or surrogacy has become an increasing business, particularly in Spain, Greece, Czech Republic and Ukraine (for the regulatory landscape within Europe, see for PGD (Council of Europe 2015) and for surrogacy (Rigon and Chateau 2016)). Documents adopted by the European Parliament with regard to oocyte trading (European Parliament 2005) and to surrogacy (European Parliament 2015) express considerable concern about practices involving third parties in reproduction within the European context. Although this effort cannot preclude an increasing market orientation of medically assisted procreation, it indicates political sensitivity towards issues of exploitation and the welfare of women and children (De Sutter 2016). Recently, also strong international opposition to any kind of surrogacy (domestic, international, altruistic, commercial) formed up (Lahl 2017). Such increased awareness about the necessity for regulating an increasingly rampant commerce of surrogacy is also reflected by legislation in Thailand and in India, as both countries recently introduced laws banning commercial surrogacy for foreigners and homosexual couples (Sándor, Chapter 3; Reddy et al., Chapter 8; Shalev, Chapter 15).

Resolutions and laws, however, are not sufficient to ensure good practice in the field of medically assisted reproduction, particularly in cross-border settings and involving third parties. In line with this, the ESHRE taskforce on CBRC implemented a good practice guide in 2011 (Shenfield et al. 2011). This is a help for clinicians but, as the authors note, the guide is still difficult to put into practice even 'in "protected" Europe' (Shenfield et al. 2011, p. 659), let alone in a global context. Balancing of competing interests for the sake of the well-being of those who are physically and emotionally involved in bringing a child into the world is particularly complex and currently far from resolved (Shalev, Chapter 15).

Marketing Desire and Hope

Already in the 1980s, surrogacy was a well-established practice that 'was not brought to us by the march of scientific progresses but rather by brokers, by people who saw a new market and went after it'

(Rothman 1989, p. 159). This also holds true for human oocytes, so-called eggs,[5] which are on the market at least since the 1990s when entrepreneurs started openly offering oocytes from fashion models or Ivy League students (Rothman 1989; for a review of the 'commerce of conception' see Spar 2006). By 1998, also commercial sperm banks 'had become a $164 million per year industry in the United States' (Rothman 1989, p. 174). However, establishing ARTs as players in the global health market, now even listing at the stock exchange and in some of its expressions showing 'the unacceptable face of fertility capitalism' (Brown 2014, p. 25) is a more recent development, triggered particularly by the possibility for genetic testing and a dynamic biotech industry merging the strongly proliferating genomic sequencing sector with ARTs (Brown 2014). In line with this, also a press release by a market analysis provider on the IVF market size stated recently: 'the availability of genomic testing enabling the prevention of the transfer of genetic disease during IVF use is further expected to drive the market demand' (Grand View Research 2016).

The so-called IVF-market was valued to be about 10 billion USD in 2014 and is expected to grow to around 27 billion USD in 2022 according to professional market analysts.[6] The global market for sperm grew to 3.51 billion USD in 2015 and is expected to be 4.96 billion USD in 2025, with a strong demand for high-priced sperm from known donors and largest revenues in the US (Grand View Research 2017).

Numbering the surrogacy market size appears to be more difficult. As a recent report on surrogacy law and policy in the US puts it, 'it is important to note that there is very limited data regarding surrogacy trends' (Finkelstein et al. 2016, p. 6). Figures that are mentioned regarding this sector are 6 billion USD annually in 2008 (Finkelstein et al. 2016), and 445 million USD alone in India for the same year with an estimated growth to 1–2 billion USD by 2012 (Sarojini and Venkatachalam 2016). With all due caution, one can therefore conclude that the volume of for-profit surrogacy arrangements in India grew at least fivefold since 2008. Similar trends possibly occur also in other destinations such as Mexico, Ukraine, Russia or the US. The International Social Service (ISS) estimates that worldwide at least 20,000 children are born per year due to international surrogacy arrangements (ISS 2016). Also the parentage/surrogacy project of *The Hague Conference on Private International Law* (HCCH) gives valuable insight to the monetary dimension of international surrogacy highlighting the big variations of the amounts intended parents (IPs) have to pay and surrogate mothers as well as egg donors will receive (HCCH 2014). It also illustrates the lucrative shares that lawyers, agencies and other intermediaries can expect. While the 'global costs' for an international surrogacy arrangement in India is according to this source all in all between 63,000 and 72,300 USD, it is considerably higher in the US and reached even 454,091 USD in a case that involved insurance problems (HCCH 2014, p. 65).[7]

Oocyte 'Donation' and Coproducing Perfection

The need and hence the market for oocytes extends the one of surrogacy by far and is also crucially linked to gestational surrogacy in case oocytes from the intended mother cannot be used or in case of surrogacy for male homosexual couples. Oocytes are kind of 'all-rounders' in the biological sense and of versatile use in medicine as they cannot only be fertilised for procreation but also accept a number of manipulations that give rise to biological specimens useful in research (see later). Due to vitrification,[8] oocytes can be banked for donor programmes, now routine in many clinics (De Munck et al. 2016). Vitrification also allows oocyte storage for later use due to fertility preservation in the case of cancer treatment (so-called medical freezing) or as extra-corporeal oocyte storage for possible later use in the case of postponing motherhood to a later stage in life (so-called elective or social freezing). Storing oocytes for non-medical reasons is a much debated addition to the ARTs basket which creates another market of hope for clients who undergo and pay for stimulation and retrieval of oocytes, storage and probably also IVF later on (for a review of various aspects to consider for prospective users and their doctors, see Petropanagos et al. 2015).

The procurement of oocytes by hormone-induced superovulation and aspiration from the ovaries is an invasive procedure with a number of risks to donors and patients such as the well-characterised iatrogenic ovarian hyper stimulation syndrome (OHSS). Risk factors for developing OHSS are, among others, young age, low body weight, number of retrieved oocytes and pregnancy (Practice Committee American Society for Reproductive Medicine 2008). In addition and despite the lack of systematic research, there is increasing evidence for long-term cancer risk (Schneider et al. 2017), particularly for donors who—reflecting the double standard of care (Gruben 2013)—often undergo stimulation protocols optimised for oocyte retrieval rather than reproductive care.

For reproduction as well as research, 'good', i.e. 'young' oocytes are the biomaterial of choice. Hence both reproduction clinics and researchers aim at young donors who are not infertility patients. Nevertheless, also women participating in 'egg-sharing' programmes, meant to help other IVF patients, contribute oocytes to research, often in exchange for reduced IVF fees, a controversial issue in itself (Haimes 2013). An expanded supply line from fertility centres to research institutions and companies in Europe and the US has been established (Braun and Schultz 2012).

An even more lucrative international business is oocytes for reproduction. Global Egg Donors based in California, for example, offer arrangements with fertility clinics in New Delhi, Johannesburg, California (near Los Angeles), Nicosia, Los Angeles, Toronto, Cancun and Tbilisi. Prices per egg donation range from 3500 USD in Tbilisi (and 37,000 USD in case of surrogacy) over 10,500 USD in Nicosia to 18,900 USD in Los Angeles but in this case preimplantation genetic screening (PGS, a controversial method thought to improve IVF success, see below) of 10 embryos is included. The agency proudly offers quality by advertising 'Fully Screened Egg Donors Worldwide' on their website (Global Egg Donors 2017). This sort of quality label was always paramount in medically assisted reproduction: From the beginning, gametes were marketed according to certain traits that are supposed to come along with them such as intelligence, personality traits or physical appearance as well as race (Rothman 1989; Spar 2006). In addition, quality control of prospective gamete donors/providers checks for 'disease' traits insinuating that this kind of tests warrant 'health' thus granting creation of a 'healthy' embryo, an ethically challenging topos of biomedicine (Nisker et al. 2010).

Oocytes are also required to further develop ARTs, to generate stem cells for developing cell replacement therapies in regenerative medicine,[9] for research and development of applications based on mitochondrial replacement,[10] and for research and development of applications for genome editing[11] with the vision to prevent genetic diseases linked to mutations in the mitochondrial or in the nuclear genome. Mitochondrial replacement and genome editing are both highly controversial biotechnologies as they are deemed experimental and unsafe and are crossing lines widely accepted so far. Nevertheless, a New York-based fertility doctor used unapproved mitochondrial replacement in 2016[12] and markets the method also for treating age-related infertility.[13] His work caused a warning by FDA and critique from peers (Lowthorpe 2017). Also experiments editing the human germ line are rapidly advancing and stocks are already traded based on the assumption that this technology 'could be used to treat and in some cases cure literally hundreds of diseases' (Divine 2017). It seems to be only a matter of time that current worldwide regulations which prohibit alterations of the germ line might be changed as is also indicated by a much-debated report from the US National Academy of Sciences (Kaiser 2017).[14] It should also be kept in mind that this research depends on ART as platform technology and women's bodies as a resource, an issue rarely discussed (Werner-Felmayer and Shalev 2015).

Conclusion

Medically assisted reproduction has become a biotechnology and a platform for many other fields of research and potential applications. As such the field is part of a proliferating bio-economy which is shaped by the globalisation and economisation waves of this era as well as by technology convergence. As can be currently observed, clinics worldwide—also in regions with regulation and sophisticated instruments of oversight—go increasingly for profit, offering unnecessary additional treatments and methods to check for 'embryo quality' through PGS or time-lapse microscopy, both methods of increasingly controversial usefulness (Heneghan et al. 2016). This indicates that business flourishes in the gaps which the healthcare system does not cover (te Velde et al. 2017). Thus also in countries with a highly regulated healthcare system that covers infertility treatment to a certain extent, the boundaries between non-profit- and profit-oriented care and services are blurring and it is getting increasingly difficult for those seeking infertility treatment to make well informed decisions (Heneghan et al. 2016). As has been emphasised previously, the 'focus on commercial returns has resulted in less academic oversight of who receives treatment and when' (Kamphuis et al. 2014, p. 1) and a general overuse of ARTs is observed (te Velde et al. 2017). This applies to IVF in general and to ICSI in particular as in some contexts more than 70% of women undergoing IVF receive ICSI, a rate that is certainly far beyond the prevalence of male infertility (Evers 2016; Xiong et al. 2017). Such practice is not only costly and reflects an unjustified 'therapeutic illusion' about the effectiveness of ICSI but is also harmful since for couples without diagnosed male factor infertility, ICSI consistently leads to fewer live births than IVF (Evers 2016) and to higher rates of birth defects (Xiong et al. 2017). As vitrification turns the *zona pellucida*, a glycoprotein layer of crucial function for fertilisation surrounding the oocyte, less penetrable and thus requires the use of ICSI (De Munck et al. 2016), the increasing use of vitrification possibly contributes to ICSI overuse. Moreover, recent work showed that recipients of embryos created by using donated oocytes have a significantly higher risk to develop high blood pressure, a risk factor for preeclampsia, than those who use their own oocytes which are syngeneic in terms of immunology (Letur et al. 2016). This raises the question whether or not the practice of oocyte donation as well as increasing use of vitrified oocytes should be further extended.

Moreover, the increasing use of ARTs is unsettling in itself as IVF and ICSI shorten the duration of pregnancy and reduce neonatal birthweight (De Geyter et al. 2006). They also cause higher rates of a number of conditions such as imprinting disorders and congenital malformations particularly of the male urogenital tract (Kamphuis et al. 2014; Hyrapetian et al. 2014; te Velde et al. 2017). ICSI may also affect cognitive development (Rumbold et al. 2017). Moreover, there is still a so-called twin epidemic in infertility care as too many embryos are routinely transferred leading to 'an unacceptable high incidence of maternal, perinatal and childhood morbidity and mortality. Healthcare costs due to infertility therapy are too high and this may lead to social and political concern' (Ombelet 2016, p. 189). A major concern of responsible scientists is that long-term studies on safety of several procedures are still lacking and not funded and that there is a lack of knowledge on best practice and long-term safety (Kamphuis et al. 2014).

To conclude, from the perspectives of science and medicine and considering several challenges of the current practice, we should realise that even though ARTs are a medical success story we still do have only very limited knowledge about several steps which are manipulated by IVF, ICSI, oocyte procurement, third party reproduction, etc. For example, there is ample evidence that the composition of embryo culture medium influences the phenotype of embryos by epigenetic mechanisms but research in this field is scarce (Sunde et al. 2016). Who would have thought that even the age of embryo culture medium affects the birthweight of new-borns (Kleijkers et al. 2015)? And who would be able to know what the whole procedure, particularly in a surrogacy setting, is doing to the child in terms of, e.g. bonding (Tieu 2009)? It is time to understand that ARTs should not be overused, their applications not be extended far beyond very well defined conditions of infertility and that profit-orientation, going global and involving third parties on a large scale are inadequate settings for a highly sophisticated field of expertise where we only start to gather experience and insight. We also should be very careful to use ARTs as just an auxiliary platform for genetic engineering of embryos. Otherwise, science and medicine will create new forms of harm instead of helping patients.

Notes

1. ARTs comprise all 'treatments or procedures that include the in vitro handling of both human oocytes and sperm or of embryos for the purpose of establishing a pregnancy. This includes, but is not limited to, in vitro fertilization and embryo transfer, gamete intrafallopian transfer, zygote intrafallopian transfer, tubal embryo transfer, gamete and

embryo cryopreservation, oocyte and embryo donation, and gestational surrogacy. ART does not include assisted insemination (artificial insemination) using sperm from either a woman's partner or a sperm donor' (Zegers-Hochschild et al. 2009, p. 1521).

2. There are several definitions of infertility according to World Health Organization (WHO 2017): clinical infertility is 'a disease of the reproductive system defined by the failure to achieve a clinical pregnancy after 12 months or more of regular unprotected sexual intercourse' or 'the inability of a sexually active, non-contracepting couple to achieve pregnancy in one year'; demographic infertility is the inability to become or remain pregnant within five years. See text for some critical issues with regard to definitions of infertility.
3. A treatment cycle refers to the time from starting the hormonal treatment for inducing superovulation to implanting the embryo.
4. Also the notation of 'clinical labour' is used in social science theorisations (Parry 2015).
5. In English, the term 'egg' for oocyte is common and I was informed that the term 'egg cell' is not idiomatic at least in the US context (M. Darnovsky, personal communication). In the European context, however, the term 'egg cell' is commonly used, e.g. also in documents by the European Parliament (European Parliament 2005). In terms of biology, 'egg cell' is more appropriate as the 'human egg' is a cell without yolk, egg white or off-white (in contrast to the hen's egg where the biological equivalent to the human egg cell is the yolk). In order to avoid this linguistic issue, I will use the scientific term 'oocyte' throughout the text unless when quoting other authors.
6. The market is calculated for IVF, intracytoplasmic sperm injection (ICSI), frozen embryo replacement (FER), using fresh or frozen donor and non-donor gametes, as well as PGD, instruments and materials required for these procedures (Grand View Research 2016).
7. Medical Travel Quality Alliance (MTQUA), a US-based International Corporation, mentions costs from 16,000 to 35,000 USD in Georgia, 28,000–35,000 USD in India and 38,000–120,000 USD in the US. Other destinations for which pricing is mentioned are Canada, Mexico, Thailand, Russia/Ukraine and Armenia (MTQUA 2014). The prices are generally lower than the ones stated in the HCCH report (2014), as not so obvious side costs for lawyers and agency fees are probably not included.

8. Oocytes are sensitive to damage by freezing procedures that can be used for other cell types. Vitrification (in which no ice crystals are formed by freezing) got increasingly adapted and used for oocyte storage since 2005 when the first protocol with a high survival rate of oocytes was established (for a review see De Munck et al. 2016).
9. Stem cells are derived from embryos (i.e. fertilised oocytes) or from embryos from parthenogenesis (i.e. unfertilised oocytes; for review see Bos-Mikich et al. 2016).
10. In mitochondrial replacement, embryos are created which contain the nuclear genomes of the patient and her partner and the mitochondrial genome from a healthy donor. The method is also termed 'three-parent IVF'.
11. For genome editing, specified sequences can be cut out by an enzyme system called CRISPR/Cas. In principle, any desired change of DNA sequence can be achieved by this method.
12. The embryo with mitochondria from a donor was created in New York but implanted in a Mexican Clinic.
13. According to current hypothesis, age-related infertility is caused by 'a mitochondrial component' (Wolf et al. 2015, p. 5).
14. Outlining the ethics of editing the human germ line is far beyond the scope of this chapter (for current news and expert debates, see Center for Genetics and Society 2017).

Bibliography

Argyle, C. E., Harper, J. C., & Davies, M. C. (2016). Oocyte cryopreservation: Where are we now? *Human Reproduction Update, 22*(4), 440–449.

Bos-Mikich, A., Bressan, F. F., Ruggeri, R. R., Watanabe, Y., & Meirelles, F. V. (2016). Parthenogenesis and human assisted reproduction. *Stem Cells International.* Available at http://dx.doi.org/10.1155/2016/1970843.

Braun, K., & Schultz, S. (2012). Oocytes for research: Inspecting the commercialization continuum. *New Genetics and Society, 31*(2), 135–157.

Brown, S. (2014, January). *Business is booming.* Focus on Reproduction, European Society for Human Reproduction and Embryology (ESHRE), pp. 24–27 [online]. Available at http://www.eshre.eu/~/media/emagic%20files/Publications/Focus/Focus%20Jan14.pdf. Accessed 25 Feb 2017.

Bumgarner, A. (2007). A right to choose? Sex selection in the international context. *Duke Journal of Gender, Law & Policy, 14,* 1289–1309.

Center for Genetics and Society. (2017). *Human genetic modification* [online]. Available at https://www.geneticsandsociety.org/topics/human-genetic-modification. Accessed 5 Apr 2017.

Chambers, G. M., Adamson, G. D., & Eijkemans, J. C. (2013). Acceptable cost for the patient and society. *Fertility and Sterility, 100*(2), 319–327.

Council of Europe. (2015). *Background document on preimplantation and prenatal genetic testing. Clinical situation, legal situation* [online]. Available at https://rm.coe.int/inf-2015-6-dpi-dpn-e/168078bad2. Accessed 14 Apr 2018.

DeCherney, A. H., & Barnett, R. L. (2016). In vitro fertilization research is translational research. *Reproductive Sciences, 23*(12), 1634–1638.

De Geyter, C., De Geyter, M., Steimann, S., Zhang, H., & Holzgreve, W. (2006). Comparative birth weights of singletons born after assisted reproduction and natural conception in previously infertile women. *Human Reproduction, 21*(3), 705–712.

De Munck, N., Belva, F., Van de Velde, H., Verheyen, G., & Stoop, D. (2016). Closed oocyte vitrification and storage in an oocyte donation programme: Obstetric and neonatal outcome. *Human Reproduction, 31*(5), 1024–1033.

De Sutter, P. (2016, September 23). *Children's rights related to surrogacy.* Council of Europe, Parliamentary Assembly, report doc. 14140 [online]. Available at http://assembly.coe.int/nw/xml/XRef/Xref-DocDetails-EN.asp?fileid=23015&lang=2. Accessed 31 Mar 2017.

Divine, J. (2017, July 12). *CRISPR stocks: How to invest in a medical miracle* [online]. Available at https://money.usnews.com/investing/articles/2017-07-05/best-crispr-stocks-to-buy. Accessed 3 Sept 2017.

Dyer, S., Chambers, G. M., de Mouzon, J., Nygren, K. G., Zegers-Hochschild, F., Mansour, R., et al. (2016). International committee for monitoring assisted reproductive technologies world report: Assisted reproductive technology 2008, 2009 and 2010. *Human Reproduction, 31*(7), 1588–1609.

ESHRE. (2018). *ART fact sheet 2018* [online]. Available at https://www.eshre.eu/~/media/sitecore-files/Guidelines/ART-fact-sheet_vFebr18_VG.pdf?la=en. Accessed 14 Apr 2018.

European Parliament. (2005). *P6 TA(2005)0074, planned egg cell trade, European Parliament resolution on the trade in human egg cells* [online]. Available at http://www.europarl.europa.eu/sides/getDoc.do?pubRef=-//EP//NONSGML+TA+P6-TA-2005-0074+0+DOC+PDF+V0//EN. Accessed 4 Apr 2017.

European Parliament. (2015). *REPORT on the annual report on human rights and democracy in the world 2014 and the European Union's policy on the matter (2015/2229(INI)* [online]. Available at http://www.europarl.europa.eu/sides/getDoc.do?pubRef=-//EP//NONSGML+REPORT+A8-2015-0344+0+DOC+PDF+V0//EN. Accessed 4 Apr 2017.

Evers, J. L. H. (2016). Santa claus in the fertility clinic. *Human Reproduction, 31*(7), 1381–1382.

Finkelstein, A., Mac Dougall, S., Kintominas, A., & Olsen, A. (2016). *Surrogacy law and policy in the U.S.: A national conversation informed by global lawmaking.* Report of the Columbia Law School, Sexuality and Gender Law Clinic [online]. Available at http://www.law.columbia.edu/media_inquiries/news_events/2016/june2016/surrogacy-law-report. Accessed 25 Mar 2017.

GBD 2015 Maternal Mortality Collaborators. (2016). Global, regional, and national levels of maternal mortality, 1990–2015: A systematic analysis for the Global Burden of Disease Study 2015. *The Lancet, 388*(10053), 1775–1812.

Global Egg Donors. (2017). *Global egg donors provides fully screened egg donors worldwide* [online]. Available at https://www.globaleggdonors.com/. Accessed 27 Mar 2017.

Grand View Research. (2016). *IVF market size projected to reach USD 27 billion by 2022* [online]. Available at http://www.grandviewresearch.com/press-release/global-ivf-market. Accessed 12 Mar 2017.

Grand View Research. (2017). *Sperm bank market analysis by donor type, by service type (sperm storage, semen analysis, genetic consultancy), by technology (donor insemination, in-vitro fertilization), and segment forecasts, 2014–2025* [online]. Available at http://www.grandviewresearch.com/industry-analysis/sperm-bank-market. Accessed 5 Apr 2017.

Gruben, V. (2013). Women as patients, not spare parts. Examining the relationship between the physician and women egg providers. *Canadian Journal of Women and the Law, 25*(2), 249–283.

Haimes, E. (2013). Juggling on a rollercoaster? Gains, loss and uncertainties in IVF patients' accounts of volunteering for a U.K. 'egg sharing for research' scheme. *Social Science and Medicine, 86,* 45–51.

HCCH, The Hague Conference on Private International Law. (2014). *A study of legal parentage and the issues arising from international surrogacy arrangements. General affairs and policy* [online]. Available at https://assets.hcch.net/upload/wop/gap2014pd03c_en.pdf. Accessed 25 Mar 2017.

Heneghan, C., Spencer, E. A., Bobrovitz, N., Collins, D. R. J., Nunan, D., Plüddemann, A., et al. (2016). Lack of evidence for interventions offered in UK fertility centres. *British Medical Journal, 355,* i6295. https://doi.org/10.1136/bmj.i6295.

Hyrapetian, M., Loucaides, E. M., & Sutcliffe, A. G. (2014). Health and disease in children born after assistive reproductive therapies (ART). *Journal of Reproductive Immunology, 106,* 21–26.

Inhorn, M. (2016). Medical cosmopolitanism in global Dubai: A twenty-first-century transnational intracytoplasmic sperm injection (ICSI) depot. *Medical Anthorpology Quarterly, 31*(1), 5–22.

Inhorn, M., & Patrizio, P. (2015). Infertility around the globe: New thinking on gender, reproductive technologies and global movements in the 21st century. *Human Reproduction Update, 21*(4), 411–426.

International Social Service (ISS). (2016). *Call for action 2016: Urgent need for regulation of international surrogacy and artificial reproductive technologies* [online]. Available at http://www.iss-ssi.org/index.php/en/what-we-do-en/surrogacy. Accessed 31 Mar 2017.

Irshad, A., & Werner-Felmayer, G. (2016). An ethical analysis of assisted reproduction providers' websites in Pakistan. *Cambridge Quarterly of Healthcare Ethics, 25*(3), 497–504.

Kaiser, J. (2017, February 14). U.S. panel gives yellow light to human embryo editing. *Science (Health Policy)* [online]. Available at http://www.sciencemag.org/news/2017/02/us-panel-gives-yellow-light-human-embryo-editing. Accessed 5 Apr 2017.

Kamphuis, E. I., Bhattacharya, S., van der Veen, F., Templeton, A., & The Evidence Based IVF Group. (2014). Are we overusing IVF? *British Medical Journal, 348,* 252.

Kleijkers, S. H., van Montfoort, A. P., Smits, L. J., Coonen, E., Derhaag, J. G., Evers, J. L., et al. (2015). Age of G-1 PLUS v5 embryo culture medium is inversely associated with birthweight of the newborn. *Human Reproduction, 30*(6), 1352–1357.

Lahl, J. (2017, March 21). *Stop this global trading on the female body.* The Center for Bioethics and Culture Network [online]. Available at http://www.cbc-network.org/2017/03/jennifer-lahl-at-the-united-nations-stop-this-global-trading-on-the-female-body/. Accessed 5 Apr 2017.

Letur, H., Peigné, M., Ohl, J., Cedrin-Durnerin, I., Mathieu-D'Argent, E., Scheffler, F., et al. (2016). Hypertensive pathologies and egg donation pregnancies: Results of a large comparative cohort study. *Fertility and Sterility, 106,* 284–290.

Lowthorpe, L. (2017, June 9). *Researchers condemn fertility doctor's rogue science* [online]. Available at https://www.geneticsandsociety.org/biopolitical-times/researchers-condemn-fertility-doctors-rogue-science. Accessed 7 Sept 2017.

Mascarenhas, M. N., Flaxman, S. R., Boerma, T., Vanderpoel, S., & Stevens, G. A. (2012). National, regional, and global trends in infertility prevalence since 1990: A systematic analysis of 277 health surveys. *PLoS Medicine, 9*(12), e1001356.

Medical Travel Quality Alliance (MTQUA). (2014). *Commercial surrogacy tourism* [online]. Available at https://www.mtqua.org/wp-content/uploads/2014/10/Commercial-Surrogacy-Tourism.png. Accessed 17 Mar 2017.

Mitra, S., & Schicktanz, S. (2016). Failed surrogate conceptions: Social and ethical aspects of preconception disruptions during commercial surrogacy in India. *Philosophy, Ethics and Humanities in Medicine, 11*(1), 9.

Nisker, J., Baylis, F., Karpin, I., McLeod, C., & Mykitiuk, R. (Eds.). (2010). *The 'healthy' embryo*. Cambridge: Cambridge University Press.

Ombelet, W. (2016). The twin epidemic in infertility care: Why do we persist in transferring too many embryos? *Facts, Views and Vision in Obstetrics and Gynecology, 8*(4), 189–191.

Parry, B. (2015). Narratives of neoliberalism: 'Clinical labour' in context. *Medical Humanities, 41,* 32–37.

Pellati, D., Mylonakis, I., Bertoloni, G., Fiore, C., Andrisani, A., Ambrosini, G., et al. (2008). Genital tract infections and infertility. *European Journal of Obstetrics and Gynecology and Reproductive Biology, 140,* 3–11.

Petropanagos, A., Cattapan, A., Baylis, F., & Leader, A. (2015). Social egg freezing: Risk, benefits and other considerations. *Canadian Medical Association Journal, 187*(9), 666–669.

Practice Committee of the American Society for Reproductive Medicine. (2008). Ovarian hyperstimulation syndrome. *Fertility and Sterility, 90*(3), 188–193.

Riben, M. (2016, October 14). American surrogate death: NOT the first. The Blog. *The Huffington Post* [online]. Available at http://www.huffingtonpost.com/mirah-riben/american-surrogate-death-_b_8298930.html. Accessed 31 Mar 2017.

Rigon, A., & Chateau, C. (2016). *Regulating international surrogacy arrangements: State of play*. European Parliament Briefing [online]. Available at http://www.europarl.europa.eu/RegData/etudes/BRIE/2016/571368/IPOL_BRI%282016%29571368_EN.pdf. Accessed 5 Apr 2017.

Rothman, B. K. (1989). *Recreating motherhood.* New Brunswick, NJ: Rutgers University Press.

Rumbold, A. R., Moore, V. M., Whitrow, M. J., Oswald, T. K., Moran, L. J., Fernandez, R. C., et al. (2017). The impact of specific fertility treatments on cognitive development in childhood and adolescence: A systematic review. *Human Reproduction, 32*(7), 1489–1507.

Sarojini, N., & Venkatachalam, K. (2016). Marketing reproduction: Assisted reproductive technologies and commercial surrogacy in India. *Indian Journal of Gender Studies, 23*(1), 87–104.

Schneider, J., Lahl, J., & Kramer, W. (2017). Long-term breast cancer risk following ovarian stimulation in young egg donors: A call for follow-up, research and informed consent. *Reproductive BioMedicine Online, 34*(5), 480–485.

Shalev, C., & Werner-Felmayer, G. (2012). Patterns of globalized reproduction: Egg cells regulation in Israel and Austria. *Israel Journal of Health Policy, 1*(15) [online]. Available at https://ijhpr.biomedcentral.com/articles/10.1186/2045-4015-1-15. Accessed 15 Mar 2017.

Shenfield, F. (2011). Implementing a good practice guide for CBRC: Perspectives from the ESHRE cross-border reproductive care taskforce. *Reproductive Biomedicine Online, 23*(5), 657–664.

Spar, D. (2006). *The baby business: How money, science and politics drive the commerce of conception*. Boston, MA: Harvard Business School Press.

Sunde, A., Brison, D., Dumoulin, J., Harper, J., Lundin, K., Magli, M. C., et al. (2016). Time to take human embryo culture seriously. *Human Reproduction, 31*(10), 2174–2182.

te Velde, E., Habbema, D., Nieschlag, E., Sobotka, T., & Burdorf, A. (2017). Ever growing demand for in vitro fertilization despite stable biological fertility: A European paradox. *European Journal of Obstetrics and Gynecology and Reproductive Biology, 214*, 204–208.

Tieu, M. M. (2009). Altruistic surrogacy: The necessary objectification of surrogate mothers. *Journal of Medical Ethics, 35*, 171–175.

Werner-Felmayer, G., & Shalev, C. (2015). Human germline modification: A missing link. *American Journal of Bioethics, 15*(12), 49–51.

Whittaker, A. (2011). Reproduction opportunists in the new global sex trade: PGD and non-medical sex selection. *Reproductive BioMedicine Online, 23*(5), 609–617.

WHO. (2017). *Infertility definitions and terminology* [online]. Available at http://www.who.int/reproductivehealth/topics/infertility/definitions/en/ Accessed 3 Apr 2017.

Wolf, D. P., Mitalipov, N., & Mitalipov, S. (2015). Mitochondrial replacement therapy in reproductive medicine. *Trends in Molecular Medicine, 21*(2), 68–76.

Xiong, X., Dickey, R. P., Buekens, P., Shaffer, J. G., & Pridjian, G. (2017). Use of Intracytoplasmic sperm injection and birth outcomes in women conceiving through in vitro fertilization. *Paediatric and Perinatal Epidemiology, 31*(2), 108–115.

Zegers-Hochschild, F. Adamson, G. D., de Mouzon, J., Ishihara, O., Mansour, R., Nygren, K., et al. (2009). International Committee for Monitoring Asssisted Reproductive Technology (ICMART) and the World Health Organization (WHO) revised glossary of ART terminology, 2009*. *Fertility and Sterility, 92*(5), 1520–1524.

3

Transnational Surrogacy: An Overview of Legal and Ethical Issues

Judit Sándor

Introduction

Surrogacy is based on a contract in which a woman, who is not the intending mother, is hired to carry a child to term and to bring it into the world. Within this broad definition, there are many forms of surrogacy and the different arrangements raise a number of legal and ethical questions: on the national level, surrogacy may be prohibited, ignored, tolerated, restricted or permitted. This diversity also depends on the culturally embedded concepts of family, maternity and paternity. Transnational practices produce an increasing number of cases where legal controversies have occurred due to the domestic recognition of family ties.

This chapter elaborates these controversies and argues for the need to formulate international norms. First, I will provide a typology of these

J. Sándor (✉)
Center for Ethics and Law in Biomedicine (CELAB), Central European University, Budapest, Hungary
e-mail: Sandorj@ceu.edu

S. Mitra et al. (eds.), *Cross-Cultural Comparisons on Surrogacy and Egg Donation*,
https://doi.org/10.1007/978-3-319-78670-4_3

legal–ethical differences by focusing on two fundamentally different forms: the local-altruistic surrogacy, on the one hand, and the transnational commercial surrogacy, on the other. In this chapter, I would particularly like to examine the key ethical and legal issues that shape the contours of the future normative instrument on transnational surrogacy. Judicial cases from different jurisdictions are used to illustrate the problems of unregulated transnational surrogacy. Due to the lack of comprehensive norms, these cases offer the best tools for highlighting the relevant ethical and legal issues. Second, I will examine what the adequate international legal framework that would better respond to the challenges of transnational surrogacy agreements would be. In addition, I argue that national and transnational surrogacy agreements also raise different kinds of legal questions.

In the following, I will distinguish between:

> 'Full surrogacy' (also known as traditional surrogacy) involves artificial insemination of the surrogate using the sperm of the intending father, or via *in vitro* fertilisation using the egg of the surrogate mother and the sperm of the intending father.
>
> 'Partial surrogacy' (also known as gestational surrogacy) involves the egg from the intending mother and the sperm from either the intending father or a sperm donor.

Further distinction is based on the nature of the contract (or personal agreement), where we differentiate between altruistic (without any payment) and commercial surrogacy. Within commercial surrogacy, we differentiate between an individual agreement and direct contact between the intending parent and a surrogate mother, and a contract drawn up by an agency. The latter perhaps not only provides better legal guarantees, but also leads to higher prices and provides a less personal agreement. By using these differences, it is possible to draw a scale from the most controversial or debated to the more acceptable practice: transnational commercial full surrogacy for same-sex couples or for a single intending parent at one end of the spectrum (more controversial) vs. national altruistic surrogacy offered to a married couple (less controversial).

In most parts of the world, surrogacy is neither prohibited nor considered a private, non-enforceable agreement.

Altruistic vs. Commercial Surrogacy

The legal and philosophical literature focuses predominantly on the commercial aspects of surrogacy. It also forms an important legal puzzle, especially in Europe, where prohibition of financial gain in relation to the human body and its parts is the cornerstone of the normative framework of bioethics. While commercial aspects and controversies around surrogacy dominate the literature, I want to emphasise that there are some cases where there are compelling and respectable reasons behind seeking a surrogate mother. For example, when the intending mother has had cancer and therefore has had to undergo a hysterectomy, but she has retained her ovaries and is thus still able be a mother both genetically and socially.[1] In such a medical case, it is very hard to see consistency in legal regulation that allows gamete donation, perhaps even mitochondrial donation—but forbids surrogacy.

Indeed, the prohibition of altruistic surrogacy is far from self-evident. The concern that a third person, other than the mother and the father, takes part in the reproductive process is no longer a sufficient ground for prohibition. Looking at the full range of reproductive procedures, we find various interventions that involve third persons such as gamete donors. Indeed, artificial insemination is performed with donated sperm, *in vitro* fertilisation is performed with donated eggs, and even embryo donation is permitted in some countries. However, certain causes of infertility seem to produce a strange contradiction: the reproductive rights of a woman who has eggs but no uterus are valued less than those of another woman who has a uterus but no eggs, or who has eggs but suffers from fertilisation issues (Sándor 2015). Therefore, the prohibition of altruistic surrogacy can be seen as a gendered and gamete-centred practice. It does not consider that a woman who donates her gametes (as is possible for men) may also become a mother. According to this legal construction, only the woman who gives birth to the child that she will consequently raise as her own is considered a true mother, even if she received the egg from another woman. In other words, under this system, a woman who only carries the child to term and gives birth to it without being its genetic mother is considered a mother, while she would not be seen as such, if she were only the child's genetic mother.

Nonetheless, with regard to altruistic surrogacy, many legal issues arise on the side of the surrogate mother (Ragoné 1999). Is it possible for a woman to offer her uterus to, for example, a sister, a sister-in-law or a cousin who is unable to give birth, but has her own eggs? Organ donation between relatives is permitted almost everywhere, even if these interventions result in irreversible changes, some of which may affect the donor's health as well. When comparing organ donation with surrogacy, it turns out that organ donation is a more serious invasion of the physical integrity of the donor than bearing a child for nine months (Steinbock 2002). Objections towards altruistic surrogacy show how the society has fundamentally different perceptions of paternity and maternity. Paternity originates primarily in marriage and in common-law partnerships, while maternity traditionally requires the physical act of giving birth.

The International Social Service (ISS)[2] estimates that, every year, more than 20,000 children are born through various types of surrogacy and that the profit is a crucial part of it (Werner-Felmayer, Chapter 2). Countries currently allowing commercial surrogacy for intending parents without the need to have a place of residence there or to be a citizen of the country are Georgia,[3] Mexico, Russia, Ukraine and several states in the US.[4] Still, in the last couple of years, the destinations for commercial surrogacy have frequently changed (Reddy et al., Chapter 8). Because of controversial cases of transnational surrogacy in Thailand and India, these are no longer among the countries that allow international surrogacy agreements within their territory. Some destinations, however, such as Georgia, are rarely mentioned in international publications and reports—Georgia is a country that specifically allows surrogacy, and here, the surrogate mother is not considered the mother of the child she has born. Ukraine is another example, where both commercial and altruistic form of surrogacy are recognised under the family law statute,[5] the act on basic health care protection and a ministerial decree (Philips 2013, p. 67). The law was amended in 2011 to exclude requests for surrogacy from same-sex couples. In Israel, surrogacy has been allowed since 1996. Here, the law prescribes that a man and a woman who are partners are entitled to finding a surrogate and to enter into a surrogacy agreement with her. The surrogacy agreement has to be submitted to the Board for Approval of Surrogacy. Most of the applications have been approved by the committee (Birenbaum-Carmeli 2009; Shalev, Chapter 15). Even though the procedure is expensive and requires the additional costs of legal services, travel and administration, in addition to the fee paid to the surrogate mother and, if applicable, to the egg donor as well, the US is often chosen as a site for surrogacy due to its high standards for organisation and its legal environment. Although India is less expensive and a number of private clinics are specialised in recruiting surrogate mothers, these women are more vulnerable and at greater risk of, for instance, miscarriage or other health complications (Mitra, Chapter 5). As they do not even speak the same language as the clients, they can more easily be deceived. In a patriarchal society, another issue frequently arising is that a woman might become a surrogate mother not from her own free will but because her family seeks to generate income in this manner. The first case of surrogacy in India was registered in 1994, when the surrogate mother was paid 50,000 rupees and she used the money to have her paralysed husband treated. Other parents travel to Thailand, Mexico or Ukraine. Low price is just one of the aspects that influences the decisions of future parents.

The main moral and legal argument against commercial surrogacy is that it can involve exploitation. Different forms of exploitation can be observed: women living in extreme poverty may even be forcefully recruited or purposefully misled about the surrogacy, their vulnerability might be used for the benefit of the reproductive industry. In the case of a voluntary and informed surrogacy, transactional exploitation takes place when only a small portion of the money paid by the intending parents is given to the surrogate mother. Similarly, there are cases in which she does not automatically receive sufficient post-natal health care and has to pay for it out of her own pocket. Even when such types of exploitation cannot be observed, some form of commodification, depersonalisation or instrumentalisation might take place. This especially happens in cases of transnational surrogacy where the surrogate mother's role is restricted and the pre-planned delivery of the baby is usually performed via caesarean section. Intending parents choose this option to lower the costs and to have as little contact with the surrogate mother as possible. A further reason for exploitation originates from the asymmetric information and negotiating capacity of intending parents and surrogate mothers. Most surrogate mothers in commercial arrangements, especially in developing countries, are relatively poor and not well educated (Deonandan et al. 2012).

Despite all this, while many countries have not adopted a specific law on surrogacy, they still have a surrogacy practice on a national and/or transnational level. Absence of any regulation or the silence of law can also be interpreted differently, depending on the legal culture or on the structure of the legal system. Usually, it means that there is no prohibition; therefore, the practice is allowed. However, it could also mean that the practice is illegal as it is not authorised, licensed or it is contrary to the general principles of the law (e.g. the prohibition of selling infants is stated generally in criminal or civil law).

Commercial surrogacy differs from altruistic or intra-family surrogacy as it is based on a financial agreement. It is usually arranged between the clients (intending parents or parent) and private, specialised clinics and only very rarely directly between the intending parents and the surrogate mother. The fees that the clients must pay differ extremely between countries (Werner-Felmayer, Chapter 2); where in some cases the profit is obvious, in others it is more of a cost-covering fee. The ratio between the payments to the agency and/or clinic and the payment received by the surrogate mother may indicate exploitation not only by the intended parents (IPs), but also often by the local agency and/or clinic. Relying on the vulnerable status or poverty of the surrogate mother often plays an important role in exploitation, treating surrogate mothers differently from other mothers who carry their own children. Furthermore, for the success of the procedure, a planned caesarean section is often performed.

Ethical controversies in commercial surrogacy cases, in particular, surface when the media focuses on scandals. A well-known scandal is the *Baby Gammy* case in 2014.[6] In this case, an Australian couple, Mr. Farnell and his wife Wendy contracted a surrogate mother in Thailand. The costs were nearly 10,000 Euros, but after the birth of the twins the parents took only the healthy baby girl Pipah back to Australia with them and left the baby boy Gammy, who had Down syndrome, behind. Later on, the surrogate mother sued the parents and requested the baby girl for herself, but the Australian Court decided in favour of the intending parents. Although the parents later on denied that they left the baby boy behind because of his illness, the case has become the basis for ethical and legal discussions worldwide and has resulted in the change of the Thai law on surrogacy. The case took a new, unexpected turn when it became public that the father, who had abandoned Gammy because of his Down syndrome, had committed sexual offences against children in the past. As the Thai surrogate mother was willing to also take care of the healthy twin sister of the child she was raising, she submitted her legal claim for the custodial right over the healthy child in Australia as well. Although the Court did not grant her this right, other disturbing facts were revealed, such as earlier crimes committed by the father. This illustrates the potential solutions the law can find in cases of assisted procreation where the couple or the person requesting such assistance has certain demands and refuses to accept the child or its sibling. It appears that many have the false belief that, if a service is based on a new technology and if the client has paid for it, this means that certain characteristics of the unborn child are guaranteed as well.

In addition to the problematic financial arrangement, the possibility of exploitation and legal complications with the recognition of the children in the parents' home country also raises broader ethical questions. One of the main issues is whether the commercial arrangement changes the perception of the parent–child relationship and includes the demand for a perfect, healthy child with certain biological or physical traits and whether the delivery can be 'demanded' within a certain time framework. Reproduction and birth are not calculable events in life. There is the risk of wrong and problematic expectations that, with the help of money and *in vitro* fertilisation services, the exact time of birth and the health of a child can be guaranteed. Such images are often supported by media reports of celebrities who have babies at an advanced age, without even interrupting their work.

According to China's long-standing one-child policy and its legislation against surrogacy, childless couples or those wishing to have more children or a child of a specific sex pay for a surrogate mother to travel out of China to a country where jurisdiction allows or at least does not forbid surrogacy. Every year, the Baby Plan Medical Technology Company assists in the birth of 300 Chinese children in this way. Client couples do not even meet the surrogate mother. Fearing condemnation, surrogate mothers also attempt to avoid publicity as much as possible (Johnson and Li 2014).

Although the mother giving birth generally takes precedence *vis-a-vis* the parties ordering the surrogacy, there are also exceptions to this. In the case of an especially wealthy Russian couple, Olga Mirimskaya and Nikolai Smirnov, who were, due to their advanced age, only capable of assisted procreation, they signed a surrogacy agreement whereby the daughter of Ms. Mirimskaya from a former marriage would be the egg donor, while the embryo was carried to term by a Russian surrogate mother. However, after the birth of the child, when the contracting mother (the child's biological grandmother) wanted to take the child home with her, the surrogate mother (presumably with the assistance of the biological father) refused to hand over the child and left the country, taking the child with her. Olga Mirimskaya took legal action against the surrogate mother before Russian courts, accusing her of human trafficking (Pravda.ru 2016). The highest court found her, i.e. the contracting mother to be the mother of the child, and not the surrogate mother.

Legal Consequences of Seeking Transnational Surrogacy Arrangements

As more and more couples or single individuals are looking for transnational surrogacy, another important legal question arises: What are the legal challenges of such a decision? Until recently, India was a favoured destination for transnational surrogacy (Ghosh 2013), but its current legal transformation changes both its context and conditions (Kotiswaran, Chapter 7). Currently, it is still unclear in which direction the Indian law will ultimately move and how it will change the practice (see Kotiswaran, chapter for understanding the legal transformation in Indian surrogacy). Intending parents might now look for new destinations, such as Laos, Nepal and Mexico.

In most of Europe, agreements whereby the birth mother bears the child of another woman on the basis of a prior agreement are either prohibited or not recognised. However, the permissive regulations in Ukraine and in Russia present the European courts and birth registry officials with serious challenges. In the UK, the cases of *X & Y*[7] together with the *A vs. P* cases[8] reflect the problems that arise in connection with Ukrainian surrogate births. From a legal perspective, it is often only after the child's birth that the real difficulties begin. The father signs a voluntary paternity acknowledgement form, and the father's wife adopts the newborn child. In a 1998 Californian case, the status of a child born for Mr. and Mrs. Buzzanca by a surrogate mother was disputed.[9] None of the two spouses in the Buzzanca case was the genetic parent of the child: The embryo carried by the surrogate had been created from both donated sperm and egg. Although, in first instance, the court held that neither spouse could be considered a parent as they had no genetic connection to the child, the California Court of Appeals later held that the husband's paternity had arisen through his consent to the sperm donation and that the wife could similarly be deemed to be the mother by her request for the *in vitro* procedure.

Legal disputes on transnational surrogacy have already appeared in front of the European Court of Human Rights. The European Court of Human Rights is an international court set up in 1959. It hears applications alleging that a contracting state has breached one or more of the human rights provisions set out in the European Convention on Human Rights and its protocols. In June 2014, the European Court of Human Rights issued rulings in two cases (*Mennesson vs. France* and *Labassée vs. France*). The cases concerned the recognition of the status of children born as a result of surrogacy agreements under family law.[10] Upon their return to France, the Labassées encountered difficulties when trying to obtain a birth certificate and citizenship for their child. The court of guardians, then the court of second instance in Lille (France) and finally the court of appeal declared the surrogacy agreement to be null and void, finding it contrary to French law.

Indeed, under French law, the carrying surrogate mother is considered the child's mother. The applicants, however, argued before the Strasbourg court that failure to recognise the child's status violated the Convention on the Rights of the Child and Article 8 of the European Convention on Human Rights. It was also argued, by referring to the *Genovese vs. Malta* case,[11] that citizenship is an integral part of personal identity. The court ruled that Article 8 had indeed been violated in both cases and concluded that the child's origin is an important element of privacy rights, which were breached by the French authorities' failure to recognise the child's relationship with the parents. On 21 July 2016, the European Court of Human Rights delivered judgment in two cases: the *Foulon vs. France* case[12] and the *Bouvet vs. France*.[13] Both cases were concerned with the recognition of the paternity of impending biological fathers of children born in India. *Foulon vs. France* was the case of a single man who wanted to have a child. He entered into a contract with an Indian surrogacy mother, who is also the genetic mother of his child. In return for a payment of 1300 Euros, the woman, who has never met Mr. Foulon personally, renounced all her rights as a mother. The outcome of the case followed the same approach as in the *Menesson*, the *Labassée* and, more recently, the *Paradiso*[14] cases, by stating that the interests of the child warrant the legal recognition of this agreement. The *Bouvet vs. France* case involved a gay couple who used the services of an Indian surrogacy mother. According to the European Court, homosexuals must have equal opportunities as other couples to have children by that procedure. Again, it stated that the duty to respect the children's privacy had been violated by the state of France. In another case, in *Laborie vs. France*,[15] the Court had to re-examine the non-recognition of Ukrainian birth certificates in France with respect to two children who were born by a surrogate mother.

Despite these rulings, surrogacy is still illegal in France and can be punished with a 7500-Euro fine and up to six months in prison. The legal situation in Germany (Dücker and Hörnle, Chapter 11; Beier, Chapter 12) is somewhat similar to France. It indicates that even in a legal environment that forbids surrogacy there might be cases when a child born as a result of transnational surrogacy agreement can be legally recognised. This is all based on the rights of the child, including the recognition of the identity rights of the child and from the right not to be discriminated based on the circumstances of birth. Still, there are legal uncertainties that may have impact on the parent–child relationship and an ad hoc, case by case assessment may result in different and inconsistent legal solutions.

In the *Paradiso and Campanelli vs. Italy*[16] case, a couple decided to use surrogacy services in Russia after eight unsuccessful *in vitro* treatments in Italy. The first applicant travelled to Moscow, transporting her husband's cryopreserved semen from Italy, which she handed over at the clinic. A surrogate mother was found and the applicants entered into a gestational surrogacy agreement with the company Rosjurconsulting. After a successful *in vitro* fertilisation on 19 May 2010, two embryos were implanted into the surrogate mother's womb on 19 June 2010. On 16 February 2011, the Russian clinic certified that the second applicant's seminal fluid had been used for the embryos to be implanted into the surrogate mother's womb. The first applicant travelled to Moscow on 26 February 2011, before the child was born in Moscow on 27 February 2011. On the same day, the surrogate mother gave her written consent to the child being registered as the applicants' son. Italian authorities requested a DNA test, which revealed that the intending parents were not the biological parents of the child. The Russian clinic did not find the person responsible for the mistake. Still, the parents wanted to adopt the child. Following the decision of the Minors Court in 2011, the child was placed in a children's home for about fifteen months in a location that was unknown to the intending parents. The applicants found that the measures adopted by the Italian authorities amounted to an interference with their family life. In their opinion, that interference had a formal basis in law, as the impugned measures had been taken under the provisions of the Italian Adoption Act. However, these measures resulted from an arbitrary analysis by the domestic courts, in so far as they had concluded that the child had been "in a state of abandonment". The applicants also submitted that although the Medically Assisted Reproduction Act, in its Sections 6 and 14, prohibited the practice of gestational surrogacy, criminal proceedings had nonetheless never been taken against surrogate mothers or IPs. In the absence of an extraterritoriality clause, gestational surrogacy arrangements entered into legally in another state could not, in their view, be prosecuted in the Italian courts. The European Court of Human Rights recognised that there was a de facto family life between the applicants and the child; therefore, Article 8 of the Convention was applicable. Nevertheless, the European Court of Human Rights held eleven votes to six that there had been no violation of Article 8 of the Convention. The Court stated that it does not 'underestimate the impact which the immediate and irreversible separation from the child must have had on the applicants' private life. While the Convention does not recognise a right to become a parent, the Court cannot ignore the emotional hardship suffered by those whose desire to become parents has not been or cannot be fulfilled'.

Main Legal–Ethical Norms for the Assessment of Domestic and Transnational Surrogacy Agreements

As argued above, in all juridical cases on surrogacy, the key factor for the decision was the assessment of the rights of the children. Lacking specific legal sources on the rights of children born as a result of transnational, assisted procreation technologies, it is necessary to rely on general sources and to interpret them. For over 25 years, the United Nations Convention on the Rights of the Child (UNCRC) has guaranteed the following rights for children:

> The right to be registered immediately after birth and the right to a name from birth, the right to acquire a nationality and, as far as possible, the right to know and be cared for by his or her parents; (Art. 7)
>
> The right to not be separated from his or her parents, and to maintain personal relations and direct contact with both parents on a regular basis, except if it is contrary to the child's best interests; (Art. 9)
>
> The right to leave and enter any country, including their own. A child whose parents reside in different states shall have the right to maintain, save in exceptional circumstances of personal relations, direct contact with both parents on a regular basis. The right to leave any country shall

> be subject only to such restrictions as are prescribed by law and which are necessary to protect the national security, public order (*ordre public*), public health or morals or the rights and freedoms of others, and are consistent with the other rights recognised in the present Convention. (Art. 10)

All the above-mentioned provisions are relevant for assessing the field of assisted reproduction, egg donation and surrogacy. The message of these sections (such as in the Thai case, mentioned above) is that irrespective of whether the child was born as a result of assisted reproduction, surrogacy or egg donation, she or he has rights similar to other children. It is also clear that this international convention does not restrict surrogacy as such. It provides important procedural requirements to make sure that the identity rights of the child are not infringed irrespective of the medical procedure.

As it has been argued here, the key moral issues in surrogacy that need to be assessed are exploitation, health risks for the surrogate or the child, and the legal status of the child 'ordered' from another country, as well as the possibility of clients changing their minds. So, if we can accept the notion of *in vitro* fertilisation, including egg donation, this logically leads to the conclusion that some forms of surrogacy must also be authorised. In the case of egg donation, the pregnant woman is in fact (genetically speaking) carrying the child of another woman, with all the related consequences for her own mental and physical health. Thus, if we can accept these procedures, why can a woman who has lost her uterus not be a mother, while a woman who does not have eggs, thereby having lost the opportunity for genetic motherhood, still has this privilege? With the increase in the number of cases of foreign surrogacy, we can expect that restrictive legislation in those countries will be relaxed as well. In fact, surrogacy is already permitted in a few European countries (e.g. altruistic surrogacy is legal in Belgium, in the Netherlands and in the UK; in Georgia and Russia, surrogacy is allowed in general). Overall, it is important that an international convention in this area is needed, as is the case for international adoption. (The Hague Convention on Protection of Children and Co-operation in Respect of Intercountry Adoption.)

There is no doubt that, among the various reproductive interventions, surrogacy agreements are the ones that run into the most legal and ethical problems, especially with international agreements. The most salient cases depend on the motivations of the involved parties for using this procedure. Some people, due to serious medical operations, illnesses or the loss of their own children, find themselves in a situation where they have no choice but to have someone else bear their children for them. Others who use the services of a surrogate do so because, due to their financial or social position, they feel that they have the right to choose between procedures and to select the child they want.

A uterus transplant would be a theoretical alternative, albeit one that is still very risky at the moment. This major operation is available to women suffering from uterine abnormalities, or who do not have a uterus, thus offering them procreative assistance without exploiting another woman. At the moment, however, there are still very few successful results. The first child to be born as a result of this procedure was delivered in 2015. Should this procedure become more widespread, regulations for reproduction would need to be combined with those for organ transplantation. Currently, there are not many internationally accepted standards on the regulation of reproductive rights. In Europe, most countries have ratified the Oviedo Convention which in its Article 18 states that 'where the law allows research on embryos *in vitro*, it shall ensure adequate protection of the embryo. The creation of human embryos for research purposes is prohibited'.

Transnational surrogacy agreements are generally used for two reasons. One is where surrogacy is prohibited locally, in all its forms, or where it is not allowed for specific groups of people (e.g. for same-sex partners or single parents). In such situations, seeking an international agreement seems to be the only available avenue for the intending parents because of the legislative restrictions. The other reason why people turn to transnational surrogacy arrangements is to save money and to find a location where the costs for such a procedure are lower—in their own country it's allowed, but too expensive for them. It is in this case that poorer countries become involved, and people living in difficult circumstances may be exploited. For a long time, this was the reason for the great number of surrogacy agreements concluded in India and now in Nepal, Mexico and Thailand as part of reproductive tourism.

To my understanding, here the most ethical and legal concerns emerge. Because of the physical and social distance between the contracting parties, births in such arrangements are frequently induced and performed by caesarean section, and there is little opportunity to have any sort of personal contact with the surrogate mother (see Beier, Chapter 12 for this point). A further risk factor in transnational surrogacy agreements is, as seen above, that the documents of the newborn infant are not accepted by the home country of the contracting parents, or that the family law status of the child becomes uncertain for this reason.

Should International Law Respond to the Problems Raised by Transnational Surrogacy?

Based on this mere cursory look at different laws, juridical cases and the legal literature, it becomes clear that there is no consensus about the moral assessment of surrogacy. One cannot find a coherent agreement that would be valid in each of these categories (protection of the child's identity, security of motherhood, prohibition of exploitation, discrimination, etc.) on why surrogacy is allowed and why it is allowed or restricted. While in most European countries the practice of surrogacy is either not regulated or specifically banned because of the fear of exploitation and commodification of women's bodies, in Asia, commodification and gender issues are taken into account less. However, these more permissive countries show certain restriction, e.g. excluding same-sex couples based on the ambition of protecting traditional views about the heterosexual family as a norm. Controversial cases about countries that are popular destinations for making transnational surrogacy agreements—like the few mentioned above—are commonly published in the Western media. As we have seen, sporadic case laws from different countries offer an unpredictable approach to surrogacy. They leave surrogate mothers, commissioning parents and children in a legally hazardous position. Surrogate motherhood raises significant challenges in legal interpretation as numerous issues arise, not only in connection with the circumstances of the agreement, but also with regard to the health of the unborn child and the surrogate mother, as well as the legal recognition of the child in the home country of the would-be parents. Commercial issues raised by the appearance of the growing transnational adoption market have been addressed by many scholars (Bhabha 2004). 'Predatory globalisation' (Falk 1999) plays a different role in the surrogacy market. While adoptive parents often aim to avoid local bureaucratic obstacles by going abroad, in transnational surrogacy, seeking cheaper alternative elsewhere raises a different kind of exploitation as it involves a non-voluntary pregnancy.

If there are so many moral and legal concerns against surrogacy, especially against its commercial and transnational form, would an overall ban be a good legal solution? I believe that with the widespread and accepted practice of various *in vitro* fertilisation procedures, gamete donation and genetic testing of the embryos it would be hard and inconsistent to argue that women who have lost their wombs would be the only group to be excluded from assisted reproduction technologies (Shevory 1990). There are also several other arguments in favour of surrogacy. In addition to the legal consistency of reproductive regulations (to allow remedy for different types of infertility, not only when gamete donation could help), genetic relation has been valued more since the Human Genome Project (Franklin 2013). The Court of Appeal in Singapore recently applied the term 'the loss of genetic affinity' in a case when, due to a mistake at the clinic, the intending parents' gametes were not used in the *in vitro* procedure.[17]

The key moral issue regarding surrogacy is that it turns motherhood into a business enterprise. Furthermore, the vulnerability of the gestational mother and how she may be deprived of her privacy, reproductive rights and health is important. With regard to such commercial agreements, complications with surrogacy can arise when the would-be parents have special requests or if they change their minds. For instance, one rich Japanese man had a number of babies from several surrogate mothers all at the same time in order to guarantee success (Hawley 2014). Currently, little legal or ethical attention is given to the absence of the birth mother after the surrogacy agreements. For psychologists, this is often the primary concern in the assessment of reproductive services (Golombok et al. 2013). The legal background is even more important to clients, as most of them only travel to distant countries because surrogacy agreements are illegal in most of Europe. Some countries are even threatening to apply criminal law to the parties involved, while the invalidity and unenforceability of the contract make it risky in other countries. Many European courts have faced this problem, and no uniform jurisprudence has emerged in this respect. The court of first instance in Valencia (Spain) stated that an agreement whereby another woman bears one's child is a violation of public order, even if the agreement was signed in another country.[18]

The proliferation of international surrogacy raises yet another legal issue: If the surrogacy agreement is concluded abroad, can the child be recognised under the law of a country that does not accept or that outright prohibits surrogacy agreements? Here, the issue cannot be settled as simply as with other unlawful actions under the law of a specific country, as the child would suffer severe legal consequences if he or she could not be granted citizenship in the parents' home country and if the parents would have to sedate the child with drugs and smuggle him or her across the border. On the other hand, the notion of surrogacy is hard to reconcile with the legal systems of many European countries. We might add that in agreements with surrogate mothers in India or Ukraine, the surrogates find themselves in a particularly disadvantaged position *vis-a-vis* the heterosexual or homosexual parents from richer countries compared to when the intervention and the raising of the child occur in the same country. If, for instance, the surrogate mother requires medical care later on, she would most likely not have access to the same level of medical care as she would have in the country of the commissioning parents. The rights of surrogate mothers may be curtailed in countries where information services and patient rights are not very developed.

Another important aspect is that same-sex couples have to go through more complicated procedures to adopt a child; accordingly, it may be that surrogacy is the easier option. Although the preparations seem complicated, it is only after the child's birth that the real legal difficulties begin. The father signs a voluntary acknowledgement of paternity form and the father's wife usually adopts the newborn child.

Even if they have a birth certificate, they are required to have their child's parentage recognised by the legal system of their home country after their return from abroad.

Conclusion

Reproductive markets transform continuously, partly in response to changing legal frameworks throughout the world. If surrogacy or egg donation is banned in one country, for everyone or for groups of individuals, another country may take its place in offering these services. For example, after legislative controls were tightened in India and Thailand, Laos became the new destination for those seeking surrogate mothers.

If a transnational ban is not a practical option, an international convention would be a solution. After studying numerous court cases as well as the literature, my opinion is that if gamete donation is allowed, surrogacy should not be banned—provided that only citizens and permanent residents are involved in the agreement and that the surrogate mother is not exploited, she receives good medical care before pregnancy, during pregnancy and after the child's delivery. She should be entitled to good health care even if the pregnancy was unsuccessful or complications occur. The agreement should not involve any form of exploitation and, in general, should not be a commercial, for-profit contract. Surrogate mothers may have a right to stay in touch with the family.

The best solution would be to offer this option to relatives or those who have friendship or other long-standing emotional ties—in other words, in cases where the surrogate mother is a (*quasi*) family member, not just a service provider. A surrogacy contract made between partners in physical, not only social proximity would further guarantee that the surrogate mother is not pressured to 'deliver' on a specific, pre-defined time, as is most often the case in transnational surrogacy agreements.

In my opinion, we have sufficient information to draft a normative instrument for international law, within the United Nations framework, that could respond to international reproductive interventions effectively, clarifying definitions of family relations in connection with surrogacy and, most importantly, providing legal protection to the children born in this manner.

Notes

1. The medical literature mentions some primary indications for gestational surrogacy: (1) not having a uterus as a result of a hysterectomy or being born without a functional uterus, (2) significant uterine abnormalities, including uterine scarring (Asherman's Syndrome) and inability to develop a thick, supportive uterine lining, (3) significant medical conditions that make carrying a pregnancy risky for a woman's health (such as hypertension), and (4) chronic reproductive loss.
2. The ISS is an international NGO founded in 1924; today a network of national entities and a General Secretariat that assist children and families confronted with complex social problems as a result of migration.
3. Article 143 of the Georgian Law No 3553 of 21 July 2010—LHG I, No 46, 4.8.2010, Art. 282 provides that 1. *in vitro* fertilisation shall be allowed:
   (a) to treat infertility, if there is a risk of transmitting a genetic disease from the wife or the husband to the child, using the gametes or embryo of the couple or a donor, if a written consent of the couple has been obtained;
   (b) if a woman does not have an uterus, by transferring the embryo obtained as a result of fertilisation to the uterus of another women ('surrogate mother') and growing it there; obtaining a written consent of the couple shall be obligatory.
4. Alabama, Arkansas, California, Connecticut, Illinois, Iowa, Maryland, Massachusetts, Minnesota, Nevada, North Dakota, Ohio, Oregon, Pennsylvania, South Carolina, Tennessee, Texas, Utah, West Virginia, Wisconsin.
5. Article 123 of The Ukrainian Family Code, amended December 22, 2006, No. 524-V.
6. Family Court of Western Australia, judgement of April 14, 2016—*Farnell & Anor and Chanbua* [2016] FCWA 17.
7. High Court of Justice, *X & Y* (Foreign Surrogacy) [EWHC 3030 (Fam)], judgement of December 9, 2008.

8. High Court of Justice, *A vs. P* (Surrogacy: Parental order: Death of Applicant) [EWHC 1738 (Fam)], judgement of July 8, 2011.
9. California Court of Appeal, Fourth District, Division 3, California, *John A. Buzzanca vs. Luanne H. Buzzanca* [61 Cal.App.4th 1410], judgement of March 10, 1998.
10. European Court of Human Rights, *Mennesson vs. France*, app. no. 65192/11, judgement of June 26, 2014 and *Labassée vs. France*, app. no. 65941/11, judgement of June 26, 2014.
11. European Court of Human Rights, *Genovese vs. Malta*, app. no. 53124/09, judgement of October 11, 2011.
12. European Court of Human Rights, *Foulon vs. France*, app. no. 9063/14, judgement of July 21, 2016.
13. European Court of Human Rights, *Bouvet vs. France*, app. no. 10410/14, judgement of July 21, 2016.
14. European Court of Human Rights, *Paradiso and Campanelli vs. Italy*, app. no. 25358/12, judgement of January 24, 2017.
15. European Court of Human Rights, *Laborie vs. France*, app. no. 44024/13, judgement of January 19, 2017.
16. European Court of Human Rights, *Paradiso and Campanelli vs. Italy*, app. no. 25358/12, judgement of January 24, 2017.
17. Court of Appeal of Singapore, *ACB vs. Thomson Medical Pte Ltd and others*, judgement of March 22, 2017.
18. Case No 193/2010, decision of the Tribunal de Primera Instancia No 15 of Valencia, September 15, 2010.

Bibliography

Bhabha, J. (2004). Moving babies: Globalization, markets, and transnational adoption. *Fletcher Forum of World Affairs, 28*(2), 181.

Birenbaum-Carmeli, D. (2009). Contested surrogacy and the gender order. An Israeli case study. In D. Birenbaum-Carmeli & M. C. Inhorn (Eds.), *Assisting reproduction, testing genes: Global encounters with new biotechnologies* (pp. 189–210). New York: Berghahn Books.

Deonandan, R., Green, S., & Van Beinum, A. (2012). Ethical concerns for maternal surrogacy and reproductive tourism. *Journal of Medical Ethics, 38*(12), 742–745.

Falk, R. (1999). *Predatory globalization: A critique*. Cambridge: Polity Press.

Franklin, S. (2013). *Biological relatives. IVF, stem cells, and the future of kinship*. Durham: Duke University Press.

Ghosh, D. (2013). Rent-a-womb service: An overview of commercial surrogacy in India. In J. Sándor (Ed.), *Studies in biopolitics*. Budapest: Center for Ethics and Law in Biomedicine.

Golombok, S., Blake, L., Casey, P., Roman, G., & Jadva, V. (2013). Children born through reproductive donation: A longitudinal study of psychological adjustment. *Journal of Child Psychology and Psychiatry, 54*(6), 653–660.

Google Baby. (2009). [Film] Directed by Zippi Brand Frank. Israel: Brandcom Productions.

Hawley, S. (2014). Japanese man fathers 16th baby via surrogate in Thailand. *ABC News* [online]. Available at http://www.abc.net.au/news/2014-09-10/japanese-surrogacy-man-has-another-baby/5732856. Accessed 3 Aug 2017.

Johnson, I., & Li, C. (2014). China experiences a booming underground market in surrogate motherhood. *The New York Times* [online]. Available at https://www.nytimes.com/2014/08/03/world/asia/china-experiences-a-booming-black-market-in-child-surrogacy.html. Accessed 3 Aug 2017.

Philips, A. (2013). *Our bodies, whose property?* Princeton: Princeton University Press.

Pravda.ru. (2016). Genetic theft failed. Olga Mirimskaya confirmed by court as lawful mother. *Pravda Report* [online]. Available at http://www.pravdareport.com/news/society/stories/02-04-2016/134034-olga_mirimskaya-0/. Accessed 3 Aug 2017.

Ragoné, H. (1999). The gift of life: Surrogate motherhood, gamete donation, and construction of altruism. In L. L. Layne (Ed.), *Transformative motherhood: On giving and getting in a consumer culture* (pp. 65–88). New York: New York University Press.

Sándor, J. (2015). Consistency of the regulation on assisted reproduction: Is it a missing element of reproductive justice? In V. Kantsa, G. Zanini, & L. Papadopoulou (Eds.), *(In)Fertile citizens: Anthropological and legal challenges of assisted reproduction technologies* (pp. 23–39). Mytilene: (In) FERCIT, University of the Aegean.

Shevory, T. C. (1990). Rethinking public and private life via the surrogate contract. *Politics and the Life Sciences, 8*(2), 173–184.

Steinbock, B. (2002). *Legal and ethical issues in human reproduction*. Dartmouth: Ashgate.

4

A Case for Restrictive Regulation of Surrogacy? An Indo-Israeli Comparison of Ethnographic Studies

Elly Teman

Introduction

The past decade has seen a huge increase in the number of surrogate motherhood arrangements as well as in the ethnographic interest in the practice of surrogacy. When I began my anthropological research on gestational surrogacy in Israel in 1998 (Teman 2010a), the only ethnography of surrogacy was Helena Ragoné's (1994) study in the late 1980s of traditional surrogacy arrangements in the US. Three smaller-scale ethnographic studies began to examine gestational surrogacy as it gained in popularity during the early to mid-1990s (Ragoné 1999; Goslinga-Roy 2000; Roberts 1998a, b), yet there was very little ethnographic interest in the topic throughout the next decade.

Qualitative studies steadily appeared in the psychological scholarship (see Teman 2008 for a review of this literature), and surrogacy was consistently theorised by philosophers, ethicists, feminist theorists and legal scholars, yet as Markens (2012) contends, there was a marked drop

E. Teman (✉)
Ruppin Academic Center, Emek Hefer, Israel

S. Mitra et al. (eds.), *Cross-Cultural Comparisons on Surrogacy and Egg Donation*,
https://doi.org/10.1007/978-3-319-78670-4_4

in media attention to the subject. It was only after surrogacy in India became available to foreigners in 2005 that it resurfaced a 'hot topic' in the international media (Markens 2012) and became a locus of scholarly attention for sociological and anthropological research. While there has been some recent ethnographic inquiry into surrogacy in the US (Berend 2016; Jacobsen 2016) and Thailand (Whittaker 2014), it has been India that has attracted the most ethnographic interest.

This heightened interest in India, rather than in other countries where surrogacy occurs, stems perhaps from what Markens (2010) suggests is the way that Indian surrogacy seems to embody the feminist dystopia of reproductive inequality and exploitation imagined by Margaret Atwood in her classic, The Handmaid's Tale. Feared by many feminists since the 1980s, the idea that poor women of colour in Third World countries would be held in 'reproductive brothels' to service the reproductive desires of white elite women seems to many to come to fruition in scenes depicting Indian surrogates laying docilely in rows of beds in Indian clinics (Markens 2010).

In the past five years alone, the number of publications based on qualitative studies of Indian surrogates and of intended parents (IPs) pursuing surrogacy in India has quickly multiplied (Mitra et al., Chapter 1). Yet despite this growing body of sociological and anthropological research on Indian surrogacy, there has been little meaningful comparative discussion of how these case studies of Indian surrogacy compare and contrast with one another, and even less attention to how they converse with studies of surrogacy in other countries. Publications sometimes do include the perfunctory citations of the relevant research, but a citation is not the same as a conversation. By this, I wish to suggest that the studies do not include direct engagement with and comparative exploration of the data emergent from one particular case study with other comparative qualitative research on surrogates, IPs, and on the practice of surrogacy more generally. An exception is France Windance Twine's (2011) comparative look at global surrogacy yet her thoughts on India were formulated before most of the more recent India-based studies were published.

In the following, I wish to begin such a conversation by drawing upon the data emergent from the India-based studies and my own ethnographic work on surrogacy in Israel. Although it is beyond the scope of this chapter to broaden this conversation to include the US, the UK, and other countries where surrogacy is currently being researched (Zandi et al. 2014 on Iran; Whitaker 2014 on Thailand), I hope that this attempt at such comparative engagement will inspire further cross-cultural discussions of surrogacy.

Moreover, it is beyond the scope of this research to compare and contrast the findings of each of the Indian studies with one another; instead, I reference the studies as a group, concentrating on the similarities that arise from them rather than the differences, and on the way these similarities contrast with the Israeli data. I anticipate that through such discussion we can begin to formulate ethnographically based conclusions as to what might be the best practices to include in formulating evidence-based local and global policies on surrogacy. In this sense, my work follows the call of other scholars who have looked at the way that comparative analysis of ethnographic research can inform legislation and lead to evidence-based policy and bioethical debates (Haimes 2002; Krones et al. 2006).

Regulation has been discussed at length in the legal scholarship, and ethics have been heatedly debated from a theoretical perspective (Kirby 2014); this chapter aims to contribute to the ongoing conversations about surrogacy policy and to explore what might make surrogacy a more ethical practice from the ground up. I ask: What are the main themes that arise from the ethnographic comparison of surrogacy research and how might these themes translate into more informed policy considerations? What can a comparative assessment of the ethnographic data on surrogacy contribute to discussions of the regulation and ethics of surrogacy?

Methods

My thoughts on this subject have developed in the framework of a larger ethnographic study of Israeli surrogacy arrangements, conducted between 1998 and 2006, which took a multi-sited approach (Teman 2010a). The study also included in-depth, open-format interviews with 26 Jewish–Israeli gestational surrogates and 35 intended mothers, as well as intended fathers, doctors and other professionals involved in these arrangements. An additional 20 interviews were conducted in the summer of 2016 with surrogates who had given birth in the past two years. The study also drew upon textual analysis of the Israeli surrogacy law, governmental debates surrounding its legislation, protocols of court cases relating to the law, as well as analysis of newspaper articles on surrogacy in the Israeli press. Fuller ethnographic findings from the study are explored elsewhere (Teman 2010a, b).

The chapter compares these findings from the Israeli study to those emergent from the growing corpus of ethnographic and qualitative studies of surrogacy in India. As someone who has been researching surrogacy for twenty years, I was fascinated by each and every ethnographic exploration of this topic. As a result, this chapter is the outcome of several years over which I have been reading the emergent data from Indian surrogacy and thinking about that data comparatively. As I read each study, I highlighted themes that compared and contrasted with those I had examined in my own anthropological study of surrogacy, including themes of regulation, protections for surrogate, contract, informed consent, surrogate's experience and surrogate–IP relationship. The studies included in this chapter are those that were part of the crystallisation of this comparative line of thinking, beginning with the first ethnographic studies of Indian surrogacy by Pande (2014) and Vora (2009, 2014) on surrogacy in Anand, and going on to include those of Deomampo (2013, 2016), Mitra (2015, 2017), Parry (2015), Rudrappa (2015), Saravanan (2013) and Tanderup et al. (2015) on surrogacy in other areas of India. These studies do not include all of those now in process on Indian surrogacy, but only those that were published before writing of this chapter was completed. I hope that the chapter will inspire further comparative thinking on this subject.

The Setting: Strict Regulation Versus Lack of Regulation

In order to contrast the picture of surrogacy emergent from India and to compare it to the Israel-based data, it is necessary to outline the crucial difference in each country's approach to regulating surrogacy. At the time of this writing, India's approach to surrogacy is in flux; a proposed surrogacy bill has been pending in the Indian parliament since 2010 which may significantly change the parameters of the practice once passed. Moreover, after issuing guidelines in 2013 banning gay couples and single persons from pursuing surrogacy in India, the Indian government has recently added guidelines barring these arrangements for foreign couples (Kotiswaran, Chapter 7).

Yet at the time when the majority of the qualitative studies on Indian surrogacy were conducted, surrogacy was available to same sex couples, single and foreign persons, and it was actively marketed as a commercial, economically lucrative offshoot of medical tourism and outsourcing with no effective regulation at all (Harrison 2014). Indeed, surrogacy was heralded at the time as a booming business for the Indian nation with some estimating up to $450 million per year in revenue (Mohapatra 2012), and the country reportedly granted tax breaks to medical institutions that provided surrogacy-related services to foreign clients (ibid).

During this time, there were only voluntary, non-binding guidelines proposed by the Indian medical council in the absence of a law. As a result, there was a wide variety of practices and policies among different clinics, with some writers estimating (although these numbers are impossible to validate as no official statistics were gathered) up to 3000 different clinics arbitrating surrogacy agreements up through 2013 (Saravanan 2013; Mohapatra 2012). This means that the 'rules' of surrogacy are widely variable, depending on the clinic's individual policy. What is most important to take into consideration in this account is that it was within this unregulated, variable context that the studies I discuss below explored the experiences of those involved in such agreements, and that this situation contrasts starkly with the highly regulated context in which Israeli surrogacy is practised. Indeed, these two countries pose what might be considered to be polar opposite approaches to surrogacy.

Israel's surrogacy law of 1996 (The Embryo Carrying Agreements Law, Agreement Authorisation & Status of the Newborn Child, 1996, Statute Book 176) not only legalised surrogacy but also called for its close regulation and monitoring. This law made Israel the first and only country in the world to implement a form of state-controlled surrogacy in which each and every contract must be approved directly by the state (Shalev, Chapter 15). The state-appointed surrogacy committee has the solitary right to permit, refuse or demand revisions of all surrogacy contracts signed in Israel, and any arrangements that do not receive the committee's approval are illegal and punishable by up to a year in prison. Transnational arrangements are completely forbidden, as are inter-religious arrangements (only Israeli citizens and permanent residents who share the same religion are permitted) and parties cannot be related to one another.

The law is also patriarchal and hetero-normative: IPs must be married and heterosexually paired, while the surrogate must be unmarried unless the committee approves otherwise, if convinced that the couple is unable to find an unmarried surrogate (Rimon-Zarfaty, Chapter 17). In practice, since 2013 the committee has accepted married women as surrogates, and the majority of surrogates are now married. Surrogacy is not a reproductive option that is officially encouraged by the state and it is permitted only as a last resort for citizens who match these criteria and who have exhausted all other avenues towards genetic parenthood. All arrangements must be medically conducted in one of 24 state-approved clinics that are subject to close government supervision. Surrogates and IPs must be interviewed by the committee and sign the contract before its members. The process of receiving committee approval can take anywhere between two to nine months, after which the approved contract is fully legal and can be upheld in a court of law (Teman 2016). In the following, I outline what each of these regulatory frameworks potentiates.

Protections for the Surrogate

From the currently available ethnographic data on surrogacy in India, it seems clear that because of the wide variability in practice and the lack of regulation, there is negligible protection for the surrogates. In contrast to the rarity of finding a candidate who can match the long list of criteria outlined by the Israeli surrogacy committee, most studies of Indian surrogacy highlight the surplus of women wishing to undertake this pursuit, resulting in the clinics treating candidates as dispensable, disposable and easily interchangeable with other women willing to be surrogates (Saravanan 2013; Pande 2014).

In many of the clinics, there seem to be no standardised criteria for screening surrogates for their medical and psychological compatibility with the surrogate role, potentiating harm to both the surrogate and the baby. With regard to medical screening, each clinic individually decides the degree and type of screening they apply. Vora (2014) writes that women in the Anand clinic are given blood tests to screen for chronic medical conditions or illnesses, and their husband's blood is tested for transmittable diseases. However, none of the women in her study had medical records or consistent medical care in their previous pregnancies, and most had previously given birth at home rather than hospital. Clinics had no way to ascertain history of miscarriage, low birth weight or health issues if the surrogate chose not to tell them in order to get the job (Mitra 2017). While some clinics prohibit women who had more than five prior births from being surrogates, Deomampo (2016) writes about a surrogate who lied that she had only two children at the intake interview because admitting that she was a mother of five would have disqualified her as a potential surrogate.

Saravanan (2013) contends that even the limited guidelines imposed by the clinics she studied could be easily manipulated: surrogates were required to obtain their husband's permission to do surrogacy, but women reportedly brought another family member or friend to pose as their husband. And although surrogates were supposed to be at least 21 years old, one surrogate in Saravanan's (2013) study was aged twenty. In lack of consistent and comprehensive screening, women who become surrogates in India may be taking chances with their own health by submitting themselves to the regimen of hormonal preparation for IVF, risking a pregnancy with multiple foetuses and surgeries such as foetal reduction and repeat caesarean operations.

In lieu of standardised psychological screening, Vora (2014, p. 69) writes that the doctor who screened candidates in the clinic she studied in Gujarat did so solely on intuition, claiming that 'you know if they are the right type'. Saravanan (2013) contends, based on her study in Gujarat, that one of the most important hidden criteria adopted by medical practitioners and surrogacy agencies for selecting surrogates was their inertness, courteousness and submissiveness; women who showed signs of aggressive or assertive behaviour were rejected on medical pretexts. Other studies suggest that psychological fragility might be grounds for recruitment at certain clinics. Parry (2015), for instance, notes that surrogates in her study had been recruited by clinic agents specifically because their life circumstances had been compromised because of a recent life crisis, such as death of a relative or marital abandonment. Saravanan's (2013) interviewees included two surrogates who each had an ill family member, one who had a child with severe disabilities and one with a sick husband; Pande (2014) writes of an interviewee who became a surrogate in order to pay for her child's heart surgery. Thus, it seems that in these circumstances surrogates are recruited based on their submissiveness and dependency or even desperateness. What directs their recruitment is not their emotional and medical stability, but how easy it might be to control them and the resultant stability and easy management of the surrogacy process.

None of the current studies report any screening of the IPs, other than the intended mother needing to show medical necessity for surrogacy in the clinic where Vora and Pande did their studies. While that clinic publicly prohibited same-sex couples from contracting a surrogate, Vora (2014) notes that the clinic was discreetly accepting single men as clients. This shows that rules can seemingly be bent for well-paying clients. While screening of IPs is controversial, it prevents situations such as the baby Gammy controversy in Thailand, in which the intended father was revealed to be a convicted paedophile in his native Australia, or another headlining case in Thailand in which a Japanese billionaire reportedly fathered fourteen children through different clinics (Whittaker 2014). Screening of IPs, although controversial, may serve as protection in these cases for the future child.

In contrast to the situation in India, all of the surrogates in Israel must undergo centralised screening. Medically, surrogates and IPs must agree to transparency of their full medical histories before being allowed to proceed and undergo blood tests to establish that they are all healthy, disease free and not substance abusers. Since Israel's social medicine system centrally computerises all medical and mental health-related data, surrogates cannot conceal relevant information that could impact upon their own health or the baby's health, nor can IPs conceal information that may affect the surrogate's health (Teman 2010b, 2016).

The surrogate's reproductive history, including past pregnancy, abortion and birth records, is closely scrutinised by the committee. She must also pass a general medical exam and a gynaecological exam, including a pelvic ultrasound, and undergo a battery of blood tests. A potential surrogate will not be approved by the committee if she has not given birth before, has already delivered four babies, has had two or more caesarean operations, past miscarriages, or past experiences with toxic pregnancy, low-birth-weight infants (under five pounds) or early deliveries (before the thirty-sixth week). Candidates can also be rejected if they are beyond the age limits of 22–38 years, if they have an elevated BMI (above 30), if they smoke, or if they have taken antidepressants or undergone gastric bypass or other cosmetic surgeries in the past (Teman 2010b, 2016).

In addition, prospective surrogates and IPs must undergo comprehensive psychological screening, including written psychological testing and interviews with a psychologist. For surrogates, this screening can last up to six hours. The approvals committee carefully interviews all candidates after psychological testing is completed, and when the surrogate has a steady boyfriend, the committee interviews him as well. Women who pass the psychological tests but have undergone divorce, separation, childbirth, a loss or a major medical event in her family, or any other loosely defined 'trauma' or 'stressful event' are asked to wait before applying to the committee until a year has passed since the incident. Unofficial committee guidelines also prevent women from becoming surrogates if one of their own children is mentally or physically disabled or ill so that surrogacy will not take away from that child's care. In general, the committee will not approve a surrogate they determine to be desperately poor or whom they believe upon interview to have unrealistic expectations for using the money, such as the belief that she will be able to buy a house outright with her fee (Teman 2010b, 2016).

Couples are also screened medically and psychologically. The intended mother must be aged 22–45 if providing the ova and 22–51 if using donor eggs. She must present convincing medical evidence that she cannot otherwise conceive or carry a baby to term, such as prolonged infertility measured by at least 8 failed attempts to become pregnant through IVF or at least seven miscarriages. Other medical evidence includes the absence of a uterus or severe risk to her health in pregnancy. The number of surrogacy contracts a couple can enter into is also strictly limited. Only childless couples or those with one child from the present marriage are approved; in rare cases, a couple will be allowed to contract a surrogate to have their third child. In practice, the surrogacy committee has added additional restrictions. For instance, when the intended mother has a life-threatening disease, such as cancer, the committee is hesitant to approve a surrogacy contract unless her prognosis is favourable or she has been in remission for a significant period. This, in turn, may be interpreted as a (controversial) attempt to protect the future child. Intended mothers who are older than the permitted age limit have been approved on a case-by-case basis, but petitions to hire a surrogate by single women or men and by same-sex male couples have consistently been rejected by the committee (Teman 2010b, 2016).

Contract and Informed Consent

In all of the Indian studies, it is reported that surrogates have no legal representation or advocate, and it is the clinic doctors who arbitrate the contract signing and control the payment scheme (Mitra, Chapter 5). This, of course, may be a direct implication of selecting submissive and thus easy to control surrogates. A common thread in the Indian studies is that surrogates are unable to advocate on their own behalf during contract signing because they fear that they will lose the contract if they try to negotiate (Saravanan 2013; Deomampo 2016). Most studies note that the Indian surrogates cannot read or understand the English language and therefore cannot read or understand the contract (Vora 2014; Deomampo 2016; Rudrappa 2015; Pande 2014).

Surrogates are generally not able to negotiate the terms of the contract because of clinic practice. Deomampo (2016) indicates this in the case of a surrogate she interviewed who was literate in English, but reported that the doctor turned the pages of the contract too fast for her to read it. This surrogate claimed that she didn't dare ask for a copy of the contract so she could read it at home. Other studies also suggest that contracts are withheld from surrogates after they have been signed (Saravanan 2013). Mitra (2015) reports that the surrogates at one of the clinics she observed were only given a contract to sign after the embryo transfer was completed and that all hormonal preparation up through the transfer is carried out with no contract at all. She adds that only certain clauses of contracts were explained to surrogates in the clinics she studied: their payment and how often they would be able to see their children.

In general, these studies all suggest a lack of transparency in medical practices and an absence of informed consent from surrogates for medical procedures, with surrogates reporting that they rarely were offered the opportunity to ask about procedures or medications (Deomampo 2016). The agent/caretakers recruiting surrogates reportedly reassure the women that there is no sexual contact involved in the process (Pande 2014), but do not prepare them for the physical and emotional challenges of the IVF and possible medical complications during pregnancy that can result from the transfer of multiple embryos, which seems to be a widespread practice in most of the clinics, along with foetal reduction (Mitra 2017). These possible medical complications are also of relevance to the lack of protection of surrogates in India.

After interviewing surrogates and physicians at 18 different clinics in Delhi, Tanderup et al. (2015) found that surrogates were not able to object to foetal reduction if more than two embryos implanted, and they were not able to negotiate the implantation of fewer embryos in order to prevent the chance of carrying multiples. Likewise, Rudrappa (2015) reports that in Bangalore, surrogates are routinely implanted with four embryos and undergo selective reduction if more than two embryos take. She notes that it is a widely prevalent practice in the fertility clinics in Bangalore to have two surrogates pregnant simultaneously so as to increase the odds of a full-term pregnancy for the IPs. This has resulted in cases of IPs receiving two sets of twins born from different women.

All of the studies report that surrogacy births are routinely and necessarily caesarean, with the added health risks that caesarean deliveries entail. Evidence from a study by Sayani Mitra (2017) strengthens the notion that informed consent is absent in these practices. Mitra found that doctors at the clinics she observed told her they would write in the surrogate's file that two embryos had been transferred even if the transfer had actually been of five embryos, and that surrogates were not informed that it had been a multiple transfer.

The Israeli surrogacy law, as well as the guidelines enforced by the surrogacy committee, maintains mechanisms meant to ensure the surrogate's informed consent and to protect her as much as possible. These include the guarantee of her payment: couples must deposit all of the surrogate's payment in advance into a trust administered by a state-approved attorney who is written into the contract and contracts are not approved unless all funds have been secured. Couples must also deposit extra funds to pay for psychological counselling for the surrogate and her children for the duration of the process and up to six months after the birth. While all Israeli citizens have mandatory national health insurance as part of the social medicine system, couples must buy the surrogate extra medical and disability insurances so as to replace her income should she be on bed rest during the pregnancy (Teman 2016).

The committee also makes every effort to ensure the surrogate's informed consent. She must be represented by her own lawyer, paid for by the IPs, who is to advocate on her behalf and negotiate any changes she requests in the contract. The committee then interviews her and asks her questions about the contract to ensure she knows exactly what she is signing. Surrogates in my study viewed this stage as a crucial selection test by the committee and spent time preparing for the interview by learning the details of the contract in order to 'pass the test' (Teman 2010a). The committee ensures that she understands the terms of the contract, including the terms of payment, the process of surrendering the baby, how many embryos will be implanted, what will happen in case of multiples and which prenatal tests she will undergo.

In all cases, the surrogate remains autonomous in regard to her body; if she chooses to terminate the pregnancy, she is requested to return the money paid to her, but no one can prohibit her from doing so or force her to return the fee. Surrogates can choose the type of delivery they want; while it was fashionable for several years among Israeli surrogates to ask for elective caesareans in order to complete the pregnancy at 38 weeks instead of waiting for birth to occur naturally, most surrogates now choose to have regular deliveries unless they are carrying twins (Teman 2016).

Relationships: Subjectivity Versus Disposability

In my study of Israeli surrogacy arrangements, the geographical proximity between surrogates and IPs enabled them to interact throughout the pregnancy far more frequently and intensely than what might be possible in transnational surrogacy agreements or even in surrogacy within the continental US, where surrogates and IPs often live in different states (Berend 2016). After meeting the couple independently or through a matching agency, surrogates were often in touch with the couple for up to a year before their contract was approved by the committee. By the time they achieved a pregnancy, after an average of three embryo transfers, a familiarity had already been established between the surrogate and couple.

This relationship usually intensified as the pregnancy progressed, involving the challenges of negotiating interpersonal boundaries, yet in the majority of cases intense bonds developed between surrogates and intended mothers during the process. Most intended mothers would speak with their surrogate at least once a week, if not every day, and would accompany the surrogate to every doctor's appointment. Intended fathers would attend the major ultrasound examinations as well. Surrogates and intended mothers often developed an intimacy that they likened to a romance, to 'falling in love' or a 'temporary marriage' that excluded the intended father, in a similar manner to that described by Berend (2012) in the US. Surrogates made creative efforts to share the pregnancy with the intended mother as much as possible, and the intended mother's constant presence during the surrogate's pregnancy gave surrogates the feeling that the intended mother was 'holding' the pregnancy with them. Surrogates expected intended mothers to be their 'partner' in this process. An intended mother's lack of participation in the pregnancy, disinterest or emotional distance was interpreted by the surrogates as hurtful and as making their own holding of the pregnancy more difficult without the intended mothers' reciprocal containment and support (Teman 2010a).

This connection between the women, as Elizabeth Roberts (1998b) has put it, 'warmed up' the technologically facilitated, contract-arbitrated procedure and supported the surrogate's efforts to renounce her potential entitlement to the social label of 'mother' to this baby by constantly being reminded of the other candidate who wished to be assigned that title. As their relationship took centre stage during the pregnancy, surrogacy became as much about the *process* as the product, as much about the intended mother's transition to motherhood through identification with her surrogate and participation in a relationship with her during this period as it was about the couple receiving a baby. The surrogates' efforts to share the process of bringing the baby into the world with the intended mother led many surrogates to describe surrogacy as 'birthing a mother' (Teman 2010a). This shift from contractual relations to human fellowship and solidarity emphasises the surrogate as an irreplaceable subject and her central role in actively ushering the intended mother into the maternal role; the surrogate is not erased, and she is not objectified as a rented womb or valued only for her reproductive labour.

In nearly all cases, both members of the couple were present during the delivery, or in some cases only the intended mother, and birth stories were often relayed as the height, or peak, of their camaraderie. The moment of relinquishment, when surrogates saw their intended mother hold the baby for the first time or when they saw the intended father's expression as he first saw the baby, was described by the surrogates as their proudest and happiest moment, and one they were crestfallen to have missed in cases of full anaesthesia during emergency caesarean delivery (in planned caesareans, intended mothers usually accompanied surrogates during the operation). In most cases, the solidarity with their intended mother was viewed by surrogates as creating a lasting bond between them, replacing the contractual relations with the indebtedness of a gift relationship. Surrogates looked for acknowledgement in the post-birth period that they still mattered to the couple after the baby was born; a heartfelt thank you or other symbolic act would signify for the surrogate that they still mattered to the couple.

In interviews up to ten years after surrogacy had ended, surrogates would speak of their bond with the intended mother nostalgically, like one speaks of a comrade in arms or best friend from high school, even if they had not seen the intended mother or spoken to her for several years. In such cases, contact often dropped off gradually during the first few months or years following the birth, but this was not interpreted as an insult because adequate acknowledgement had already been secured (Teman 2010a). None of the surrogates reported bonding with the baby or at any time reported feeling that they were the mother of the baby; all noted that the woman with whom they were in contact was the 'only' mother of the baby they were carrying.

In India, conversely, there is little contact between surrogates and IPs. Rudrappa (2015), in her study of Bangalore surrogates, suggests that most surrogates did not even know whom they were being contracted by: a heterosexual couple, same-sex couple or single man. Deomampo (2013, p. 523) notes that Mumbai surrogates knew only that their IPs were 'foreign', a label they used to denote that they were not from India. All of the studies report that surrogates rarely spoke English and clients rarely spoke Hindi or the local language, so any interactions that did occur necessitated an interpreter. Vora (2014) reports that most IPs leave India after the collection of eggs and semen for the embryo transfer. She notes that when IPs were non-resident Indians who did speak one of the local languages, they sometimes did have long-distance phone communication with surrogates, but this type of communication was rare, leaving very little opportunity for emotional bonding between them (Vora 2009, p. 9).

Beyond the language barrier, several studies note that clinics actively discouraged contact and deliberately prevented the surrogates and IPs from ever meeting one another at all, or restrict their interaction to one or two supervised meetings (Deomampo 2013; Vora 2009). Vora (2009) and Deomampo (2013) both suggest that the limited interactions between surrogates and IPs were almost always mediated or monitored by the clinic staff as intermediaries, often serving as interpreters as well. Deomampo (2013) also contends that when surrogates did meet the IPs, they were unable to have a dialogue with them not only because of language but also because of cultural barriers. She relays the case of a surrogate whom the IPs visited repeatedly but the surrogate reportedly did not utter a word.

Along these lines, Harrison (2014) notes that even when they do meet one another, the surrogate–IP relationship is often unable to grow because of deep stereotypes that the foreign couples sometimes hold of their Indian surrogate. Many foreign IPs reportedly imagine their surrogate in line with the imagery promoted in the US media as desperately poor, and IPs see themselves as 'saving' the surrogate through their financial contribution (Vora 2014; Pande 2014). They see the surrogate as an exotic other and worry about her conduct, the food she eats, and many choose clinics that offer surveillance in surrogacy hostels because of this sense of deep distrust.

Donahue-Singh (2014) adds that in surrogacy arrangements that occur with Indian citizens or Indian nationals living abroad, other factors may hinder communication, including huge gaps in socio-economic status (Bollywood stars and desperately poor surrogates), caste differences (Brahman couples and low-caste surrogates) and religion (Muslims, Hindus and Christians). Accordingly, Mitra (2015) reports that in several cases she observed Indian IPs present within the same room as their surrogate during the embryo transfer but that they did not introduce themselves to her even after Mitra's encouragement to do so because they told Mitra they had no interest in meeting her.

These impediments to a relationship don't mean, however, that surrogates do not hope for a relationship with their IPs. Pande (2014) suggests that surrogates in Anand seemed to hope for a relationship with their intended mother, even imagining one to be true. Their hopes for a close and lasting relationship with the intended mother seemed to be reinforced by the rare instances of surrogates who did have IPs who called them, wrote to them, or visited them and told the other hostel residents of their experience (Pande 2014).

The surrogacy agreement, in the absence of an emotional bond or personal relationship between surrogate and IPs, remains a contractual exchange of money for commodified goods rather than a gift relationship. Surrogacy becomes a service arbitrated by an array of middlemen; surrogates in India are usually closely accompanied by 'agent-caretakers' or 'hostel matrons' whose job is to accompany them to medical appointments, dispense medications and 'monitor' their behaviour (Deomampo 2016; Vora 2009). In this context, surrogacy in India is often theorised as work or 'reproductive labour' rather than as relations based on trust, solidarity and human fellowship. Surrogates are constructed as 'easily disposable' and as 'perfect mother workers' whose disposability and interchangeability are compared to other forms of ‘temporary work’ such as employment in call centres (Vora 2009), garment factory sweatshops (Rudrappa 2015) and domestic help (Pande 2014). In the absence of a meaningful relationship with the recipients of her gestational efforts, the surrogate remains a faceless, anonymous and dehumanised worker, rather than a valued subject.

Conditions and Cultural Acceptance

Since participants in Israeli surrogacy arrangements share cultural knowledge, they are privy to the same basic cultural attitudes and understandings of motherhood and childbearing. In Israel, motherhood is a basic status and role for women (Kahn 2000). Surrogates and intended mothers alike had been socialised into a cultural realm often characterised as obsessed with motherhood and fertility, where non-natalist voices are seldom heard in public and private arenas, even as being ‘childfree by choice’ has become a legitimate social option in many other societies (Donath 2010). Intended mothers all agreed that they would do anything to have a child, and some had gone through up to 30 IVF attempts before turning to surrogacy. Surrogates in my study understood what was at stake for the intended mother within this pronatalist society; soon after meeting the intended mother and hearing her story, surrogates came to realise how ‘priceless’ this pregnancy was to her and how important it was for her to be a part of the process of bringing her child into the world rather than to receive the baby as a ‘finished product’ at the end of that process.

In line with this cultural attitude towards childbearing, surrogates did not usually encounter stigmatic or critical attitudes regarding their role from their families and friends. In the early years of the study, surrogates did report stigmatic questions and comments they had received from family and friends who were unfamiliar with surrogacy. However, for the most part, Israeli surrogates took pride in their role and told everyone they knew about their surrogate role. There was no question in their eyes that what they were doing—making another woman into a mother—was just about the greatest *mitzvah*, or moral deed, a woman could do. They spoke of giving the IPs a continuing generation, a notion that reveals the loaded sociopolitical meaning underlying their contribution in a society where the lingering effects of the Holocaust and of continual war endow children with national, and not just personal, value.

As the surrogates' relationship progressed with the IPs, their investment in what many spoke of as their 'mission' came to a height and many recounted their stories in retrospective interviews as heroic narratives relayed with pride. They spoke of themselves as angels or messengers of God, believing they had done what God, nature, technology and even the best physicians could not do—they gave this couple a baby, and even more significant in Israel's highly familial society, they made this other woman into a mother and this childless couple into parents. These surrogates described the feeling in the weeks following delivery as a 'high' from realising the magnitude of their actions. As one surrogate who gave birth to twins put it: 'No one knows what it is to give, to give life to someone else. I mean, it is like donating your organs while you are still alive'. In cases where the surrogate did not receive acknowledgement for the many 'gifts' she had given the couple, this heroic narrative was replaced by one of grief and betrayal; nevertheless, the majority of surrogates in my study relayed a narrative of empowerment that was in part contingent on the relationship with their couple (Teman 2010a, p. 267).

Whereas Israeli surrogates are vocal and openly proud of their role in surrogacy, Indian surrogates live in a culture where surrogacy is smeared by stigma and women who become surrogates are wary of being accused of infidelity or prostitution. As a result, most ethnographers report that Indian surrogates keep their pregnancy a secret from their extended families in order to escape the social stigma imposed by community members and to avoid the potential of being shamed (Vora 2014; Pande 2014; Deomampo 2013).

From the moment they become pregnant, surrogates are highly encouraged, and sometimes coerced, to leave their own homes and live in hostels with shared dormitory-like spaces or in apartments rented by the clinics. Studies consistently report that surrogates are under constant surveillance by the clinic staff in these maternity homes, where, as Deomampo (2013, p. 526) writes, their mobility is restricted, their daily activities are 'tightly controlled' and their daily timetables are 'monotonous and unpleasant'. All of the studies report that surrogates are expected to be on bed rest most of the time; one such maternity home filmed in the documentary *Google Baby* (2009) depicts a crowded row of cots in a dormitory style room with women laying on the cots in various stages of the process, some following the embryo transfer and others pregnant.

While Pande (2014) suggests that the surrogates in Anand seemed to enjoy living in the hostel and formed ties of sisterhood with the other surrogates, Deomampo (2013) suggests that most of the surrogates in the maternity homes she studied in Mumbai remained cautious and avoided interaction with other surrogates in the building. Deomampo further contends that the surrogates shared stories of loneliness, isolation, 'spatial imprisonment' and powerlessness. Far from being restful, they experienced 'high levels of stress and anxiety because of the restrictions on their mobility and the separation from their families' (p. 526).

After being subjected to extreme vigilance during the surrogate's pregnancy, all of the studies imply that surrogates have limited medical care post-delivery and no medical insurance to cover any medical issues that might arise in the period following delivery. This means that they may experience post-partum health problems, such as infections, haemorrhoids, depression, pain or other complications without receiving any treatment (Makherjee and Sekher 2015). The studies also report that there is no standard psychological counselling for surrogates except by the clinic doctors, by agent-caretakers who recruit surrogates and receive their commission upon successful relinquishment, and by hostel matrons who are former surrogates who have been employed by the clinics to serve as housemothers in the maternity homes and to ensure that surrogates conform to clinic rules (Saravanan 2013; Pande 2014; Deomampo 2013). As such, it seems clear that those in charge of 'counselling' the surrogate are all parties with clear monetary incentives to make sure that the surrogates 'do their job' rather than offering unbiased professional counselling and support.

Most studies note that it was common among the surrogates they interviewed to have been sterilised before becoming surrogates as part of India's anti-natalist population control policy (Rudrappa 2015). Surrogates' own children and families were thus devalued, while their reproductive potential was harnessed to produce children for foreign nationals and upper-class Indians (Deomampo 2013). In this sense, Vora (2009, p. 9) suggests that 'the surrogate is doing work that enhances and actually reproduces the life of that person and family, and as such she is contributing to the enrichment and longevity of the U.S. economy and society', while her own country is valuing her only for the potential marketability of their 'otherwise unused uterus'—as a body part—but not as a person.

Conclusion

In view of the picture emergent from India, I suggest that the unique legislation regulating surrogacy in Israel, the small geographical scope of the practice and its specific cultural and social milieu have imbued Israeli surrogacy with a distinct character, with critical implications when juxtaposed with India. I argue that the Israeli model produces the structural framework for more protections for both the surrogate and child and potentiates frequent interaction, intimacy and human fellowship between surrogates and IPs.

In contrast to India's transnational, cross-border arrangements, the Israeli law eliminates the option of crossing any borders at all: national, religious or geographical. The Israeli law replaces the 'reproductive exile' (Inhorn and Patrizio 2009, p. 904) that has been suggested as characteristic of cross-border reproduction with a very local, nationally bound version of the practice: Jewish–Israeli 'kosher' surrogacy. Elsewhere (Teman 2009), I have critiqued this law as a symbolic act of the state to preserve the national, natural, nuclear family in face of the culturally anomalous situations surrogacy can introduce and as a mechanism of symbolic control (Teman 2016). There, I criticised the law's restrictiveness and argued that legislators passed the law so as to ensure that local surrogacy will only produce Jewish–Israeli citizens to hetero-normative, nuclear families; that it is strongly patriarchal, paternalistic, controlling and invasive into the surrogate's body and the private realm of individual families (for relevant discussion, see Rimon-Zarfaty, Chapter 17).

Yet, in the light of these comparisons to the unregulated surrogacy industry in India, the Israeli surrogacy law appears to have structured the practice in a way that constructs the surrogate as a valued 'subject'. By limiting the arrangements nationally and geographically, the Israeli law potentiates relationships between surrogates and IPs based on shared language and cultural understandings as well as frequent personal interaction that opens up the space for intimacy and human fellowship to develop between surrogates and IPs. While these relationships are often fraught with challenges, boundary negotiations and misunderstandings, the direct interpersonal relationship is necessarily conducted between the persons immediately involved in the process.

As a result, the surrogate cannot be erased, exoticised or kept at an anonymous distance as in India; the Israeli surrogate is constructed as an important, valued national subject, rather than an alienated service worker (Vora 2014) or docile, disposable 'perfect mother worker' (Pande 2014, p. 970), as in India. Instead of hiding her behind an array of middlemen under the auspices of a 'one stop package deal' as offered through Indian clinics, the Israeli surrogate forms a relationship with the IPs. IPs can, and sometimes do, intend at the outset to keep their relationship with their surrogate as strictly business, but in my experience, they ultimately fail because they are thrown together in a process where they ultimately enter into emotional bonds of intimacy, caregiving, solidarity and gift with the surrogate.

This leads me to conclude that a case may be made for restrictive regulation in this case as grounds for a more ethical and humanistic surrogacy. In as much as many legislators and laypersons in other countries might be less open to state intervention in the private sphere (see Berend 2016 on US surrogates' ideas about privacy and regulation), the Israeli case demonstrates that a patriarchal, paternalistic law may actually prevent the furthering of potential inequalities and exploitation from the process. Banning surrogacy only sends IPs across national borders to undertake transnational arrangements that may be morally and ethically harmful to all those involved (Mitra, Chapter 5). Ignoring surrogacy also leads to legal and ethical quagmires as discussed above. Legislating surrogacy nationally, while paying close attention to the cultural context and to empirical research, may prevent further harm.

Finally, I end this chapter with the call for further comparative analysis of findings from ethnographic and qualitative studies of surrogacy. It is only through comparative examination of empirical findings that we can look beyond the many cultural assumptions that inform both the theoretical and psychological scholarship on surrogacy as well as public debate (Teman 2008). Such comparative study can elicit the many implications of different regulatory frameworks, as well as the ethical and social issues involved in shaping the surrogacy process. Evidence-based policy may have critical outcomes for the surrogates' emotional and physical well-being, as well as serious implications for the protection of the future child.

Acknowledgement I would like to thank my husband, Avi Solomon, for his comments and suggestions on multiple versions of this manuscript. I would also like to thank Diane L. Wolf, who first inspired me to think about surrogacy in Israel and India comparatively at a conference she organised at UC Davis in 2009. I also thank Zsuzsa Berend for always inspiring me to think about surrogacy in a comparative framework and the anonymous reviewer of this chapter for the helpful suggestions and comments.

Bibliography

Berend, Z. (2012). The romance of surrogacy. *Sociological Forum, 27*(4), 913–936.

Berend, Z. (2016). *The online world of surrogacy*. London: Berghan Books.

Deomampo, D. (2013). Gendered geographies of reproductive tourism. *Gender & Society, 27*(4), 514–537.

Deomampo, D. (2016). *Transnational reproduction: Race, kinship and commercial surrogacy in India*. New York: New York University Press.

Donath, O. (2010). Pro-natalism and its 'cracks': Narratives of reproduction and childfree lifestyles in Israel. *Israeli Sociology, 11*(2), 417–439 [Hebrew].

Google Baby. (2009). [Film] Directed by Zippi Brand Frank. Israel: Brandcom Productions.

Goslinga-Roy, G. (2000). Body boundaries, fictions of the female self: An ethnographic perspective on power, feminism and the reproductive technologies. *Feminist Studies, 26*(1), 113–140.

Haimes, E. (2002). What can the social sciences contribute to the study of ethics? Theoretical, empirical and substantive considerations. *Bioethics, 16*(2), 89–113.

Harrison, L. (2014). 'I am the baby's real mother': Reproductive tourism, race, and the transnational construction of kinship. *Women's Studies International Forum, 47*, 145–156.

Inhorn, M. C., & Patrizio, P. (2009). Rethinking reproductive 'tourism' as reproductive 'exile'. *Fertility and Sterility, 92*, 904–906.

Jacobsen, H. (2016). *Labor of love: Gestational surrogacy and the work of making babies*. New Brunswick, NJ: Rutgers University Press.

Kahn, S. M. (2000). *Reproducing Jews: A cultural account of assisted conception in Israel*. Durham, NC: Duke University Press.

Kirby, J. (2014). Transnational gestational surrogacy: Does it have to be exploitative? *The American Journal of Bioethics, 14*(5), 24–32.

Krones, T., Schlüter, E., Neuwohner, E., El Ansari, S., Wissner, T., & Richter, G. (2006). What is the preimplantation embryo? *Social Science & Medicine, 63*(1), 1–20.

Makherjee, R., & Sekher, T. (2015). Women at risk in the unregulated surrogacy 'industry'. *Economic & Political Weekly, 1*(26–27), 83–89.

Markens, S. (2010). Interrogating narratives about the global surrogacy market. *The Scholar and Feminist Online*, Special Issue, *9*(1–2). http://sfonline.barnard.edu/reprotech/markens_01.htm. Accessed on April 4, 2017.

Markens, S. (2012). The global reproductive health market: U.S. media framings and public discourses about transnational surrogacy. *Social Science & Medicine, 74*, 1745–1753.

Mitra, S. (2015, September 9–11). *Reproductive disruptions during surrogacy: End of a beginning*. Magic 2015 Anthropology Conference: Session on reproductive disruptions & flows: Surrogacy & obstetric care in India and the US, University of Sussex, Sussex, UK.

Mitra, S. (2017). *Disruptive embodiments: An ethnography of risks and failures during commercial surrogacy in India*. Ph.D. dissertation, Social Science Faculty, University of Göttingen, Göttingen, Germany.

Mohapatra, S. (2012). Stateless babies and adoption scams: A bioethical analysis of international commercial surrogacy. *Berkeley Journal of International Law, 30*(2), 412–450.

Pande, A. (2014). *Wombs in labor: Transnational commercial surrogacy in India*. New York: Columbia University Press.

Parry, B. (2015). Narratives of neoliberalism: 'Clinical labour' in context. *Critical Medical Humanities, 41*, 32–37.

Ragoné, H. (1994). *Surrogate motherhood: Conception in the heart*. Boulder, CO: Westview Press.

Ragoné, H. (1999). Surrogate motherhood, gamete donation, and constructions of altruism. In L. L. Layne (Ed.), *Transformative motherhood: On giving and getting in a consumer culture* (pp. 65–88). New York: New York University Press.

Roberts, E. F. S. (1998a). Examining surrogacy discourses: Between feminine power and exploitation. In N. Scheper-Hughes & C. F. Sargent (Eds.), *Small wars: The cultural politics of childhood* (pp. 93–110). Los Angeles: University of California Press.

Roberts, E. F. S. (1998b). Native narratives of connectedness: Surrogate motherhood and technology. In J. Dumit and R. Davis-Floyd (Eds.), *Cyborg babies: From techno-sex to rechno-tots* (pp. 193–211). New York: Routledge.

Rudrappa, S. (2015). *Discounted life: The price of global surrogacy in India*. New York: New York University Press.

Saravanan, S. (2013). An ethnomethodological approach to examine exploitation in the context of capacity, trust and experience of commercial surrogacy in India. *Philosophy, Ethics, and Humanities in Medicine, 8*(10), 1–12.

Singh, H. D. (2014). "The world's back womb?": Commercial surrogacy and infertility inequalities in India. *American Anthropologist, 116*(4), 824–828.

Tanderup, M., Reddy, S., Patel, T., & Nielsen, B. B. (2015). Informed consent in medical decision-making in commercial gestational surrogacy: A mixed methods study in New Delhi, India. *Acta Obstetricia et Gynecologica Scandinavica, 94*(5), 465–472.

Teman, E. (2008). The social construction of surrogacy research: An anthropological critique of the psychosocial scholarship on surrogate motherhood. *Social Science and Medicine, 67*(7), 1104–1112.

Teman, E. (2009). The last outpost of the nuclear family: A cultural critique of Israeli surrogacy policy. In D. Birenbaum-Carmeli & Y. Carmeli (Eds.), *Kin, gene, community reproductive technologies among Jewish Israelis* (pp. 107–126). Oxford: Berghahn Books.

Teman, E. (2010a). *Birthing a mother: The surrogate body and the pregnant self*. Berkeley: University of California Press.

Teman, E. (2010b). The last outpost of the nuclear family: A cultural critique of Israeli surrogacy policy. In D. Birenbaum-Carmeli & Y. Carmeli (Eds.), *Kin, gene, community: Reproductive technology among Jewish Israelis* (pp. 107–126). Oxford: Berghahn Books.

Teman, E. (2016). Surrogacy in Israel: State-controlled surrogacy as a mechanism of symbolic control. In E. S. Sills (Ed.), *Handbook of gestational surrogacy: International clinical practice and policy issues*. Cambridge: Cambridge University Press.

Twine, F. W. (2011). *Outsourcing the womb: Race, class and gestational surrogacy in a global market. Framing 21st century social issues*. New York: Routledge.

Vora, K. (2009). Indian transnational surrogacy and the disaggregation of mothering work. *Anthropology News*, p. 9.

Vora, K. (2014). Experimental sociality and gestational surrogacy in the Indian ART clinic. *Ethnos, 79*(1), 63–83.

Whittaker, A. (2014). Merit and money: The situated ethics of transnational commercial surrogacy in Thailand. *The International Journal of Feminist Approaches to Bioethics, 7*(2), 100–120.

Zandi, M., Vanaki, Z., Shiva, M., & Mohammadi, E. (2014). Nine centuries waiting: The experiences of Iranian surrogacy commissioning mothers. *Iranian Journal of Nursing and Midwifery Research, 19*(3), 224–232.

5

Cross-Border Reproflows: Comparing the Cases of India, Germany, and Israel

Sayani Mitra

Introduction

Commercial surrogacy and egg donation have become popular fertility options within the artificial reproductive technologies (ART) discourse in various parts of the world. Yet, several countries like Germany, Italy, Austria, and Switzerland strictly ban the use of such reproductive technologies. However, instead of preventing its actors from getting involved in practices that these legislations deem morally unacceptable, such bans push the citizens to cross borders in search of the same technologies elsewhere. Egg donation and surrogacy are amongst the top reasons why people traverse national borders (Nahman 2016, p. 418). Together, the pursuit of technology, available biogenetic material, and accommodating regulatory regimes provide the ground for what is termed as 'cross-border reproductive care' (CBRC) or 'reproductive travel'.

S. Mitra (✉)
Medical Ethics & History of Medicine, University Medical Center Göttingen, Göttingen, Germany
e-mail: sayani.mitra@medizin.uni-goettingen.de

S. Mitra et al. (eds.), *Cross-Cultural Comparisons on Surrogacy and Egg Donation*,
https://doi.org/10.1007/978-3-319-78670-4_5

The three countries (Germany, Israel, and India) discussed in this volume present three completely different national contexts of ART and yet, each country plays a significant role in cross-border reproductive travel. The case of Germany serves as a rather interesting example in this regard. The country is known for its restrictive ART legislations while Germans constitute a substantive population of commissioners crossing borders in search for egg donation and/or surrogates (Sheinfield et al. 2010). In contrast to the absolute ban on surrogacy and egg donation in Germany, cross-border reproductive travels also result due to partial prohibition or systematic discrimination of actors based on sexuality, marital status, nationality, etc. A prominent example of this is the case of Israel: the first country in the world to regulate surrogacy and to make it available exclusively for its heterosexual citizens and hence preventing its homosexual and single citizens from accessing the same (Rimon-Zarfaty, Chapter 17). As a result, Israeli gay couples form a large community within the cross-border surrogacy commissioners. Furthermore, egg donation has been allowed in Israel since 2010 but the restrictions on donor categories based on religion, age, marital status, and the strict limitation on the number of donation cycles create a shortage of oocytes (Birenbaum-Carmeli 2016). This prompts Israelis to either physically cross borders in search of egg donors or to commission egg donation services across the border without physically travelling there. The latter allows them to obtain frozen embryos with donor eggs, collected from egg donors and fertilised with their own sperm samples at clinics in a different country like Romania (Nahman, Chapter 16). Such flows of technologies, services, and biogenetic materials like eggs or embryos are made possible by the free flow of knowledge and expertise, accompanied by a fertile line of women making their bodies and biogenetic materials 'bioavailable' (Cohen 2005, p. 83). Therefore, the bioavailability of fertile bodies willing to labour in the reproductive bioeconomies plays a crucial role in giving rise to transnational reproductive centres or 'reprohubs' (Inhorn 2016, p. 2). One of the earliest and most well-known transnational reprohub was the surrogacy industry in India.

The aim of this comparative analytical chapter is to capture the cross-border 'reproflows' (Inhorn 2010, p. 181) of actors, technologies, desires, hopes, gametes, embryos, etc., to and from Germany, Israel, and India. The concept of reproflows as coined by Inhorn (2010) refers to the global circulation of fertility in the form of the movement of bodies, gametes, and technologies within the global cross-border reproductive industry (Inhorn 2010). I use this conceptual frame to trace how each of these three different national legislations along with the actors' social and cultural attitudes contribute to creating various routes and patterns of reproduction around the world. I show how each national ART regime gives rise to various choices, forms of consumerism, and variations in standards of reproductive services leading to the production of values and labour relations around surrogacy and egg donation. It is these aspects that I refer to as 'reproductive bioeconomies'. I want to show how the cross-border movement of actors from the three countries discussed shape the global reproductive terrain. Following Appadurai's (1996) conceptualisation of global scapes, Inhorn (2010) developed the notion of 'reproscapes' (Inhorn 2010, p. 181) to describe the ways in which bodies are located within the global reproductive terrain. This concept helps in identifying the various local forces at play that place actors within the global field, while at the same time explaining the conditions of their local positioning. I show how the varying local contexts of Germany, Israel, and India shape the cross-border reproflows and constitute the global reproscape. Doing so helps in explaining the diverse and often stratified, hierarchical positioning of the actors in the reproscapes.

In the following section, I discuss the case of each of the mentioned countries, i.e. Germany, Israel, and India. Based on the findings from ethnographic studies as well as chapters in this volume from the fields of law, anthropology, and sociology, I identify the specific legal and sociocultural contexts that further stratify reproduction. I also show how reproductive pathways of surrogates, egg donors, intended parents (IPs), and agencies from India, Germany, and Israel cross in the global reproscapes, revealing how interconnected the transnational reproduction network actually is.

Movement from Germany

Germany is known to have a rather conservative and restrictive take on any form of ART that tampers with embryos and can be seen as disintegrating motherhood. The German law considers it illegal to interfere with natural motherhood (also Schicktanz, Chapter 6). According to the

Embryo Protection Act (*Embryonenschutzgesetz*, ESchG),[1] egg donation and surrogacy[2] are considered illegal in Germany, although the use of in vitro fertilisation (IVF) with one's own eggs as well as sperm donation is allowed (Dücker and Hörnle, Chapter 11). It is legally punishable to advertise for surrogacy or egg donation or to act as an agent or solicitor for these practices. However, the law in Germany is rather vague and it allows people to seek third-party eggs and embryos by travelling to a different country (Dücker and Hörnle, Chapter 11). As a result, the country does not punish its citizens if they engage in cross-border egg donation or surrogacy (be it altruistic or commercial). According to a study conducted during the mid-2000s, the public attitude in Germany seemed to be largely supportive of both egg donation and surrogacy (Stöbel-Richter 2005). Yet, the German media discourse on surrogacy and egg donation has been highly critical of both practices (Beier, Chapter 12; König, Chapter 13). Hence, German heterosexual as well as homosexual couples regularly cross borders in search of egg donation and surrogacy. The German legal and ethical discourse thus influence the reproflow of actors, desires, services, and technologies and substantially contribute towards shaping the global reproscape and the bioeconomies around cross-border reproduction.

Unlike Israel, where cross-border travel for surrogacy and egg donation takes place within a religiously guided moral ideal of the Jewish family and the nation state, in Germany, that same discourse is influenced by a strong sense of bioethics emerging from a widespread scepticism towards the (too liberal) use of biotechnology—partly due to the historical experience of the Holocaust. The German ethical discourse criticises cross-border travel on grounds of the exploitation of women under unorganised and unregulated healthcare systems abroad and for converting babies, egg, or embryos into 'sellable commodities' (Beier, Chapter 12). This can be contrasted with Israel, where ART as well as selective technologies are embraced due to pronatalist and pro-market leanings. Thus, the fear of exploitation, baby-selling, and degradation of the human life by its commodification constructs the German ethical discourse as fundamentally opposed to any form of bioeconomies that could commodify motherhood or the reproductive capacities of women or involve the use of technologies to manipulate eggs or embryos.

Thus, when German citizens travel to other European countries like Spain, they fear a strong moral opposition at home and prepare themselves to face possible social stigma or shame.

Travelling to gain access to ART is financially demanding and not everyone can afford it. Therefore, costs become a deciding factor in the choice of destinations. As for egg donation, some might prefer South Africa over Spain or the Czech Republic, while in the case of surrogacy, Germans for instance are known to prefer Ukraine and previously India over the US (see Teschlade, Chapter 14 for more details on cross-border egg donation commissioned by German citizens). Apart from the costs, couples are also reported to be opting for destinations that are likely to provide more transparent facilities and a larger range of technological possibilities. Thus, some couples prefer the US, especially California, over other destinations, like Ukraine, due to their trust in the quality of clinical services, in clauses of anonymity, or the opportunity to meet the surrogates as well as the routine use of preimplantation genetic screening or diagnosis to ensure the birth of a 'healthy' child (König, Chapter 13). Yet, since the cost of surrogacy in the US is much higher (around USD 150,000) than in India or in Ukraine (USD 40,000–50,000), not everyone can afford the luxury of a choice. The choice of destination in itself thus reflects the economic capacities of these couples, creating the very first level of stratification.

Cross-border travel requires a lot of information and the couples or individuals need to become familiar with a considerable amount of knowledge regarding the scopes and technical opportunities in the global reproscape. In the absence of any agencies in Germany to provide interpersonal counselling and handholding to IPs (like in Israel) and without doctors offering medical advice on the topic (since doing the same is legally punishable and could lead to cancellation of one's medical licence), German IPs find themselves at a comparatively disadvantaged position. Those couples which are unable to research or comprehend such complex sets of information from the internet seldom manage to cross borders (Teschlade, Chapter 14). Hence, it is important to note that the internet plays a major part in the recruitment of the German IPs (König, Chapter 13). Online forums, virtual platforms, advertisements in German language, and Skype calls guide the prospective parents across unknown territory. These virtual communications set the flow of information, hopes, desires, and bodies and direct the reproflows from Germany to other areas. They play a major part in assimilating Germans into the global reproductive bioeconomy. Additionally, agencies employ English or German-speaking doctors or agents especially to guide an entire community of prospective German couples.

German citizens travel extensively in search of egg donors as well. Unlike Israel, where the act of egg donation faces religious opposition for disintegrating the unity of the Jewish womb and egg, in Germany, egg donation poses a threat to the common moral notion of motherhood and naturalness. Hence, German couples often prefer places like Spain and the Czech Republic that ensure donor anonymity and spare them from social stigma and social disapproval (Bergman 2011). However, the choice of destinations for egg donation is determined largely by the profile of the donors available, as the study by Sven Bergmann (2011) reveals. For instance, some German couples prefer to travel to the Czech Republic over Spain since the former still counts as 'Central Europe'. Hence, IPs expect a donor with a close phenotypic match to themselves. Spain, on the other hand, is known to have a diverse range of egg donors and greater choices of clinics and services to choose from and those preferring diversity often travel there (Bergman 2011). As heterosexual couples opting for IVF with egg donation are able to pass on their genes from at least one side (sperm from the male partner), they often emphasise a 'correct' genetic egg donor profile. Such genetic profiling becomes necessary for those planning to conceal their participation in gamete donation. Concealment may or may not also be a requirement for citizens from other countries like the UK or Australia, but the social, ethical, and, most importantly, legal codes of the German society set particular prerequisites for this selective choice. It becomes the constituting feature of their reproflows. By strategic matching of the phenotypes of the donors with themselves, couples thus hope to attain social legitimacy and satisfy the social (or even internalised) expectations towards kinship. Therefore, concealment becomes important for these couples in order to prevent the social stigmatisation of their child and also to escape the moral burden of being known as someone participating in an illegitimate procedure (Bergman 2011).

Pressure to conceal one's participation and the fear of disclosure also play a vital part in German parent's experience during surrogacy. These needs could previously be met by the surrogacy services in India. India's surrogacy clinics guaranteed complete anonymity and the severing of all future ties through a legally binding contract. While it is difficult for gay couples to conceal their parenthood through surrogacy, heterosexual couples are often faced with the dilemma of being socially rejected or being labelled as having been a part of an illegal practice. In fact, the fear of some German IPs that their child will be taken away shapes their expectations, the services they choose, and the destinations they go to (König, Chapter 13). A major trouble typical for the Germans' surrogacy experience is the fear of not being able to bring the child to Germany. Hence, legal concerns and securing the right documentation become a part of their surrogacy journey right from its onset. Such fear prompts IPs to choose destinations where the law meets their own legal requirements for taking the newborn child home without any trouble. Therefore, reproflows consist not only of technical and medical services, but is also shaped by the availabilty of legal services in the destination countries. The specific situation in Germany thus opens up the scope for legal attorneys specialising in the German legal situation to find employment in the global reproductive industry.

To circumvent legal uncertainties, German IPs in particular choose destination like the US, because it allows them to bring their child to Germany on a US passport, which ensures the child's US citizenship and allows visa-free travel to Europe (Dücker and Hörnle, Chapter 11). However, not many German couples can afford travelling to the US. Thus, the reproflow of IPs towards countries like Ukraine or India segregates and stratifies them based on their socio-economic status. For those children born via surrogacy in Ukraine, or previously in India, the process is much more complicated. It requires either concealing the arrangement from the government officials or undergoing months of legal ordeals in a foreign country. Such varied ways of coping with the legal regime of the destination countries diversify the position and experience of the German IPs within the global reproscapes. The famous Jan Balaz case in India, which involved a German couple and their child being stuck in India for a year, revealed the ordeals faced by German citizens due to such a prohibitionist law in Germany (Mohapatra 2010;

Kotiswaran, Chapter 7). Although the landmark judgement of the German Federal Court of Justice in 2014 has now made it possible for German citizens to bring their child home legally by establishing genetic link of the child to one of the parents, the law continues to be vague and each case needs to be seen as a one-off experience (Dücker and Hörnle, Chapter 11).

The legal non-permissibility not only creates internal stratifications between Germans based on class and education background but also places them at a disadvantaged position (in the case of surrogacy) compared to IPs from other countries like Israel or Spain, who can return to their own countries relatively easily after a paternity test.

Movement from Israel

Israel is known to be a pronatalist and pro-technology country. As per the Jewish religious law, the pressure to reproduce is so strong that the Israeli state is one of the most permissive states in the world in terms of accepting and promoting all forms of ART (Shalev, Chapter 15). Interestingly, ART in Israel has developed as part of a privatised healthcare sector, which only later came to be controlled and monitored by the state to ensure the propagation of a certain acceptable kind of family (Birenbaum-Carmeli 2016). However, the state policies in Israel regarding the use of ART are not the same towards all its citizens and can, in fact, be seen as discriminatory (Rimon-Zarfaty, Chapter 17). Israel was the first country to regulate surrogacy under its Embryo Carrying Agreement Law in 1996—mainly to prevent cross-border travelling and to also regulate the birth of its citizens. The Israeli law only allows gestational surrogacy and does not allow surrogacy for gay couples, even though same-sex unions are legal in the country. By doing so, it places surrogacy within the bounds of the Israeli Jewish ideologies of the heteronormative family (Birenbaum-Carmeli 2016). However, the state grants citizenship to children born out of surrogacy abroad who are then converted to Judaism on their return (if the parents are Jewish). This leads Israeli gay men to cross borders and travel to other countries to gain access to surrogacy. Heterosexual couples who wish to distance themselves from the surrogates still travel overseas, as the Israeli surrogacy norms that have developed under the watchful eyes of the state demand close involvement between the intended mother and the surrogate. Thus, the surrogacy law and practice in Israel become a part of the state's nationalism project. By doing so, the state, first, honours the intention of all (including gay) citizens to propagate the Jewish race without encouraging limitless use of technology that could be contradictory to the state's religious ideologies. Second, the state directly feeds into stratifying reproduction by selectively pushing same-sex couples towards a financially demanding procedure while heterosexual couples find the option of surrogacy easily available within their own country. Third, it produces a specific group of potential cross-border commissioners of surrogacy as the exclusionary law prompts the cross-border reproflows of its gay IPs and contributes towards the global reproductive bioeconomies.

Since gay men are not allowed to adopt in Israel, cross-border surrogacy becomes their sole option to attain parenthood. While this gives rise to new types of family by breaking the Jewish notion of the heteronormative family, ethnographic studies have reported that the Israeli gay men do in fact apply commercial surrogacy as a means to emulate the Jewish heterosexual ideology of a nuclear family (Moreno 2016). They often see parenthood as a means to reintegrate themselves back into the Jewish society. The Israeli notion of a healthy, desirable, and Jewish biogenetic citizen appeals to these gay parents as well. Although some IPs are indifferent to biogenetic relatedness, they, however, claim to choose surrogacy due to the lack of certainty at receiving healthy children through transnational adoption (Farber 2014). In fact, unlike their heterosexual counterparts, most Israeli gay couples reportedly intend to raise their children—whether or not born out of surrogacy—in a conservative manner. They choose this as a way to fulfil the normative kinship expectations of the society by compensating for being non-normative in certain other ways (Seidman 2001). However, since cross-border surrogacy is costly and much more expensive than surrogacy in Israel (Birenbaum-Carmeli 2016), this still creates stratification between the heterosexual and homosexual couples/singles as well as amongst the homosexual couples/singles themselves as not everyone can afford to travel.

Yet, since forming families through surrogacy is popular and morally acceptable in Israel (unlike in Germany), surrogacy—even though cross-border—has become a desirable option for the Israeli gay couples as well. Their demand for cross-border surrogacy is met by Israeli surrogacy agencies with offices in both Israel and other reprohubs like India or Thailand (Moreno 2016). Unlike Germany, in Israel it is legal to advertise for surrogacy or to mediate such arrangements. These agencies thus offer counselling to prospective gay parents and support their arrangements at the destination country. They play a crucial role in persuading people to be a part of the transnational bioeconomies by connecting IPs with clinics or agencies abroad and by being their constant source of support and mediation while their pregnancies progress in a different country.

Empirical studies have shown that most Israeli men report frustration and anxiety due to being part of a long-distance surrogacy arrangement. They struggle to connect with the child-to-be that is growing in the womb of an unknown woman in another part of the world (Ziv and Freund-Escher 2015). As a result, ultrasound reports and knowledge about the sex of the baby become significant moments in their surrogacy journey. Such flow of affective information takes place with the help of the global surrogacy network and its line of agents, clinic staff, and patient coordinators. All these are ensured and coordinated by specialised Israeli surrogacy agencies like *Tammuz*.[3] Thus, while the Israeli state, through its exclusionary legal policies, create a subset of its population as potential commissioners of cross-border surrogacy, through its privatised healthcare policies, it paves the way for surrogacy agencies and their cross-border business.

Still, the surrogacy experiences of Israeli gay men are often marked with insecurities due to the changing legal landscape of commercial surrogacy around the world. The sudden and unexpected ban on same-sex and single parents' surrogacy in India in 2012 (Kotiswaran, Chapter 7) left several Israeli commissioning parents in a legal limbo. They were stuck in India for months, trying to fight a legal ordeal in an unknown country by either waiting for their child to be born amidst a legal disapproval, or by fighting to get their child back to Israel, or to get their embryos shipped to another surrogacy destination (Mishra 2010). While this ban diverted the global reproflows and took Israeli gay couples first to Thailand (Fiske 2014), then Nepal, and finally Cambodia, the same tale of untimely ban was repeated in each of these destinations.

Nevertheless, even though the Israeli state prevents gay parents from accessing surrogacy on its own soil, it still made an exception when several children born through the biogenetic material of Israeli citizens were stuck in Nepal. The earthquake, which struck Nepal in 2015, exposed the surrogacy network of Israeli gay couples in Nepal that was run by Indian doctors and Indian surrogates (see Mitra, under review). The earthquake led the Israeli state to fly out 29 newborn 'Israeli' surrogate babies along with their commissioning parents from Nepal. The state even offered to fly out all those Indian surrogates who were pregnant with 'Israeli babies' (Kamin 2015). While this event was applauded as an instance of solidarity by the Israeli state towards its citizen irrespective of their sexuality, the fact remains that every time Israeli gay couples are stuck abroad due to legal or natural emergence, the overtly discriminatory policy of the state, that urges them to use cross-border services, becomes evident. Again, cross-border surrogacy is often marked by other forms of disruptive events leading to custodial disputes over the child. News of unintended gamete mixing or low-quality controls in surrogacy clinics hit the headlines when an Israeli same-sex couple was denied entry to Israel with their child after their child's paternity test failed. The test indicated that the surrogate girl child whom they 'picked-up' from Nepal was mistakenly not born out of the semen samples provided by them (Surkes 2017). Thus, cross-border surrogacy, on the one hand, allows Israeli gay couples to find their place within the heterosexual realm of the Jewish family, on the other hand, it also makes this particular set of its citizens precarious and highly disadvantaged as compared to their heterosexual counterparts, who enjoy the privilege of being a part of a highly regulated, morally regarded, and well-organised surrogacy arrangement at home.

Unlike surrogacy, egg donation was not legally allowed in Israel until 2010 (Birenbaum-Carmeli 2016). IVF in Israel was only allowed with the egg of the IP's mother in order to maintain the unity of the Jewish womb and egg in the creation of a Jewish citizen. This often led couples to travel to countries like Romania in search of egg donors. In 2010, the new law legalised egg donation in order to prevent people from crossing borders. However, it still prohibited cross-religious egg donation. By doing so, the state attempted to harden the borders of the Jewish state by trying to keep its citizens within. Needless to say, the shortage of eggs continue to take Israelis abroad for a procedure which is much more expensive than those at home. Again, transnational egg donation also takes place through a process that Nahman describes as 'reverse traffic' (2011). This means that sperm samples are flown to countries like Romania and fertilised embryos that are created using donor eggs are sent back to Israel (Nahman, Chapter 16). In this cross-border arrangement, the price of a Jewish egg is much more expensive than a non-Jewish egg. Such hierarchies of the pricing of eggs enhance the weight of the genetic material as part of the Jewish Israeli identity (Nahman 2006) and creates stratification amongst the procurers of jewish and non-jewish egg recipients.

Although some Israelis utilise eggs from non-Jewish donors as well and thereby break the Jewish notion of the union of the womb and egg and instead place the main focus on the womb, they often try and ensure that their embryos are made of the 'right mixture' (Nahman 2006, p. 206). So when looking for non-Jewish egg donation, Israelis have been reported to opt for Romanian donors, but to strictly reject Palestinian donors. As a result, the process of cross-border egg donation continues to carry the ideologies of the Israeli nation-building project. Importantly, whether at home or abroad, the privatised fertility clinics work to assist Israeli couples to produce an Israeli child and hence encourage them to participate in the imaginings of national bodies through a privatised, globalised healthcare platform (Nahman 2010). The privatised nature of the practice that makes the act of procuring eggs a part of the global neoliberal economy along with the procurement of egg from abroad only shows how the national body is made transnationally (Nahman 2010). While transnational egg donation in a way bridges inequalities at home by allowing people to choose non-Jewish donors or to procure eggs at all, it creates new forms of economic, racial, gendered, and class-based inequalities.

Attracting the Cross-Border Reproflows: The Reprohub of India

ART in India gained its popularity under the umbrella of medical tourism (Sarojini et al. 2011). Although IVF developed in India in 1978 with the birth of the world's second IVF 'Baby Durga' just two months after the birth of Louise Brown (Smerdon 2008), India only received global attention for its advances in reproductive medicine after the birth of its first transnational surrogate child through IVF in 2004. This incident soon converted India into a popular destination for commercial surrogacy or a reprohub and attracted commissioners from all around the world, including Germany and Israel. Therefore, commercial surrogacy dominates the ART discourse in India and often overshadows the need for attention towards other forms of ART, such as donor-assisted IVF. Both surrogacy and gamete donation are legal in India, in the sense that they are not forbidden and hence declared legal by both the Supreme Court and the Law Commission of India (Kotiswaran, Chapter 7). However, both practices are not legally regulated (Smerdon 2008). The various versions of the Draft ART Bills 2008, 2010, 2013, and 2015 (Kotiswaran, Chapter 7; Reddy et al., Chapter 8) attempted to regulate the use of every form of ART, but the sudden surfacing of the Draft Surrogacy Bill 2016 (Kotiswaran, Chapter 7; Reddy et al., Chapter 8) once again leaves the future and the rights of the gamete donors in a legal grey zone. Nonetheless, since cross-border surrogacy often requires donor gametes, especially eggs,[4] and since most surrogates in India are erstwhile egg donors (Mitra 2017), it is important to discuss the two together.

The reason for India's popularity in commercial surrogacy rested upon the availability of the latest world-class medical technologies and infrastructures that attracted private and corporate investment in health care (Qadeer 2010). Other factors included low costs, the availability of English-speaking doctors, and, most importantly, the easy availability of potential surrogates without a history of drugs or alcohol abuse (Kirby 2014). Another significant factor which attracted IPs to India was its clinics' ability to ensure that the surrogate has no legal right over the foetus or the child-to-be and does not have the flexibility to change their mind midway through the arrangement. These conditions, under which the surrogates labour in India, have often been termed exploitative because it further their already disadvantaged position in society (Saravanan 2015). Nonetheless, ethnographic studies have reported that the surrogates considered surrogacy to be 'labour' and find it life-reaffirming, without misreading the exploitation involved in such a labour (Rudrappa 2015). They are vocal about their financial motivations behind choosing surrogacy and derive a moral worth from their labour by considering it a gift—which until recently made them easy recruits for the transnational surrogacy agencies.

The surrogacy contracts in India are known to allow limited rights for the surrogates and instead to ensure their complete submission to the authority of the doctors and the IPs. The surrogates have no scope to negotiate the clauses of the contract or the financial compensation offered to them (Sama Women's Health 2012). In fact, by signing the contracts (which none of the surrogates can read as they are written in English (Sama Women's Health 2012), the surrogates and their husbands give up their right and legal claims over the 'child-to-be'. As a result, the surrogates do not have a say over or knowledge about the number of embryos transferred to their uterus (Tanderup et al. 2015; Mitra and Schicktanz 2016) and do not have the right to decide on a foetal reduction or abortion (Mitra 2017). Such conditions stem from lack of proper communication regarding the uncertainties and risks involved during surrogacy (Mitra 2017). Limited knowledge of the surrogates and the prioritisation of IPs' success over surrogates' rights by the clinics became India's defining feature within the global reproscape. As a result, the doctors could decide on the line of technological interventions without much effective opposition from the surrogates. The surety that the surrogates and their husbands would certainly sign off their rights and custody over the child attracted IPs and attracted the reproflows to India. This sense of assurance and control offered by the Indian surrogacy market brought IPs from all over the world to India, be from countries like Israel, where children born through cross-border surrogacy are legally recognised, or from Germany, where children born through cross-border surrogacy are not legally recognised.

A key reason why the surrogates in India worked with limited knowledge is their lack of contact with the IPs (Mitra and Schicktanz 2016). The absence of the IPs from the clinical spaces, the linguistic and often cultural barriers between the surrogates and their foreign commissioner placed surrogates in an agent-mediated relationship. In fact, the local agents and doctors counselled the surrogates to be good 'mother-workers' (Pande 2010). They promoted the idea that the surrogates should take on the role of a responsible worker and of a good mother, while at the same time distancing themselves from the child-to-be. In order to ensure a smooth separation and to contribute to the good reputation of the clinics and agencies, the clinics also discouraged their IPs—foreigners or Indians—to be in regular contact with their surrogates. They often created horror narratives regarding the latter taking advantage of the relative social affluence of the former (Mitra 2017). It is on such class-centric and racialised biases that the market of commercial surrogacy in India rested. It created stratifications by placing the surrogates in a very disadvantaged position in comparison to their counterparts in other countries, e.g. in the US. Furthermore, in order to save the IPs from the fear of a complicated cross-cultural kinship relation and to make the services of the Indian surrogates attractive, the agencies and doctors attempted to invisibilise the bodies and backgrounds of the Indian surrogates (Majumdar 2017). Therefore, transnational surrogacy in India flourished by distancing the surrogates and the IPs and by serving the interests of each party in isolation from the other (Majumdar 2017; Mitra 2017).

All in all, the role of the agencies as mediators was extremely important during commercial surrogacy in India. Both the surrogates and the IPs depended on them. They counselled the surrogates, mediated between the surrogates and the IPs as well as the surrogates and the doctors, monitored and supervised their pregnancies and prepared them to give up the baby. They helped IPs get access to clinics and doctors in a foreign land by selecting their surrogates and egg donors, if needed, by flying their embryo or sperm samples from either their home countries or another reproductive hub, drafting their contract as well as assisting them in attaining a clearance from the foreign office, acquiring passport for the newborn and also an exit visa. Thus, despite the fact that surrogacy in India is legally unregulated, the business rested upon a meticulous process of internal regulations and cooperation between a chain of actors who connected the flow of hopes, desires, and biogenetic materials in the global reproscapes.

Nonetheless, surrogacy arrangements do not always guarantee a successful or planned outcome. There have been several cases of custodial disputes due to conflicting laws between host and destination countries (Kotiswaran, Chapter 7). International criticism over the dangers of cross-border surrogacy and the unethical nature of the practice in India (Sándor, Chapter 3) eventually led the Indian government to close down cross-border surrogacy for foreigners (Kotiswaran, Chapter 7). However, it is difficult to arrest the reproflows of the cross-border bio-economies and closing down routes has only led to the opening of new routes. The banning of same-sex couples and singles from accessing surrogacy in India only pushed its clinics to set up parallel bases of operation in Thailand, then Nepal, and finally in Cambodia and to offer their services at these new hubs through a 'fly by the night' mechanism (Mitra, under review). The last two destinations gave rise to the phenomena of 'travelling Indian surrogates' (Mitra, under review). The embryos were transferred in India and the surrogates were then sent to Nepal or Cambodia to legally carry out the pregnancies and deliver the child. Thus, while legal restrictions and the ban on transnational surrogacy for foreigners averted the legal disputes and disruptive journeys associated with cross-border surrogacy to India, the new routes and flows of actors and networks brought along its own set of risks for the travelling surrogates by fragmenting the site of their labour and making them much more precarious (Mitra, under review).

Moreover, most surrogates in India start off as egg donors and some egg donors also become surrogacy agents (Mitra 2017). The fact that many egg donors become surrogates after having undergone multiple cycles of donation and some even go back to donation after surrogacy raises serious questions about the health risks for these women. The clause of anonymous egg donation often leaves IPs unaware of the 'source' of their child's biogenetic material. The clinics and agencies fulfil the need for Caucasian, Asian, or Black donors by either importing already fertilised embryos with the desired phenotypic egg from a repro-hub like South Africa or recruiting travelling egg donors from countries like South Africa, who undergo hormonal cycles and donate eggs in India under no legal guidelines or protection (Smerdon 2008). In the case of a medical injury, these travelling donors are left on their own to recover. Therefore, while the transnational reproductive networks connect the Indian reproductive market with the hopes of actors from all over the world, it also produces a group of precarious egg donors. Thus, the reproflows to India are marked by extreme inequalities and precarities that stratify each set of actors and their rights. In India, egg donation is strictly anonymous and in the name of anonymity, the clinics refuse to hand over medical records or proofs of donation to the donors themselves IPs or the future children (Majumdar 2017). This makes the transnational kinship relations or the urge to reconnect in future seemingly unattainable within the global reproscapes.

Conclusion

I discussed three completely different legal, social, and moral contexts of ART in Germany, Israel and India which, in their own typical ways, shape the global reproflows during cross-border surrogacy and egg donation. Cross-border reproductive travel is known to create stratified reproduction between populations and between countries due to sexualised, racialised, and economic inequalities. The cross-border reproflows from Germany give rise to inequalities amongst its IPs based on class, while it stratifies the German IPs in comparison with the Israeli IPs on the basis of their nationality due to the restrictive German laws. The reproflows from Israel, on the other hand, result from unequal access on the grounds of its citizens' sexualities and hence make their relative disadvantaged position in comparison to the heterosexual Israeli IPs evident. The reproflows to India are marked by inequalities between the surrogates and the IPs on the grounds of both race and class. A closer look at the bioeconomies of reproductive travel and the services that are on offer or have developed to suit the specific needs of a set of people reflects their local sociocultural norms and national legal agendas. In the process, the global reproflows produce new forms of inequalities and stratifications amongst both the workers in the fertility industry and the commissioners.

Notes

1. The ESchG is a German criminal code for regulating IVF. Its aim is, above all, to protect human embryos and stem cells against exploitative interference via misuse of IVF techniques.
2. Surrogacy through IVF becomes illegal in Germany since it breaks the unity of motherhood. Since there is no law directly opposing surrogacy,

traditional surrogacy, which does not involve IVF and can be done simply through the process of assisted insemination directly in the woman who then conceives a pregnancy and carries the child, is allowed.
3. *Tammuz* is a surrogacy agency based in Israel with its centres at various surrogacy destination across the world. Website [online]. Available at http://www.tammuz.com/en/.
4. Women opting for surrogacy often lose their capacity to generate fertility eggs due to age or often fear to consequence of undertaking the physical stress of further oocyte retrieval after having undergone several failed cycles of IVF.

Bibliography

Appadurai, A. (1996). *Modernity at large: Cultural dimensions of globalization* (Vol. 1). Minneapolis: University of Minnesota Press.

Bergmann, S. (2011). Reproductive agency and projects: Germans searching for egg donation in Spain and the Czech Republic. *Reproductive Biomedicine Online, 23*(5), 600–608.

Birenbaum-Carmeli, D. (2016). Thirty-five years of assisted reproductive technologies in Israel. *Reproductive Biomedicine & Society Online, 2,* 16–23.

Cohen, L. (2005). Operability, bioavailability, and exception. In A. Ong & S. J. Collier (Eds.), *Global assemblages: Technology, politics, and ethics as anthropological problems* (pp. 79–90). Hoboken, NJ: Blackwell.

Farber, A. R. (2014). *Surrogacy and fatherhood among homosexual couples in Israel.* MA, University of Haifa, Haifa, Israel.

Fiske, G. (2014). 65 surrogate babies born to Israeli gay couples stuck in Thailand. *The Times of Israel* [online]. Available at https://www.timesofisrael.com/65-surrogate-babies-born-to-israeli-gay-couples-stuck-in-thailand/. Accessed 15 Apr 2018.

Inhorn, M. (2010). Assisted motherhood in global Dubai: Reproductive tourists and their helpers. In W. Wendy Chavkin & M. J. Jane Maree Maher (Eds.), *The globalization of motherhood: Deconstructions and reconstructions of biology and care* (pp. 180–202). New York: Routledge.

Inhorn, M. C. (2016). Cosmopolitan conceptions in global Dubai? The emiratization of IVF and its consequences. *Reproductive Biomedicine & Society Online, 2,* 24–31.

Inhorn, M. C., & Patrizio, P. (2015). Infertility around the globe: New thinking on gender, reproductive technologies and global movements in the 21st century. *Human Reproduction Update, 21*(4), 411–426.
Kamin, D. (2015). Israel evacuates surrogate babies from Nepal but leaves the mothers behind. *Time* [online]. Available at http://time.com/3838319/israel-nepal-surrogates/. Accessed 1 Oct 2017.
Kirby, J. (2014). Transnational gestational surrogacy: Does it exploitative? *The American Journal of Bioethics, 14*(5), 24–32.
Mahapatra, D. (2010). German surrogate twins to go home. *The Times of India* [online]. Available at https://timesofindia.indiatimes.com/india/German-surrogate-twins-to-go-home/articleshow/5978925.cms. Accessed 9 Jan 2017.
Majumdar, A. (2017). *Transnational commercial surrogacy and the (un)making of kin in India*. Oxford: Oxford University Press.
Mishra, H. (2010). Israeli gay couple to take surrogate twins home. *The Indian Express* [online]. Available at http://archive.indianexpress.com/news/israeli-gay-couple-to-take-surrogate-twins-home/624650/. Accessed 11 Feb 2016.
Mitra, S. (2017). *Disruptive embodiments: An ethnography of risks and failures during commercial surrogacy in India*. Ph.D., Social Science Faculty, University of Göttingen, Göttingen, Germany.
Mitra, S. (Under review). Mobile bodies, mobile markets: The shifting base of commercial surrogacy from India. Special Issue. *Body and Society*.
Mitra, S., & Schicktanz, S. (2016). Failed surrogate conceptions: Social and ethical aspects of preconception disruptions during commercial surrogacy in India. *Philosophy, Ethics, and Humanities in Medicine, 11*(9), 1–6.
Moreno, A. (2016). *Crossing borders: Remaking gay fatherhood in the global market*. Ph.D., Sociology Department, University of Manchester, Manchester, UK.
Nahman, M. (2006). Materializing Israeliness: Difference and mixture in transnational ova donation. *Science as Culture, 15*(3), 199–213.
Nahman, M. (2010). "Embryos are our baby": Abridging hope, body and nation in transnational ova donation. In J. Edwards, P. Harvey, & P. Wade (Eds.), *Technologized images, technologized bodies* (pp. 185–210). Oxford: Berghahn Books.
Nahman, M. (2011). Reverse traffic: Intersecting inequalities in human egg donation. *Reproductive Biomedicine Online, 23*(5), 626–633.
Nahman, M. R. (2016). Reproductive tourism: Through the anthropological 'reproscope'. *Annual Review of Anthropology, 45*, 417–432.
Pande, A. (2010). Commercial surrogacy in India: Manufacturing a perfect mother-worker. *Signs: Journal of Women in Culture and Society, 35*(4), 969–992.

Qadeer, I. (2010). Benefits and threats of international trade in health: A case of surrogacy in India. *Global Social Policy, 10*(3), 303–305.
Raz, A. E., & Schicktanz, S. (2016). *Comparative empirical bioethics: Dilemmas of genetic testing and euthanasia in Israel and Germany*. Basel: Springer.
Rudrappa, S. (2015). *Discounted life: The price of global surrogacy in India*. New York: NYU Press.
Sama Women's Health. (2012). Birthing a market: A study on commercial surrogacy [online]. Available at http://www.communityhealth.in/~commun26/wiki/images/e/e8/Sama_Birthing_A_Market.pdf. Accessed 9 Dec 2017.
Saravanan, S. (2015). Global justice, capabilities approach and commercial surrogacy in India. *Medicine, Health Care and Philosophy, 18*(3), 295–307.
Sarojini, N., Marwah, V., & Shenoi, A. (2011). Globalisation of birth markets: A case study of assisted reproductive technologies in India. *Globalization and health, 7*(1), 27.
Seidman, S. (2001). From identity to queer politics: Shifts in normative heterosexuality and the meaning of citizenship. *Citizenship Studies, 5*(3), 321–328.
Shenfield, F., De Mouzon, J., Pennings, G., Ferraretti, A. P., Nyboe Andersen, A., De Wert, G., et al. (2010). Cross border reproductive care in six European countries. *Human Reproduction, 25*(6), 1361–1368.
Smerdon, U. R. (2008). Crossing bodies, crossing borders: International surrogacy between the United States and India. *Cumberland Law Review, 39,* 15.
Stöbel-Richter, Y., Beutel, M. E., Finck, C., & Brähler, E. (2005). The 'wish to have a child', childlessness and infertility in Germany. *Human Reproduction, 20*(10), 2850–2857.
Surkes, S. (2017). Israeli gay couple discovers surrogate baby is not their child. *The Times of Israel* [online]. Available at https://www.timesofisrael.com/gay-couple-discovers-surrogate-baby-is-not-their-child/. Accessed 11 Feb 2016.
Tanderup, M., Reddy, S., Patel, T., & Nielsen, B. B. (2015). Reproductive ethics in commercial surrogacy: Decision-making in IVF clinics in New Delhi, India. *Journal of Bioethical Inquiry, 12*(3), 491–501.
Ziv, I., & Freund-Eschar, Y. (2015). The pregnancy experience of gay couples expecting a child through overseas surrogacy. *The Family Journal, 23*(2), 158–166.

6

Beyond Relativism: Comparing the Practice and Norms of Surrogacy in India, Israel, and Germany

Silke Schicktanz

Introduction

More than a decade ago, I interviewed the German film-maker and journalist Frauke Sandig about her movie 'Frozen Angels' (D/US; Piffl Medien 2005) in Berlin. This was the first German movie dealing with egg donation and surrogacy as it was practised back then in the US, especially in California. From my perspective, it is an outstanding documentary based on its artistic framing and the immediacy of its moral messages. The documentary promulgates a strong dystopic message by criticising the increasing liberalisation and capitalisation of human reproductive behaviour. It uses visual references to Ridley Scott's original movie 'Blade Runner' (1982) and combines these with interviews of visionary researchers and businessmen to underline the extremes of urbanity, body culture, and the techno-hype in Southern California. Furthermore, emotional portraits of desperate intended parents (IPs), enthusiastic egg

S. Schicktanz (✉)
University Medical Center Göttingen, Göttingen, Germany
e-mail: silke.schicktanz@medizin.uni-goettingen.de

S. Mitra et al. (eds.), *Cross-Cultural Comparisons on Surrogacy and Egg Donation*,
https://doi.org/10.1007/978-3-319-78670-4_6

donors, tight-lipped or sometimes disappointed surrogates give the movie an affective touch. Interviewing Frauke Sandig triggered two leading research questions which still absorb a large amount of energy within my scholarly work: 'How can we understand and describe the social practices of modern medicine in a particular cultural context in an appropriate way?' And: 'What can, in terms of moral thinking and ethical assessment, be learned from looking beyond one's own (cultural) nose?'

In the following, I try to answer these two questions by, for the first time, comparing comprehensive studies from law, medical sociology, anthropology, and ethics about surrogacy from three different national contexts: Germany, Israel, and India. They provide an excellent framework for a reflective analytical juxtaposition as they represent three extremes: India is an example of a low-regulated, commercialised context; Germany is an example of an extremely restrictive, banning regulation; and Israel is an example of a highly regulated, but nonetheless permissive context (Mitra et al., Chapter 1). Moreover, all three contexts are not hermetically closed, but are interlinked by cross-national surrogacy activities (Mitra, Chapter 5). Also, their regulations are not set in stone—many regulative changes are currently being discussed or already in the making. As this volume illustrates, the contextual–national analysis of modern technology needs a critical reflection with regard to its embedding in politics, economy, and morality. However, what makes the analysis even more complex is the dialectic relationship between these contexts: the globalisation of medicine, which indicates the 'translocation' of ideas (such as the idea to use a third party's body for one's own reproduction), techniques (incl. in vitro-fertilisation (IVF), egg extraction, medical care standards), and even the parties involved (travelling IPs, medical staff, companies), produces connectivity and relatedness. This translocation creates a dialectic connectivity in terms of what is seen as morally acceptable as well as medically possible, and in terms of how national ethico-legal discourses are impacted by transnational surrogacy and vice versa.

Here, I am aiming at an ethical reflection of the 'technoscapes' in artificial reproductive technologies (ART) (also Inhorn 2011) and their global flows based on empirical insights from ethnographical studies as well as legal and ethical overviews of single contexts. 'Technoscape'

(Appadurai 1996) means the global configuration of technologies including the economical and legal regimes in their connectivity as well as in their disjunctions—distinguished from traditional geo-political landscapes. According to Appadurai, globalisation should be analysed in its building of new 'scapes' and the challenges related to the dissolution of traditional geographical territories, such as the national state. As this volume indicates, 'technoscapes' in ART are both, territorial and transnational. They are therefore an excellent example of how the globalisation of technology, on the one hand, transcends national boundaries by creating flows of people, biomaterial, and ideas, but is also able to reinforce national identity, local governance, and territorial boundaries, on the other.

My comparative ethical analysis focuses on three lines of thought: first, I identify similarities in ethical concerns that are visible despite the different regimes and practices; I will call these the 'global flow of ethical concerns'. Second, I uncover the concept of naturalised parenthood that still works as a strong medical-cultural imperative across regulative differences. Third, I argue in an ethical-reflective way why we need a transnational discourse that goes beyond normative and anthropological relativism to answer local limitations.

The Global Flow of Ethical Concerns

Ethical concerns are arguments or positions put forward within a public, legal, or ethical discourse in order to address critique, worries, and limitations of a particular issue. In modern ethics, giving priority to concerns (or conflicts) is strongly connected to the modern liberal ideal of the freedom and self-fulfilment of the individual. Only if there is evidence or strong plausibility of serious harm or disadvantages for others (individuals or collectives) can we justify restrictions or limitations of personal freedom. In our case, this indicates that only if ART or surrogacy harms others (directly or indirectly, physically, mentally, or in their self-fulfilment) do we have good reasons to limit individuals access to ART or to restrict the ways in which it is applied (also Beier, Chapter 12). Consequently, I argue for the need for an extensive and serious assessment of concerns related to modern medicine which I see as an important task for bioethics.

Comparing the provided ethnographic studies on surrogates and egg donation in India and Israel (Teman, Chapter 4; Nahman, Chapter 16), on IPs in Germany (König, Chapter 13; Teschlade, Chapter 14), and on expert and legal discourses in India (Kotiswaran, Chapter 7; Reddy et al., Chapter 8; Marwah and Nadimpally, Chapter 10), Israel (Shalev, Chapter 15; Rimon-Zarfaty, Chapter 17), and Germany (Dücker and Hörnle, Chapter 11; Beier, Chapter 12), three main concerns have been reiterated across the different contexts. These consist of the (potential) exploitation of the surrogate, the disregard of the child's best interest, and the unethical motives of (some) IPs. In the following sections, I will discuss these three themes and outline implications for future discussions.

The 'potential exploitation of surrogates' is a concern that has been raised in various nuances and variations. There are two main differences between how the concern is expressed, also depending on the normative subsets of what 'exploitation' means. Exploitation provides both consequentialist (result-oriented) as well as deontological (process and actor-oriented) connotations. Both types can occur separately. I am here referring to Alan Wertheimer's helpful definition of exploitation (Orth and Schicktanz 2017). Wertheimer's assumption (1996, p. 16) starts with 'the claim that A exploits B when A takes unfair advantage of B'. He then explains that taking unfair advantage of a transaction is indicated either by the result or the process itself. The 'result' of a transaction can be unfair either if A profits and B is harmed, or if A's profit is a lot higher than B's. The 'process' of a transaction can lead to taking unfair advantage if B is coerced or manipulated by A. The concept of 'result exploitation' is often applied within the context of transnational surrogacy when the economical gap between the surrogate, on the one side, and the IPs (as well as the involved medical industry), on the other side, seems to be too extreme, i.e. the surrogate earns much less money in comparison with what the IPs could afford or what the involved professionals (medicine, surrogacy agencies) earn. The proposal for the current shift from commercial to altruistic, intra-family surrogacy in India (Kotiswaran, Chapter 7; Reddy et al., Chapter 8) is seen as a way to avoid such forms of 'result exploitation'. Other countries (such as the US and Israel) have rather strict policies ensuring a sufficient financial compensation for the surrogate (Teman 2010). These strategies minimise the problem of financial exploitation by ensuring fair standards regarding the resulting outcome. 'Result exploitation' takes place in many areas of global industry (e.g. textile industry, food production), and global consumers tend to suppress their concerns about exploitation because of their personal benefit to purchase cheap products as well as the large spatial distance to those places where exploitation occurs. Even though there is a rising awareness to these forms of injustice amongst consumers, it remains a tricky question whether those delivering the work in question need to be aware of their exploitation, for example by being dissatisfied with the financial outcome themselves. The majority of surrogates interviewed conducted by Lamba and Jadva (see Chapter 9) claimed to be satisfied with their financial outcome; however, the larger picture of scholarly work provides a more differentiated view of this issue (Saravanan 2015; Mitra 2017). There is, for example, evidence that in cases of miscarriage or abortions, Indian surrogates did not receive the (full) payment, which can be seen as an instance of result exploitation (Mitra 2017; Teman, Chapter 4). Avoiding types of result exploitation that occur due to the nature of the economy of the social market requires a strong state as well as efficient mechanisms for regulation and enforcement (Teman, Chapter 4). This condition is still often overlooked by medical ethicists, lawyers, and psychologists.

The issue of 'process exploitation' is even more complex: as defined above, exploitation during a process occurs when one participant in the transaction is deprived of important information, is coerced or manipulated. In this case, exploitation is inherently linked to the freedom of decision-making and voluntariness, which are both basic conditions for the self-determination and autonomy of the involved party. Whether financial incentives/payment should automatically be seen as coercion (or corruption) of the free will is a crucial, but also ongoing battlefield for (post)modern ideologies (e.g. rational choice vs. Marxist psychoanalysis). Empirical research in that field is complex and ranges from neuroeconomics and decision psychology to happiness studies. While 'common' cases of labour payments are seen as fair (Marwah and Nadimpally, Chapter 10), it remains theoretically and practically contested through history and across cultures whether the use and exchange of body parts from third parties can and should be perceived as 'just' labour and as an unproblematic exchange of commodities. Indeed, for some, such body 'shopping' conflicts with our conceptions of identity, self-determination, and social recognition, because the human body has various meanings (Schicktanz 2007). Whether there is already a siege of biocapitalism over 'pagan' views of the holiness of the human body must be left unsolved. Organ trading is one practice which is unanimously rejected by the majority of experts, policy documents, and the majority of lay persons (Hoeyer et al. 2013). However, payments for donating blood, eggs, sperm, for giving one's dead body to science, or for participating in research are quite common, as is other bodily labour (such as sex work). Critical reflection acknowledges that this is an area of ongoing dispute which needs to take into account several aspects: cultural pragmatism, philosophical–anthropological as well as normative consideration about the body and about how we determine agency and autonomy (Schicktanz 2007).

To identify 'process exploitation' we need detailed, practical knowledge about the exact process of a particular transaction, while the analysis of 'result exploitation' uses manifest and evidence-based numbers of financial outcomes. For this reason, whenever exploitation is suspected by the media or in popular discourses, they are often referring to the latter. Stereotypical ideas of 'poor' lower class, lower caste women as surrogates, and 'rich' middle-upper class IPs fuel expectations and sensitivities to such exploitation (see also the critique by Kotiswaran, Chapter 7). However, empirical research is needed to provide us with detailed insights that go beyond clichés. Ethnographic and practical legal studies of real consultations, deals, and transactions in the context of India (Sandor, Chapter 3, Teman, Chapter 4; Mitra 2017) have, however, revealed that the basic conditions for good (in the sense of ethically acceptable) decision-making were indeed rarely found in India. The infringement on a surrogate's autonomy by providing her with incomplete or inadequate information (her literacy or education might be limited) about the process of surrogacy, medical risks, and her rights to withdraw is then directly linked to 'process exploitation'. In Israel, strong regulation has led to profound and robust procedures to ensure high information levels for all parties (Teman, Chapter 4). For this reason, avoiding 'process exploitation' calls for strict requirements for the decision-making process, including balanced, written, and oral information, time for consideration, and the opportunity for personal questions.

For 'result exploitation', the normative responsibility lies mainly on the IPs's side, additional to the state's side that has to ensure and enforce social market mechanisms—similar to the 'responsible' consumer situation. In the case of 'process exploitation', the responsibility lies mainly on the medical professional's and the state's side. Medical standards, based on transnational conventions and enforced by legal and internal quality controls, are needed to ensure fair procedures. Failures and irresponsibility of medical professionals, for example misinformation or manipulation, should always have legal consequences. 'Soft law', as a framework built upon various ethical guidelines, is often not sufficient in this regard. In addition to the current focus on the rights of patients (here, mainly the IPs), normative conditions for ensuring the autonomy of third 'service' parties, such as gamete donors or surrogates, also need to be specified. They are not patients, but are still also embedded in a caring relationship with the medical professionals; surrogates or egg donors must receive particular care and attention regarding their educational, social, or psychological background. As they are often deprived of access to public education, social security, information, or to legal enforcement, they require more than the minimum standards for ensuring their ability to make self-determined decisions.

Another main concern refers to 'violations or disregards of the best interest of the child (BIC)' in the context of surrogacy. This argument is very present in legal discussions (Sándor, Chapter 3; Dücker and Hörnle, Chapter 11; Beier, Chapter 12) where it is taken up from long-standing debates over adoption and custody battles (Shalev, Chapter 15). Traditional understanding of BIC tends to give biological parenthood (here, the surrogate) considerable (or even paramount) weight regarding the child's interest over the IPs. More recent court decisions in favour of IPs's right of custody for the surrogacy child, however, indicate that here, 'nurture' (in the sense of care) overrides 'nature' (here, in the sense of biology) (Sándor, Chapter 3). Also, arguments referring to the psychological well-being and good development of children provided by some psychological studies in the context of adoption, single parenthood, gamete donation, or surrogacy (Lamba and Jadva, Chapter 9; Golombok et al. 2006; Jadva and Imrie 2013) indicate that nurturing relationships are of more importance than 'just biology'. Such a nurturing view can, however, conflict with the way parenthood is still naturalised by some IPs or the medical profession, as I will argue below.

What, from an ethical point of view, is implied by BIC is not easy to answer. First, who is meant by 'child' (Archard 2016): Is it the individual child or is it the group of children ('surrogacy children')? Psychological studies tend to summarise and average their empirical data across a group. Therefore, they do not allow statements about single cases, except when an individual case assessment is undertaken. Legal regulations and court decisions regarding custody and care rights for surrogacy children (e.g. Germany, Israel, or the European Human Rights Court) focus instead on single case assessments (Sándor, Chapter 3). This legal practice counters traditional, stereotypical assumptions—such as that a mother might always be better in taking care of the child and having custody than a father. From an ethical and legal justice perspective, individual assessment appears to be more adequate than referring to statistical reports of psychological studies. Therefore, caution is required when the social practice of surrogacy is justified on the empirical basis of statistical data regarding children's well-being, as some scholars tend to do. Second, what is meant by 'best': Is it the interest paramount over all other interests, or do we assume it is the child's interest in basic living conditions, it's bodily and material welfare? And what if we assume that there are several, equal interests that a child has that need to be balanced—how do we know what is really the best for this particular child? Third, the term 'interest' also requires our normative attention: while adults and grown-up children can (and should) be asked about their own interests, we cannot automatically assume that adults (here, as lawyers, ethicists, or parents) have an objective insight into what is in the best interest of a particular child. Regarding small children and newborns, this remains a 'hypothetical choice' interpretation: what *we* assume this particular child would choose (Archard and Skivenes 2009). In some areas of bioethics such hypothetical choices are unavoidable, e.g., when decisions need to be made about persons in an unconscious state. However, in many cases we can assembly evidence from biographical events or former statements. With newborn children, on the other hand, while this is an irresolvable problem, it should not tempt us to ignore their right for moral respect (Wiesemann 2016). One would assume that establishing specific conditions is therefore in the best interest of a child, allowing children to decide for themselves as they mature. Then, practices of secrecy and lack of transparency in surrogacy, which are often mandated by the IPs (Reddy et al., Chapter 8; Beier, Chapter 12; König, Chapter 13), become problematic, because they deprive the children from their right of self-determination.

As there is a large cultural variety as well as societal pluralism in regard to what is good or necessary for each human life, the answer to what BIC is remains very difficult. The Israeli model of protecting future surrogacy children from the stigma of being a 'mamzer' (being classified as a 'bastard'), or from having conflicts regarding their religious identity (in traditional Judaism it is the gestational mother's religion that determines the child's religion) by regulating who is allowed to be a surrogate (unmarried and same religion as IPs), is a good example of a culture-specific interpretation of BIC (Rimon-Zarfaty, Chapter 17). However, it remains questionable whether such an approach does not end up reproducing cultural-religious interests rather than ensuring BIC. The international debate over the best interest of a child born via surrogacy focuses mainly on the average satisfaction with the psychological or cognitive development of such children. This empirical evidence is undoubtedly needed and more studies with different, especially culturally adopted instruments are necessary to include the long-term perspectives of these children. At the same time, the future debate would benefit from overcoming too simplified assumptions that all children are 'wanted' by their IPs and are 'voluntarily handed over' by the surrogates, and that therefore everything will (automatically) be in their 'best interest'. Elly Teman's comparison (Chapter 4) of the restrictive screening program for IPs and surrogates in Israel to the 'laissez faire' model in India indicates that there are also some good arguments in favour of such screening methods in the sake of BIC. Such screening practice may help to avoid cases of abandoned or ultimately unwanted children born via surrogacy. Apart from all doubts in regard to what BIC means, we can assume that it is not in the best interest of any child to end up as unwanted because of IPs's particular expectations, psychological problems, or relationship break-ups. While maximising the welfare of a child might be a high normative claim, it should be paramount to ensure a minimal standard for those children.

A third concern mentioned by several authors in this volume refers to doubtful motives for surrogacy that some IPs may have. This critique particularly arises when the IPs request for surrogacy is not indicated medically, but instead is driven by vanity or power (Sándor, Chapter 3; Reddy et al., Chapter 8). There are cases of rich or celebrity IPs who have no medical reason for surrogacy, but want to avoid the physical discomfort or aesthetic impairments resulting from pregnancy and birth. The above-mentioned movie 'Frozen Angels' already problematised 'wrong' motives by drawing a connection between the selection of gamete donors and surrogates on the one side and ethnic or aesthetic criteria (Teschlade, Chapter 14) as well as the racist presumptions of some white, American IPs on the other side. A different problematic motive is raised by cases such as that of *Baby Gammy* (Sándor, Chapter 3), when IPs abandon newborns because of disabilities. Another precarious motive surfaces when surrogates are forced into selective abortion by the IPs because they have induced more than one pregnancy at the same time 'to be on the safe side', but, in the end, do not want to take care of all the resulting babies (Mitra 2017). Hopefully, we can assume that such cases are very rare exceptions. Still, they require our particular normative attention. It is not the practical role of good ethical guidance and plausible regulation to avoid any and all wrongdoings (that would be unrealistic), but to reduce these events as well as to do justice if such rare cases do happen.

The normative rationale behind such concerns of wrong motives for using surrogacy is the critique of 'instrumentalisation': the surrogate (and sometimes also the child) is simply seen as pure mean for another person's interest (IPs or medical professionals), but not as an end in itself (also Beier, Chapter 12). Such a Kantian position substantiates very basic conditions of how human interaction should be guided by recognition and mutual respect. However, the argument falls short in practice if it relies mainly on stereotypical presumptions about motives without any evidence for misuse in a particular case. Concerns about such instrumentalisation are also expressed in regard to non-commercial, intra-familial surrogacy because there hierarchy and patriarchy play a stronger role (Lamba and Jadva, Chapter 9). In fact, such a concern is not out of place. Still, relevant insights can be drawn from the last three decades of living organ donation (which is mainly practised intra-familially) and which can be instructive for the context of surrogacy (also Beier 2015). It has been shown that strong and efficient regulations of psychological pre-screening, psychological post-care, transparent and balanced information have helped to protect potential donors as well as potential recipients from the practical implications of such instrumentalisation.[1] Lessons learned from both the rare cases of unethical behaviour of IPs in commercial, cross-national surrogacy as well as from practices of intra-familial surrogacy and also living organ donation provide us with good arguments for stronger regulations and for the provision of psycho-social screening and care before and after entering such arrangements.

Naturalised Parenthood as a Strong Medico-Cultural Imperative

An often-cited motivation for surrogacy is the strong wish for motherhood/fatherhood/parenthood[2]: this motive is complemented by the modern rationality of wanting to apply efficient means of ART. Both ideologies are often presented as universal values and are therefore rarely questioned—apart from some radical feminist points of view (Jagger 1983; Teman 2010). The wish for parenthood is rarely questioned because it is 'naturalised' (Strathern 1992) by using explicit or implicit evolutionary arguments, e.g. to reproduce is described as the 'basic instinct' of every living being (including members of the human species). For many, not to reproduce means to behave pathologically (Boorse 1975). And even if non-reproduction is voluntary, some might find it suspect. Though the World Medical Association (2006) has lately stated that childlessness must not automatically be seen as 'illness', it still recommends that the treatment of childlessness is 'clearly medical'. This indicates how medicalisation of childlessness is replacing the former pathologisation—however, it is still problematic as it confines alternative views on childlessness.

Despite almost four decades of intellectual constructivist or feminist critiques of such simplistic views, the ethnographies in this volume involving IPs or surrogates (Teman, Chapter 4; König, Chapter 13; Teschlade, Chapter 14; Nahman, Chapter 16) support the impression that this ideology is still influential in everyday life. Against this background, surrogacy appears as an irritating practice. Specifically, there is increasing evidence from sociological and psychological studies that surrogates can distance themselves from such 'natural instincts' as bonding (Lamba and Jadva, Chapter 9). The 'naturalised' relationship between a woman and 'her' child—often underlined by the (nowadays contested) idea of unconditional maternal love to her baby (Badinter 1981)—is replaced by (cognitive or social) constructions of a 'working' or 'gift giving' relationship. However, such ideological transformation processes—from nature to culture—may not be feasible for everybody. At least some surrogates seem to suffer from depression and emotional problems due to this ambivalence. Also, on side of the IPs, the 'basic instinct' for parenthood causes psychological and social suffering in the case of involuntarily childlessness. This pain and suffering justify their endurance through the odyssey of unsuccessful IVF treatments or failed adoption attempts (Mitra and Schicktanz 2016). Furthermore, ethico-legal critiques of surrogacy partly rely on such unquestioned naturalistic imperatives of parenthood (Sándor, Chapter 3; Dücker and Hörnle, Chapter 11; Beier, Chapter 12): it is assumed that surrogacy in general and giving away one's own baby specifically is 'unnatural'. The practice of surrogacy illuminates—even more than azny other ART before (Rapp 2001)—that there are paradoxes in the ideological relationship between modern medicine and modern motherhood: On the one hand, it consolidates bio-genetic relatedness by employing widely gamete donation of the IPs in surrogacy; on the other hand, it undermines it by expecting that biological connectedness via external pregnancy or using third-party gametes does not matter. Furthermore, ART-parenthood and the selection of egg donors or of surrogates are often interwoven with nationalistic or religious ideologies and power relations (Rimon-Zarfaty, Chapter 17). Selective practice along religious or class/caste criteria indicates how sociocultural criteria can overrule biological assumptions, but at the same time surrogacy can strengthen (perhaps only indirectly) the biological assumption of the social—incl. eugenic or racist assumptions by selecting along physiological or genetic criteria. This paradox might therefore result in self-contradictory practices as well as in strong social conflicts over its acceptance. The latter is clearly illustrated by the solid coexistence of opposing moral stances towards surrogacy and related practices.

Social practices of traditional surrogacy, adoption, or priority of social over biological fatherhood (for German legal practice, see Dücker and Hörnle, Chapter 11) as well as voluntary childlessness demonstrate that we can already refer to helpful cultural resources to rethink parenthood not as natural, but as socially construed. Such an insight would help to see surrogacy in its real complexity and connectivity and to avoid more inconsistencies. However, such a view is normally not advocated by the main driving forces behind surrogacy which is the medical-industrial complex offering this service on local (Shalev, Chapter 15) or transnational level (Werner-Felmayer, Chapter 2). Framing childlessness as a naturalistic disease does not only help physicians and people that are involuntarily childless but also sustains economic interest in these treatments.

Studies providing the exclusive narratives of a few of those affected (whether they are IPs, surrogates, or experts) suppress the connectivity of the medical-industrial complex behind this practice (Werner-Felmayer, Chapter 2). What therefore is needed is the joint study and juxtaposition of various perspectives and different actors in their interactions to detect potential conflicts and one-sidedness (see Mitra 2017 for studying the interactions between these different spheres). Sándor's analysis (Chapter 3) of international legal developments reveals that there is a trend of 'renaturalising' in local court decisions by appreciating 'genetic relationships' as legal claim for ART. This turn should be seen in connection to problematic moves to allow surrogacy only for heteronormative couples, like in India (Kotiswaran, Chapter 7; Reddy et al., Chapter 8) and Israel (Shalev, Chapter 15; Rimon-Zarfaty, Chapter 17). To normatively value genetic affinity, naturalness, and biological relationship will consistently lead to restrictions and limitations of equal rights for non-heteronormative families. This is because there is no consistent way of giving special credit to biological forms of reproduction without, at the same time, degrading alternative family forms (e.g. the lesbian, gay, bisexual, transgender, queer, intersex (LGBTQI) family). In contrast, concepts of reproductive autonomy that are free from anthropological or cultural assumptions of naturalness or the necessity of biological reproduction can consistently argue for equal rights regarding family life for all sexual orientations. Understanding such views of parenthood as socially construed would allow for accepting the connectivity produced by ART and for a more explicit discussion of the norms and responsibilities related to ART and new forms of kinship resulting thereof (Beier, Chapter 12). It forces proponents as well as critiques to argue more consistently why one practice should be allowed or provided whereas another should be prohibited. Overall, the ethico-legal problem of surrogacy and gamete donation might not be solved satisfactorily on a transnational level without critically addressing the paradoxical concept of 'naturalised by ART'-parenthood.

Transnational Discourses as Answers to Local Disjunctions

Technoscapes of surrogacy indicate how this technology spreads globally and adapts locally to fit into a respective cultural context. This process can also be understood as 'glocalisation'. Here, glocalisation is understood as a concept to analyse the process of negotiation, refraction, and mimicry between globalisation and localisation (Mitra et al., Chapter 1). In general, ethical debates focusing on 'pro' and 'contra' argumentations of surrogacy are mainly abstract, but sound balanced; while this volume highlights the peculiarities of national, local discourses and therefore single contributions seem to be dominated by few and selective arguments. Such cultural taming of the ethical debate must be understood as the result of a negotiation between the local and global context in which such debates are evolving and—at least for a particular moment in time—become fixed in a culturally 'significant' way. Applying a comparative view makes some country-specific features of the debates on surrogacy visible and more understandable: for instance, Indian sociologists have specifically pointed out that despite large economic and caste disparities, the agency and opportunities for self-determination of surrogates, even in commercial surrogacy, should not be overlooked (Teman, Chapter 4; Marwah and Nadimpally, Chapter 10). In Germany, however, the agency of surrogates is rarely discussed; instead, concerns are raised in more general terms such as violations of human dignity or the 'best interest of the child' standard (Dücker and Hörnle, Chapter 11; Beier, Chapter 12). In Israel, it is the fulfilment of religious conditions for surrogacy that has led to debates amongst social scientists and ethicists because these requirements are perceived as being too restrictive or arbitrary (Shalev, Chapter 15; Rimon-Zarfaty, Chapter 17; Teman 2010).

While none of the aforementioned issues is raised exclusively in these three national discourses, it is striking how certain lines of argument prevail in each context. We need more and detailed studies to better understand how the bioethical discourse is embedded culturally (Raz and Schicktanz 2016). Such a descriptive-analytical reflection, however, does not solve the quest for a more global and mutual understanding of moral norms—a project that has to be seen as an ideal orientation, not as a simple solution. The cultural framing of differences in the discourse[3] has, however, led to some misunderstanding within normative, conventional ethics. This misunderstanding is based on the assumption that the sociocultural study of differences in bioethical discourses would result in normative relativism. Nevertheless, it is important to differentiate between three types of relativism, i.e. descriptive (anthropological or cultural), epistemic, and normative relativism (Brandt 1967; Schaber 2008). In short, 'descriptive relativism' refers to statements such as 'members of culture X_1 believe that they should follow rule Y_1, and members of culture X_2 believe that they should follow the rule Y_2, and so on'. 'Epistemic relativism' means that 'if you are a member of culture X_1 then you should follow rule Y_1 and if you are a member of culture X_2 you should follow rule Y_2, etc.'. Finally, 'normative relativism' implies the understanding, that 'rules Y_1 are only valid and legitimate for members of culture X_1; rules Y_2 only for members of culture X_2, and so.'

The distinction between these types of relativism allows for a more differentiated understanding of the moral implications of comparison in the bioethical discourse. This is because it is quite uncontroversial that descriptive relativism, in the sense of cultural relativism, has no direct normative implication[4]: It is just a description of the social fact of global or local pluralism. Bioethical research on this descriptive level reveals that various issues are rarely discussed homogeneously. As it should have become clear by the prior comparison, today's global flows normally do not allow for simplistic assumptions about separated cultures of strong inner homogeneity. Instead, fractures, intersectionality, or inconsistencies are in many cases the standard.

Epistemic relativism confers a normative meaning to such differences. It expresses tolerance for the fact that issues can be handled differently in different cultures as long as we aim for inner-cultural consistency and moral coherence in each practice. An example for this is the way we accept the ethico-legal differences regarding pre-implantation diagnosis for late-onset diseases such as breast cancer. Some countries allow doctors to assist, others do not. This pluralism can be defended as long as the practice in question does not threaten the very basic concepts of morality, such as the moral recognition of the other, as, for example, expressed in terms of *basic* human rights. This is why practices such as killing people (as in the case of genocides or death penalty) or extreme discrimination (e.g. the exclusion of people from work or access to social goods based on gender, ethnicity, or disability) typically evoke harsh criticism of relativism. As a consequence, the tolerance across cultural contexts diminishes.

Normative relativism, however, even goes one step further. By claiming that all moral considerations are only valid locally, it disallows any questioning of local norms regarding their validity or priority. In this way, it contradicts any tolerance claims and ultimately leads to ethical nihilism, i.e. the abolishment of any belief in ethical standards or rules of argumentation (Schaber 2008; Gottschalk-Mazouz 2008). In an extremely neoliberal argumentative context, such relativism is indeed used to justify an entire permissive stance towards new technologies: because nation A, B, or C (e.g. the US, India, or the UK) is doing X (e.g. surrogacy), it would be also acceptable in D (e.g. Germany) to do X. The underlying premise then is that there exists no universally valid norm to forbid it. It is this kind of 'nihilistic' (denying all moral standards) normative relativism which is falsely concluded from descriptive relativism.

In fact, the way we deal with cultural diversity or even relativism is a crucial question for modern bioethics. I am relying here on the idea of post-conventional[5] bioethics' interest in cultural and social studies of medical practice. This requires to contextualise the leading moral justification by taking the 'real perceived' social conditions of the respective agents (e.g. patients, doctors, citizens) into account who live in certain conditions as well as the interpretation given to these living conditions. Cross-cultural comparison of moral claims and ethically relevant practices can provide reflection on three levels: first, it allows for a 'translation' of claims of validity into a joint language which leads to the identification of common concerns, as the example of the shared ethical concerns regarding surrogacy illustrates. Second, it offers 'decentration', which is a critical self-location in the relation to one's own and to others' cultural belonging (Raz and Schicktanz 2016). Third, it provides the basis for 'deliberation' by including affected, culturally embedded persons and facilitating the argumentative exchange between different opinions to reach a consensus or to identify areas of joint validity. It is therefore important to not conflate post-conventional bioethics with postmodern approaches, as the analytical focus on non-relativistic normative traditions is still viable. The search for reasons for differences or for similarities in arguments and norms and, by that, the transgression of geopolitical boundaries serve as part of a rationalisation of each claim made. This is the main condition for the 'elective affinity' of applied ethics and cultural studies (Schicktanz 2018), which means that the two social systems or mentalities (bioethics and cultural studies) are related to each other, even though there is no simple causality or natural logic for such a relationship.

Conclusion

What can therefore be learned from comparative bioethics: first, we need a critical assessment of the construction of 'culture', which, for example, indicates a differentiation between hegemonic expert discussions and lay people's perceptions of such technoscapes—including an increased sensitivity for counter-publics (according to Nancy Fraser's idea (1990)). Such counterpublics consist of social groups unlinked or excluded from common public spheres because of social injustice. This requires a critical reflection of the descriptive analysis regarding the quest of representation: to who do we have access to when conducting interviews? How far does our discourse analysis represent also vulnerable, non-hegemonic voices? Who is allowed to speak up and who is not? Second, from the example of ART technoscapes we learn that intracultural pluralism in bioethical issues sometimes exceeds intercultural differences, going beyond a simplistic cultural relativism. In cases in which we can identify strong local differences, what is needed for a re-assessment of the various levels of moral justification: Is it the validity of core moral principles? Is it about the justification of general rules? Or is it about the application to individual cases?

As this comparative book teaches us, surrogacy and gamete donation do not seem to conflict with strong core moral principles, or to contradict basic human rights. But the practice as well as the objections that are raised against the practice rest on a rather traditional understandings of parenthood as a naturalised, biological relationship. Although this conception evokes strong moral feelings, it does not, according to many (post)modern ethicists, belong to the core principles of morality. However, general doubts that remain may contest our tolerance with regard to the potential instrumentalisation of surrogates or the children involved. To dispel these doubts, one should argue strictly for best standards for protection, information, and self-determination of the surrogates—and always do this in a culturally sensitive form. Learning from other medical practices (such as living organ donation) indicates that psychological and ethical counselling are necessary conditions to avoid 'wrong' expectation. Furthermore, we need to strengthen the social recognition of what the donors or surrogates have done for others. The 'best interest of the child'-argument does not seem to provide a fundamental objection to surrogacy either, as long as minimum standards for children's protection are implemented practically. However, in many places this is not yet the case, but should become a basic claim from bioethics.

Overall, the systematic comparison of medical practices, the moral concerns expressed by various parties, and of the ethico-legal problems involved will allow us to open up our discourses regarding moral orientation and regulation on a general as well as on a concrete level. By this, we can identify alternatives and better practice examples than by only focusing on a cultural-centric perspective.

Notes

1. Ethnographic studies of living organ donation indicate that intrafamily donations create complex family dynamics evolving around power relations between spouses and parents vs. children. However, via psycho-social screening and care, extreme cases of dependency and instrumentalisation can be avoided (Wöhlke 2015).
2. In the following, I use the term 'parenthood' as an umbrella term to also include single parent and parents of all sexual orientations/gender identity.
3. Differences in practices are often explained by national law and local regulation, although law depends mutually on cultural accounts of what is seen as ethically acceptable or not. According to many ethicists, the imperative is 'law follows ethics', but in practice, legal regulation is sometimes faster and ahead of ethical discourses. Therefore, I assume a rather complex interplay between law and ethics instead of a one-way road.
4. Of course, descriptions can also be morally loaded—I am here analytically referring to the ideal forms.
5. 'Post' because conventional bioethics neglects any cultural embedding or social factors such as gender, ethnicity, and class to influence bioethical positions. A distinction between conventional and post-conventional bioethics might provoke objections because of its generalisation which can be viewed as doing injustice to individual scholars already open to interdisciplinary exchanges with sociology and cultural studies. Here, it is used to mark a more general shift.

Bibliography

Appadurai, A. (1996). *Modernity at large. Cultural dimensions of globalization.* Minneapolis, MN: University of Minnesota Press.

Archard, D., & Skivenes, M. (2009). Balancing a child's best interests and a child's views. *International Journal of Children's Rights, 17*(1), 1–21.

Archard, D. W. (2016). *Children's rights.* The Stanford encyclopedia of philosophy (Summer 2016 Edition) [online]. Available at https://plato.stanford.edu/archives/sum2016/entries/rights-children/. Accessed 22 Dec 2017.

Badinter, E. (1981). *The myth of motherhood: An historical view of the maternal instinct.* London: Souvenir Press.

Beier, K. (2015). Surrogate motherhood: A trust-based approach. *Journal of Medicine and Philosophy, 40*(6), 631–652.

Blade Runner. (1982). [Film] Directed by Ridley Scott. USA: Warner Bros.

Boorse, C. (1975). On the distinction between disease and illness. *Philosophy & Public Affairs, 5,* 49–68.

Brandt, R. (1967). Moral relativism. In P. Edwards (Ed.), *The encyclopedia of philosophy* (pp. 3–75). New York: Macmillan.

Fraser, N. (1990). Rethinking the public sphere: A contribution to the critique of actually existing democracy. *Social Text,* (25/26), 56–80.

Golombok, S., Murray, C., Jadva, V., Lycett, E., MacCallum, F., & Rust, J. (2006). Non-genetic and non-gestational parenthood: Consequences for parent–child relationships and the psychological well-being of mothers, fathers and children at age 3. *Human Reproduction, 21*(7), 1918–1924.

Gottschalk-Mazouz, N. (2008). Wann ist eine Ethik kulturübergreifend gültig? In N. Biller-Adorno, P. Schaber, & A. Schulz-Baldes (Eds.), *Zur Möglichkeit einer kulturübergreifenden Bioethik* (pp. 253–268). Mentis: Paderborn.

Hoeyer, K., Schicktanz, S., & Deleuran, I. (2013). Public attitudes to financial procurement models for organs: A literature review suggests that it is time to shift the focus from 'financial incentives' to 'reciprocity'. *Transplantation International, 26*(4), 350–357.

Inhorn, M. (2011). Globalization and gametes: Reproductive 'tourism', Islamic bioethics, and Middle Eastern modernity. *Anthropology & Medicine, 18*(1), 87–103.

Jadva, V., & Imrie, S. (2013). Children of surrogate mothers: Psychological well-being, family relationships and experiences of surrogacy. *Human Reproduction, 29*(1), 90–96.

Jagger, A. (1983). *Feminist politics and human nature.* New York: World Press.

Mitra, S. (2017). *Disruptive embodiments: An ethnography of risks and failures during commercial surrogacy in India.* Ph.D., Social Science Faculty, Georg-August University Göttingen.

Mitra, S., & Schicktanz, S. (2016, September 19). Failed surrogate conceptions: Social and ethical aspects of preconception disruptions during commercial surrogacy in India. *Philosophy, Ethics, and Humanities in Medicine, 11*(1), 9. https://doi.org/10.1186/s13010-016-0040-6.

Orth, H. G., & Schicktanz, S. (2017). The vulnerability of study participants in the context of transnational biomedical research: From conceptual considerations to practical implications. *Developing World Bioethics, 17*(2), 121–133.

Piffl Medien. (2005). *Frozen Angels. Synopsis* [online]. Available at http://www.frozen-angels-der-film.de/html/synopsis.html. Accessed 20 Nov 2017.

Rapp, R. (2001). Gender, body, biomedicine: How some feminist concerns dragged reproduction to the center of social theory. *Medical Anthropology Quarterly, 15*(4), 466–477.

Raz, A., & Schicktanz, S. (2016). *Comparative empirical bioethics: Dilemmas of genetic testing and euthanasia in Israel and Germany*. Basel: Springer.

Saravanan, S. (2015). Global justice, capabilities approach and commercial surrogacy in India. *Medicine, Health Care and Philosophy, 18*(3), 295–307.

Schaber, P. (2008). Ethischer Relativismus: Eine kohärente Doktrin? In N. Biller-Andorno, P. Schaber, & A. Schulz-Baldes (Eds.), *Gibt es eine universal Bioethik?* (pp. 159–186). Münster: Mentis.

Schicktanz, S. (2007). Why the way we consider the body matters: Reflection on four bioethical perspectives on the human body. *Philosophy, Ethics, Humanities in Medicine, 2*(30), 1–12.

Schicktanz, S. (2018). Normativity and culture in the context of modern medicine: An elective affinity. In D. Bachmann-Medick, J. Kugele, & A. Nünning (Eds.), *Futures of the study of culture* (Vol. 8). Concepts for the Study of Culture. Berlin and Boston: De Gruyter [in print].

Strathern, M. (1992). *After nature: English kinship in the late twentieth century*. Cambridge: Cambridge University Press.

Teman, E. (2010). *Birthing a mother: The surrogate body and the pregnant self*. California: University of California Press.

Wertheimer, A. (1996). *Exploitation*. Princeton: Princeton University Press.

Wiesemann, C. (2016). *Child, moral equity and bioethics*. Basel: Springer.

Wöhlke, S. (2015). *Geschenkte Organe? Ethische und kulturelle Herausforderungen bei der familiären Lebendnierenspende*. Frankfurt a.M, New York: Campus.

World Medical Association. (2006). *WMA statement on assisted reproductive technologies* [online]. Available at https://www.wma.net/policies-post/wma-statement-on-assisted-reproductive-technologies/. Accessed 20 Dec 2017.

Part II

A System Under Transition: The Case of India

7

Law's Paradoxes: Governing Surrogacy in India

Prabha Kotiswaran

Introduction

As this volume shows, the emergence of a complex transnational market in commercial surrogacy is one of the most vexed issues facing feminists today. Just as there is a range of feminist positions in relation to surrogacy, we can correspondingly identify a typology of legal regimes that states adopt in relation to reproductive technologies in general, and towards surrogacy, in particular (Fenton-Glynn et al. 2018). India is unique in that the state has adopted a range of policy positions towards surrogacy from a liberal, contract-based model in the late 1990s to a prohibitionist, carceral model in 2016. Law has thus been a site of intense political, social and economic contestation over the status of women's reproductive labour. This chapter maps these legal shifts through a socio-legal understanding of the law against the backdrop of

P. Kotiswaran (✉)
Dickson Poon School of Law, King's College London,
London, UK
e-mail: prabha.kotiswaran@kcl.ac.uk

S. Mitra et al. (eds.), *Cross-Cultural Comparisons on Surrogacy and Egg Donation*,
https://doi.org/10.1007/978-3-319-78670-4_7

the fast-changing political economy of the surrogacy sector. Towards that end, this chapter outlines the political economy of the surrogacy sector and normative approaches (primarily of feminists) to theorising it. Next, it outlines the current legal status of surrogacy. It then elaborates on the evolution of the law of surrogacy in India over the past two decades before reflecting on the continued lack of settlement of legal norms regulating the surrogacy sector.

The Indian Surrogacy Market: The Socio-economic Backdrop

Commercial surrogacy was thought to generate USD 2.3 billion annually by 2012 (Rudrappa 2015, p. 5) with the Indian Council for Medical Research (ICMR) predicting its growth into a USD 6 billion-per-year industry (Chang 2009, pp. 11–12). However, the size of the sector is likely to be more modest. Dr. Rishma Pai of the Federation of Obstetric and Gynaecological Societies of India (FOGSI) recently estimated that a mere 320 babies were born a year from surrogacy arrangements (Rajya Sabha 2017, pp. 16–17).

Surrogates in India are working-class women in prime reproductive age (20s and 30s) and married with children (Rudrappa 2015, p. 42; Pande 2014, p. 20). They are typically married because surrogacy clinics require proven fertility and having children outside of marriage is rare in India (Deomampo 2016, p. 236). Although surrogates are not poverty-stricken or destitute, they are financially desperate (Rudrappa 2015, p. 78; Deomampo 2016, p. 236). For many, income from surrogacy equals approximately five years' worth of the total family income (Pande 2014, p. 20). Little is known about the caste profile of surrogates although 30% of the surrogates that Deomampo interviewed were Buddhist and possibly Dalit (ibid., p. 237); caste was certainly a consideration for certain commissioning parents (Sangari 2015, p. 198, n. 28). Journalistic accounts and empirical data suggest that remuneration for commercial surrogacy ranges from USD 1500–2300 (ibid., p. 205, n. 68) in smaller cities to USD 8000 per delivery in Anand.

Feminist Normative Landscape on Commercial Surrogacy

According to Alison Bailey (Bailey 2014, p. 24), Western feminists theorised contract pregnancy in two phases: an intensely normative phase in the 1980s when they offered liberal, Marxist and radical feminist analyses of commercial surrogacy and since the mid-1990s when they took a biomedical ethnographic turn to understand how surrogacy work is lived, embodied and negotiated heralding a move from moral certainty to moral ambivalence. Indian feminism encompasses both normative and ethnographic strains of theorising on surrogacy. In normative terms, there are a few liberal feminists (Aravamudan 2014), radical feminist (DasGupta and Das Dasgupta 2014, p. 185) and Marxist feminist (Rao 2012) voices on surrogacy. Most Indian feminists however pursue a materialist feminist analysis of surrogacy with exploitation lying at the intersection of capitalism and patriarchy (See Marwah and Nadimpally, Chapter 10, for feminist views on surrogacy). As Teman notes in this volume (Chapter 4), several recent ethnographies view surrogates as performing reproductive labour[1]: labour which is highly gendered, exceptionally corporeal, stigmatised (Pande 2014, p. 6) and performed against the backdrop of structural inequalities, an aggressively anti-natalist state and a larger project of neo-eugenics (Pande 2014, p. 23). Materialist feminists adopt a pragmatic view to regulation opposing a ban on commercial surrogacy, which in their view will drive the market underground (Reddy et al., Chapter 8).

The Existing Legal Framework on Surrogacy

There is no specific federal or provincial statute on surrogacy as of April 2018. The Surrogacy Regulation Bill (hereafter SRB) was introduced in Parliament in late 2016 (Parliament of India 2016). The SRB being controversial was referred to a Parliamentary Standing Committee in January 2017, which submitted its report in August 2017. A revised version is likely to be submitted in the coming months.

In the absence of any statutory framework, two sets of guidelines and rules apply to surrogacy in India. The first are the 'National Guidelines for Accreditation, Supervision and Regulation of ART Clinics in India', published by the Ministry of Health and Family Welfare and drafted by the ICMR in 2005 (hereafter ICMR Guidelines) (ICMR 2005). Up until recently, these guidelines were not considered binding although several well-known Indian fertility clinics claimed to abide by them. The ICMR Guidelines did envisage the creation of state accreditation authorities to supervise and regulate assisted reproductive technologies (ART) clinics and appropriate authorities to implement its Guidelines, but these bodies were not set up. ART clinics have therefore looked to the ICMR as 'the apex body in India for the formulation, coordination and promotion of biomedical research', for guidance on ART (ICMR 2017). Apart from the ICMR Guidelines, various federal ministries have issued executive orders. The most recent of these was issued by the Ministries of Home Affairs and Health[2] clarifying that until the SRB comes into force, the ICMR Guidelines continue to apply to domestic commissioning parents. I elaborate on these orders in the following sections.

There is no explicit right to reproduction protected under the Indian Constitution. However, in the *B.K. Parthasarthi v. Government of Andhra Pradesh* case,[3] the Andhra Pradesh High Court upheld the 'right of reproductive autonomy' of an individual as a facet of his 'right to privacy', which is protected under Article 21 of the Indian Constitution. This reading was reiterated by the Supreme Court in the right to privacy judgement *Puttaswamy v. Union of India*,[4] where it held that 'a woman's freedom of choice whether to bear a child or abort her pregnancy are areas which fall in the realm of privacy' (Justice Chelamaswar, para. 38). The majority also noted that 'the sanctity of marriage, the liberty of procreation, the choice of a family life and the dignity of being are matters which concern every individual irrespective of social strata or economic well-being. The pursuit of happiness is founded upon autonomy and dignity. Both are essential attributes of privacy, which makes no distinction between the birth marks of individuals' (majority, para. 157).[5]

Another constitutional provision implicated by the practice of commercial surrogacy is Article 23, which prohibits 'traffic in human beings and other similar forms of forced labour' rendering its contravention punishable. Section 370 of the Indian Penal Code, 1860 prohibits trafficking, including for the purpose of 'sexual exploitation'. Although Section 370 is yet to be invoked against commercial surrogacy, in the absence of a definition for 'sexual exploitation', stakeholders coercing a woman into commercial surrogacy can be prosecuted for trafficking. Indeed, the Trafficking of Persons (Prevention, Protection and Rehabilitation) Bill (Ministry of Women and Child Development 2016) scheduled to be tabled before Parliament in 2018 criminalises the trafficking of a woman for purposes of bearing a child by natural means or through ART for commercial purposes. This is punishable by minimum imprisonment for ten years and can extend up to life.

As of April 2018, surrogacy contracts are legal in India (Law Commission of India 2009, p. 21). Under Section 10 of the Indian Contract Act 1872 (ICA), 'all agreements are contracts if they are made by the free consent of parties competent to contract, for a lawful consideration and with a lawful object, and are not hereby expressly declared to be void'. So far no Indian court has declared a contract for commercial surrogacy void. The Gujarat High Court held that 'commercial surrogacy is never considered to be illegal in India'[6] and the Indian Supreme upheld commercial surrogacy as a 'medical procedure [that] is legal in several countries including in India where due to excellent medical infrastructure, high international demand and ready availability of poor surrogates it is reaching industry proportions'.[7] The Law Commission of India has also commented that surrogacy is legitimate in India as no law prohibits it (Law Commission of India 2009, p. 21). A 2015 public interest litigation by Jayashree Wad argues that commercial surrogacy contracts are void under Section 23 of the ICA for being immoral and against public policy. The Supreme Court is yet to decide on this question and the petition.

In the absence of a specific statute on surrogacy, medical professional bodies specify non-binding ethical guidelines for their members to follow. The Indian Society for Third Party Assisted Reproduction (INSTAR 2017) recommends including certain contractual terms between the commissioning parents and surrogates, prioritising the 'welfare and well-being' of surrogates, including by providing for their health care and setting minimum rates of compensation. In addition, there are numerous laws governing medical practice (the qualifications and conduct of professionals, storage and sale of drugs, law governing the management of patients, laws governing the safety of patients, public and staff within hospitals), environmental protection, employment matters, professional training and research and the business aspects of the hospital. There are also laws on abortion (the Medical Termination of Pregnancy Act, 1971) and the Pre-Natal Diagnostic Techniques (Regulation and Prevention of Misuse) Act, 1994, as amended in 2003 which ART clinics have to comply with.

Regulating the Surrogacy Market 2000-Date

ART in India developed in an anti-natalist context. After all, the Indian state long-controlled women's reproductive choices in its pursuit of population control and an aggressive two-child policy, even leading to forced sterilisation during the emergency years of the 1970s.[8] Ironically, the government developed surrogacy in order to reverse the effects of sterilisation. Private fertility clinics however began to offer commercial surrogacy services only in the late 1990s. The practice grew in leaps and bounds partly due to the entrepreneurial efforts of Dr. Nayna Patel of the Akanksha Fertility Clinic in Anand, Gujarat, who in 2003 assisted a grandmother to deliver the children of her infertile UK-based daughter. As the commercial surrogacy sector has developed over the past 15 years, it has primarily taken the form of gestational surrogacy. The 2005 ICMR Guidelines do not permit the egg donor to act as surrogate and thus rule out traditional surrogacy.

Although there is no statute on surrogacy to date, several Bills have been proposed over the past 15 years corresponding to three phases of the development of the surrogacy sector and the Indian state's attempts to regulate it. These include an emergent medico-liberal phase from the late 1990s up to 2008; a highly contested phase between 2008 and 2012 and a contracting and normative phase between 2012 and 2017. These laws 'proposed' between the late 1990s and 2017 have gone from being highly favourable to fertility clinics (and less so to surrogates) to severely restricting actors who can avail of ART on the basis of marital status, sexual orientation and nationality/citizenship (correspondingly increasing levels of protection to surrogates).

Medico-liberal Phase Late 1990s–2008

The ICMR sought to regulate surrogacy from the late 1990s to 2008; I term this the medico-liberal phase. In 1999, the ICMR constituted a committee of experts to develop national standards for ART clinics. After extensive national consultations and feedback from medical professionals, patients, bureaucrats and the public, the ICMR drafted in 2005 the 'National Guidelines for Accreditation, Supervision and Regulation of ART Clinics in India'.

The ICMR Guidelines were significant for two reasons: first, they envisaged the institution of administrative bodies; second, they set the terms for interactions between stakeholders in the ART industry, thus effectively shaping its political economy in the initial years. The Guidelines also formed the basis for subsequent legislative proposals. They provided for the accreditation of ART clinics and offered guidelines on informed consent (including model forms), confidentiality, counselling, transparency around fertility treatments and maintaining information about donors and recipients.

The ICMR Guidelines permitted commercial gestational surrogacy and allowed for the compensation of gamete donors to facilitate ART. The ICMR Guidelines (Clauses 3.9.2, 3.10.3 and 3.5.3) forbade the ART Bank and clinic from facilitating monetary aspects of the surrogacy transaction leaving this negotiation to the surrogate and commissioning parents. Although the leading ART clinics claimed compliance with the ICMR Guidelines, they in fact determined levels of remuneration and matched the surrogate with the commissioning parents based on the former's financial needs.

The ICMR Guidelines (Clause 3.10.5) provided that a surrogate could not be over 45 years. Clause 3.10.6 specified that she could be a relative, a known person or a person unknown to the commissioning parents. Where a relative acted as a surrogate, she had to belong to the same generation as the commissioning mother. The ART clinic had to ascertain her medical fitness to undergo surrogacy, and the surrogate had to provide a written certificate with medical representations and future undertakings regarding her health. Finally, Clause 3.10.8 required that no woman act as a surrogate more than thrice in her lifetime.

The ICMR Guidelines required that 'surrogacy by assisted conception should normally be considered only for patients for whom it would be physically or medically impossible/undesirable to carry a baby to term' (Clause 3.10.2). Clause 3.14.1 specified that the minimum age for ART for a woman was 20 years. A woman between 20 and 30 years could avail of ART services only after failing to conceive after two years in cohabitation and/or marriage without use of a contraceptive; for a woman above 30 years, this would be reduced to one year. The two-year period did not apply where the man was infertile or the woman could not physiologically conceive.

Gametes produced by a person under the age of 18 could not be used in ART treatments (Clause 3.5.8). The accepted age for a sperm donor was between 21 and 45 years and for the egg donor between 18 and 35 years. Further, Clause 3.2.7 required that no more than three embryos be placed in a surrogate in any one cycle barring under exceptional circumstances and that no woman be treated with embryos derived from the gametes of more than one man or woman during any one treatment cycle.

The ICMR Guidelines did not permit pre-existing relationships between the gamete donors and the commissioning parents; donations had to be anonymous and facilitated by ART Banks. Donors could be compensated by the ART Bank (Clauses 3.9.1.3 and 3.9.2) and the ART Bank by the commissioning parents. Under Clauses 3.5.13 and 3.5.14, the use of sperm or eggs donated by a relative or a known friend of the wife or the husband was not permitted. Under Clause 3.5.4, an oocyte donor could not act as a surrogate. The ICMR Guidelines do not require the child to be genetically related to his or her parents; in fact, commissioning parents had to adopt the child resulting from ART unless they could prove that they were genetically related. Mention of a genetic link emerges later, in the ART Bill 2008

(Parliament of India 2008) and was recommend by the Law Commission of India (2009). The ICMR Guidelines under Clause 3.5.4 required that the birth certificate be in the name of the commissioning parents. A child born through ART was also presumed to be the legitimate child of the couple under Clause 3.12.1. Clause 3.16.1 clarified that the sperm and egg donor had no parental right or duties in relation to the child and required third-party donors and surrogates to relinquish in writing all parental rights concerning the child.

The commissioning couple was allowed to obtain information from the ART Bank on the gamete donor except for his or her identity and address. Clause 3.3.6 required information about the gamete donor to be kept confidential only to be revealed to the child upon his/her request on attaining the age of majority or as and when specified for legal purposes. It could never be revealed to the parents (except when directed by a court). The ICMR Guidelines made no mention of agents, intermediaries or middlemen. The only penalties are those that an accreditation authority imposes on clinics for failing to satisfy ethics and operating procedure guidelines. Provisions criminalising intermediaries therefore emerged only in the ART Bill 2010 (Ministry of Health and Family Welfare 2010).

The ICMR Guidelines were thus informed by a public health view of infertility, which was liberal in permitting the sale of female reproductive labour for the market. Notably, Chapter 7 also spoke of providing ART to economically weaker sections of society, including through clinics in the public sector and the need to address the high cost of ovarian stimulation hormones and reduce dependence on multi-national corporations for these drugs.

The Contested Phase 2008–2012

Between 2008 and 2012, various business models emerged to provide surrogacy services primarily to international commissioning parents. There were full-service clinics like the Anand-based Akanksha Fertility Clinic, Delhi-based SCI Healthcare and the Hyderabad-based Kiran Infertility Centre, which offered comprehensive package deals—from identifying the egg donor and surrogate to performing in vitro-fertilisation (IVF) treatment, to maintaining a surrogate hostel, to delivering the baby to facilitating local travel and sometimes even obtaining the exit visa. Then, there were agencies, which liaised with the commissioning parents, the egg donor and surrogate and looked after the surrogates (by maintaining a hostel or otherwise) while parsing out the medical treatment to a fertility clinic (for IVF) and to a hospital (for delivery). There were intermediate models as well wherein an agency run by a medical professional liaised with various stakeholders to provide comprehensive services to the commissioning parents.

Intra-industry disputes arose around 2008. Unlike in the US or UK, where surrogates litigated to keep the babies, in India, surrogates had a collective action problem even in places like Anand where the sector assumed small industrial proportions. Access to formal state institutions was a huge barrier for individual surrogates. Due to the highly medically mediated nature of the sector and surrogates' own desire to access clinics for a repeat transaction, they were loath to antagonise clinics or approach courts when their surrogacy agreements were violated.

Instead, it was the foreign commissioning parents who went to court as in the cases discussed below. As the surrogacy sector developed, the embassies of foreign commissioning parents in India had refined procedures for ascertaining parentage before issuing the necessary citizenship and travel documents. These procedures varied for different countries; the US and Australian embassies were considered less cumbersome by legal practitioners than the UK due to the six-week waiting period required under the Human Fertilisation and Embryology Act, 2008. Complications however arose where the destination country banned surrogacy, recognised only the birth mother as the legal mother, or where commissioning parents experienced a change in personal circumstances (e.g. divorce). As Justice Hedley has observed in the context of English surrogacy cases, 'the developing law of surrogacy is almost entirely made up of first instance decisions'.[9] India was no different as courts resolved these complex cases in the face of legislative inaction.

The well-known *Baby Manji* case arose in 2008.[10] Here, the Japanese commissioning parents got divorced after commissioning the surrogate to carry the baby. The wife did not want the baby. Since Japanese law only recognises the birth mother as the legal mother, the baby could not obtain a Japanese passport; under Indian law, the husband could not adopt the baby as he was single. Indian authorities refused to issue her an Indian passport; the birth certificate was finally issued with the name of the father only, which posed obstacles to his ability to take the baby back to Japan. The baby's grandmother finally approached the Indian Supreme Court and took temporary custody of her to travel back to Japan.

Another cross-border surrogacy under litigation is the *Jan Balaz* case.[11] Here, the Registrar of Births and Deaths registered the twins born to German commissioning parents with the commissioning mother as the mother on the birth certificate. When they revised the certificate to correct the birthdate, they inserted the name of the surrogate instead. Indian passports were issued to the twins on this basis. However, the Regional Passport Office (RPO) pending litigation before the Gujarat High Court in the matter requested the surrender of the passports. The father complied with this request. However, since the commissioning parents desired a UK visa, Balaz and his lawyer Dushyant Dave appealed the decision of the RPO and approached the Gujarat High Court seeking Indian passports for the twins. The High Court decided, in the absence of any legislation to the contrary, to recognise the gestational surrogate as the natural mother.

Since the surrogate was the legal mother, the children were Indian citizens under Section 3(1)(c) of the Indian Citizenship Act, 1956 and entitled to Indian passports. The Government of India appealed this decision. The Supreme Court while awaiting legislative action vigorously sought a pragmatic solution to the predicament of the Balaz twins. As the German embassy would not issue passports to the twins due to the German law prohibiting surrogacy, the Solicitor-General and Supreme Court pursued an inter-country adoption. The Central Adoption Regulation Authority issued a no-objection certificate. The Indian government provided identity documents and the Ministry of External Affairs issued an exit visa while the German embassy issued the twins with visas to enable travel to Germany. Balaz initiated proceedings in Germany to adopt them so that they could get German passports. He undertook to update the Supreme Court on progress in the case and the children's welfare.

Meanwhile, the *Baby Manji* case caused the Law Commission of India to *suo motu* submit Report 228 in August 2009 noting that:

> prohibition on vague moral grounds without a proper assessment of social ends and purposes which surrogacy can serve would be irrational. Active legislative intervention is required to facilitate correct uses of the new technologies i.e. ART and relinquish the cocooned approach to legalisation of surrogacy adopted hitherto. The need of the hour is to adopt a pragmatic approach by legalising altruistic surrogacy arrangements and prohibit commercial ones.

The Law Commission also recommended protecting the rights of the surrogate and the child and requiring a genetic link between at least one commissioning parent and the child. It suggested legislation to ensure that the birth certificate contains the names only of the commissioning parents and to clarify that no adoption or declaration of guardianship was necessary to establish parentage.

As business boomed, the ICMR went back to draft the ART Bill. The ART Bill 2008 (Parliament of India 2008) was very similar to the ICMR Guidelines. The Bill introduced offences, placed restrictions on egg donation and reiterated that ART was available to all, irrespective of marital status. It introduced stringent requirements that the baby be accepted, that a local guardian be appointed by foreign commissioning parents and that they establish that the baby could be taken home. The sample surrogacy agreement indicated that the surrogate be paid 75% of her remuneration at the time of the embryo transfer and remaining payments later.

The ART Bill 2008 was followed by the ART Bill 2010 (Ministry of Health and Welfare 2010), whose drafting was influenced by the surrogacy sector. The 2010 Bill changed the definition of the term 'couple' to cover two persons living in India and having a sexual relationship that was legal in India. Surrogacy was available to all single persons, married couples and unmarried couples. The Bill added requirements for foreign commissioning parents to ensure that there were no stateless babies; they needed documentation that their home country permitted surrogacy and that the baby could enter their country. If they failed to take the baby, the local guardian had to take delivery and hand the child to adoption within a month or keep the baby, who was entitled to Indian citizenship.

For the first time, individual brokers and paid intermediaries obtaining gamete donors or surrogates were punishable by imprisonment, up to three years and fine. The model surrogacy agreement modified the payment schedule specified in the ART Bill 2008 so that the surrogate would now be paid 75% of the remuneration only after delivery.

The Contracting and Normative Phase 2012–2017

As the surrogacy sector became well developed, but courts' repeated exhortations on the legislature to enact a specific law on surrogacy fell on deaf ears, the government used administrative procedures to regulate the sector. In 2012, the Ministry of Home Affairs (Foreigners Division) issued circulars directed at foreign commissioning parents who now needed a surrogacy visa.[12] Eligibility was restricted on grounds of marital status and sexual orientation. Only a foreign man and woman married for at least two years could commission surrogacy. They required a letter from their embassy or foreign minister that their country recognised surrogacy and that the child would be permitted entry as a biological child. The couple had to furnish an undertaking for care of the child. They could only be treated at an ICMR-registered ART Clinic and had to produce a notarised agreement between themselves and the surrogate. They could travel on a tourist visa to conclude an agreement with the surrogate but give no biological samples on that visit. When the baby was born, they needed exit visas from the Foreigners Regional Registration Officer (FRRO) or Foreigners Registration Officer (FRO); this required a certificate from the ART clinic that the baby had been duly taken custody of and that the parents' contractual liabilities towards the surrogate had been fully discharged. A copy of the birth certificate and passports would be retained by FRO or FRRO. Non-resident Indians (NRIs) and Persons of Indian Origin (PIOs) did not need a medical visa and only had to obtain special permission from the concerned FRO or FRRO on arrival provided they satisfied all the conditions applicable to foreign commissioning parents. They needed to approach the FRO/FRRO before exiting the country.[13] These circulars were upheld as binding law by the Punjab and Haryana High Court (Rajya Sabha 2017, p. 9).

The 2012 immigration orders brought some order to the sector. After years of working in the shadows, fertility clinics finally felt secure that they were not doing anything illegal and that the government was monitoring the sector, which offered them security against recalcitrant commissioning parents. The legislative fine-tuning on a surrogacy-specific statute meanwhile continued. A 2013 version of the ART Bill restricted the number of surrogacies to three live births including the surrogate's own children with two-year intervals between deliveries. Children had to be insured up to the age of 21 and at least one commissioning parent had to be genetically linked to the child. The commissioning parents had to insure the surrogate's health until the child was handed over and she was free of complications.

The ART Bill 2014 (Ministry of Health and Family Welfare 2014) confirmed earlier restrictions by explicitly excluding LGBT commissioning parents. It also excluded foreign commissioning parents. The stringent requirements under the 2012 circulars of the Ministry of Home Affairs were now applicable to NRIs, Overseas Citizens of India (OCIs), PIOs and foreigners married to an Indian citizen who commissioned surrogacy. Differential compensation rates were proposed based on whether the commissioning parents were domestic or foreign. The surrogate was required to be ever married, be between the ages of 23–35 with at least one live child (at least three years old) and could only have one live surrogate baby. She could only be subject to three IVF cycles. Several provisions protected the surrogate including insurance in the event of her death, medical emergency or medical complications for life. The ART Bank was to act as her legal representative free of cost. At delivery, her life was to be prioritised over that of the baby and her compensation was due irrespective of the birth outcome. Where she died or became disabled, the ART clinic was presumed to be negligent. Egg donors had similar protections.

Even as the federal government refined its proposed Bill, Jayasree Wad, a lawyer, filed a public interest petition in 2015 in the Supreme Court[14] praying that it prohibit commercial surrogacy and the import of embryos into the country, given the rampant subordination, exploitation and commodification of women and children. However, on the appointed day for answering questions in court, the government decided to ban commercial surrogacy for foreigners. The ICMR (as part of the Department of Health Research, Ministry of Health and Family Welfare) sent out a circular to ART clinics that surrogacy would be limited to Indian married couples; they were 'requested not to entertain any foreigners for availing surrogacy services in India'.[15]

The import of human embryos except for research (on which guidelines were said to be forthcoming) was prohibited under the 26 October 2015 notification of the Ministry of Commerce and Industry.[16] Earlier such import was permitted subject to procuring a no-objection certificate from the ICMR. The Ministry of Home Affairs issued instructions on 3 November 2015 which withdrew the 2012 circulars and set forth that no visas were to be issued to foreign nationals to commission surrogacy; no permission was to be issued by FRROs/FROs to OCIs; and no exit permission was to be given to foreign nationals except for those who had already commissioned surrogacy and whose matters would be decided on a case-by-case basis.[17] Any visas or permissions granted after the circular was issued, but before its receipt by the concerned officials were to be cancelled. This was followed by a circular dated 4 November 2015 from the Director of the Department of Health Research, which specified that until the ART Bill was passed, the instructions listed therein as well as the ICMR Guidelines would prevail (unless contrary to the circular) and be applicable to all clinics offering surrogacy and ART services.[18] The circular clarified that cases where medical treatment had already commenced (i.e. where embryos had already been imported and implantation had occurred) could proceed to completion with permission from state health authorities to be given on a case-by-case basis.

The Prime Minister appointed a working group of federal ministers to propose law reform. The ART Bill was split—one Bill dealt with ART and the other with surrogacy. In a highly publicised media event in August 2016, the Minister for External Affairs Sushma Swaraj upheld what she called the 'ethos of the Indian people' to explain that the Bill on surrogacy would ban commercial surrogacy altogether whether undertaken by domestic or non-resident Indian parents. Only altruistic surrogacy that too performed by a close relative of the commissioning couple would be permitted; only medical expenses and insurance costs were reimbursable. Commissioning parents had to be Indian citizens (OCIs were excluded) and married for at least 5 years with a medical indication of infertility. "Social" surrogacy was impermissible (see also Reddy et al., Chapter 8 for a response to the Surrogacy (Regulation) Bill, 2016).

The SRB was then drafted and introduced in Parliament in November 2016. It has an entire chapter VII creating several offences with steep punishments. Section 35 is a comprehensive section prohibiting commercial surrogacy, including where an actor might 'run a racket or an organised group to empanel or select surrogate mothers or use individual brokers or intermediaries to arrange for surrogate mothers and for surrogacy procedures' (35(1)(a)). The section also prohibits advertisement (35(1)(b)), abandoning, disowning or exploiting a child born out of surrogacy (35(1)(c)), exploiting the surrogate or the child (35(1)(d)), selling or buying embryos or gametes for surrogacy (35(1)(e)) and importing embryos or gametes for surrogacy (35(1)(f)). Violations are punishable with minimum imprisonment for 10 years with a fine of up to 10 lakh rupees. Medical professionals who contravene the Act's provisions (other than Section 35) are subject to minimum imprisonment for five years and a fine extending up to 10 lakh rupees. Commissioning parents who initiate commercial surrogacy are also liable for minimum imprisonment of five years and subject to a fine of five lakh rupees with higher penalties for subsequent offences (Section 37). Section 2(f) defines 'commercial surrogacy' as the commercialisation of surrogacy services or procedures including the sale and purchase of embryos, gametes or services of a surrogate by giving payment in cash or kind except for the payment of medical expenses and the insurance coverage for the surrogate. Section 47 however allows the central government to specify insurance coverage for the surrogate. There is no corresponding provision on medical expenses. Section 39 creates a rebuttable presumption that a surrogate was compelled by her husband, the commissioning parents or any other relative to render surrogacy services; these actors are seen as abetting surrogacy which is punishable under Section 37. All these offences are cognisable, non-bailable, non-compoundable and do not apply extra-territorially.

Section 2(ze) defines the 'surrogate mother' as the woman bearing the child that is genetically linked to the commissioning couple. No provision elaborates on this genetic link. Under Section 4(iii)(b)(II) of the SRB, only a close relative can act as a surrogate but the term 'close relative' is not defined. The surrogate must be between 25 and 35 years on the day of implantation; she must be ever married, have had a child of her own and cannot act as a surrogate more than once in her lifetime. She is required to be issued a certificate of medical and psychological fitness for surrogacy from a registered medical practitioner.

The SRB has stringent requirements for prospective commissioning parents. Section 2(g) defines a 'couple' as a legally married Indian man and woman above the age of 21 years and 18 years, respectively. This excludes any gay, lesbian, or transgender commissioning parents. Section 2(p) further defines 'infertility' as the inability to conceive after five years of unprotected coitus or other proven medical condition preventing a couple from conception. Section 4(ii)(a) specifies that surrogacy can only be undertaken when either or both members of a couple is suffering from proven infertility. Further, as per Section 4(iii)(a)(I), the intending couple needs to obtain a certificate of essentiality, which in turn requires a certificate of infertility in favour of either or both members from a District Medical Board. The commissioning parents are required under Section 4(iii)(c) to obtain an eligibility certificate from the appropriate authority satisfying certain conditions, as to the female being between 23 and 50 years and the male being between 26 and 55 years on the day of certification, that the couple are married for at least five years, are Indian citizens and that they have not had any surviving child biologically, through adoption or through surrogacy. The only exception is where the commissioning parents have a child that is mentally or physically challenged or suffers from a life-threatening disorder or fatal illness with no permanent cure as approved by the appropriate authority based on a medical certificate from a District Medical Board.

On the sale and purchase of gametes, although Section 35(e) prohibits the sale of gametes for purposes of surrogacy, Section 4(iii)(b)(III) specifies that no woman 'shall act as a surrogate mother or help in surrogacy in any way, by providing gametes or by carrying the pregnancy, more than once in her lifetime' which implies that a woman other than the surrogate could potentially donate an egg. However, the term 'altruistic surrogacy' defined in Section 2(b) includes a surrogacy where only medical expenses and insurance costs are paid; 'surrogacy' in turn is defined under Section 2(zb) as a practice whereby one woman bears and gives birth to a child for an intending couple with the intention of handing over such child to the intending couple after the birth; the altruistic donation of gametes is not included here. This is compounded by the lack of clarity on the genetic link between the commissioning parents and the child.

The SRB attempts to ameliorate the uncertain legal environment relating to parentage. A prerequisite for a certificate of essentiality is an order concerning parentage and custody from a Magistrate of the first class on an application by the commissioning parents and the surrogate. Section 7 further provides that the child born out of surrogacy shall be deemed to be a biological child of the commissioning parents with all the rights and privileges available to a natural child.

The SRB also envisages an elaborate institutional framework for regulating altruistic surrogacy. Sections 14 through 31 specify the creation of a National Surrogacy Board and State Surrogacy Boards. The main purpose of the National Surrogacy Board is to advise the central government on policy matters relating to surrogacy, review the implementation of the SRB, lay down a code of conduct for surrogacy clinics, set out minimum standards in terms of physical infrastructure and oversee the performance of the various bodies constituted under the SRB, including the state surrogacy boards. There are mirroring provisions for state surrogacy boards which under Section 23 will review the activities of the appropriate authorities set up under the SRB, monitor the implementation of the SRB and send consolidated reports about activities in the state to the central government. Chapter VI of the SRB also requires the constitution of 'appropriate authorities' in all the states and union territories, whose primary role is to manage the registration of surrogacy clinics, enforce standards within them, investigate complaints of the breach of the SRB and undertake legal action (including *suo motu*) under the SRB, recommend any modifications to the rules and regulations and make decisions on applications for certificates of essentiality and eligibility from commissioning parents and surrogates.

As mentioned earlier, the SRB was referred to a Parliamentary Standing Committee in January 2017. Based on extensive consultations with numerous stakeholders (governmental and otherwise), the Committee in its August 2017 report reversed literally every key feature of the SRB. In an extraordinary analysis, the Committee stated that imposing altruism on women was a paternalistic form of exploitation and that the 'real' exploitation that surrogates suffered arose from the lack of proper regulation of the surrogacy sector and from the poor and even lower-paying alternatives they had to commercial surrogacy. It instead proposed a "compensated" model for surrogacy whereby women's reproductive labour would be compensated at a rate fixed by the state taking into consideration numerous factors (Rajya Sabha 2017, p. 30). Their exploitation under this model could be 'minimized through adequate legislative norm-setting and robust regulatory oversight' (ibid. 2017, p. 27).

The Committee confirmed that only gestational surrogacy should be permitted. The Committee opined that the SRB's requirement for a surrogate to be a close relative would translate into high levels of coercion inflicted by relatives. Hence, the surrogate need not be related to the commissioning parents. Surrogates instead ought to be empanelled. The surrogate could bear only one live child through surrogacy. The Committee proposed strengthening the informed consent provisions of the SRB so that the surrogate received extensive counselling from a competent authority consisting of independent functionaries including NGOs before she gave consent. The Committee also recommended comprehensive insurance cover for the surrogate for one year from the date the surrogacy procedure starts and for six years from the date of confirmation of pregnancy. Life insurance must be provided for the surrogate and child. It also advocated for maternity benefits for the commissioning mother and the surrogate. It favourably referred back to the 2014 Bill on this issue and other drafting matters (including the definition of surrogacy, storage of embryos, composition of the National Surrogacy Board and the registry of ART clinics).

The Committee liberalised the eligibility criteria for commissioning parents extending the surrogacy option to live-in couples, divorced women, widows, NRIs, PIOs and OCIs but not other foreign nationals. Also one year of proven infertility was sufficient to be able to avail of surrogacy; where the need for surrogacy was absolute due to medical reasons, even the one-year period was to be waived (ibid., p. 38). Medical documentation was sufficient for proceeding with surrogacy, and a certificate of infertility was not required. The Committee suggested that the SRB offers clarification on the role of gamete donors and includes a provision on birth certificates. Furthermore, a parentage order was inadequate to spell out the obligations of the various stakeholders; a comprehensive, legally binding agreement between the commissioning parents, the surrogate and the clinic, registered with the state was necessary. Babies born from surrogacy had to be provided breast milk through banks for six months.

Significantly, the Committee advised reworking the highly carceral framework of the SRB calling instead for punishments to be proportionate to the nature of the infraction, always keeping in mind the best interests of the child born through surrogacy. Finally, it called for setting up an independent agency with quasi-judicial powers to resolve disputes between parties and for the ART Bill to be passed prior to the SRB.

Conclusion

The policy framework for surrogacy in India has gone from a medico-liberal model in 2005 to a socially conservative prohibitionist model in 2016. Through these years, the categories of who could avail of commercial surrogacy progressively narrowed excluding gay, lesbian and transgender individuals and couples in 2012, then foreigners, including even those of Indian origin in 2015. The 2016 SRB went a step further to prohibit commercial surrogacy and replace it with a familial model of altruistic surrogacy, wherein the surrogate had to be a close relative of the commissioning parents. The practice of surrogacy is being brought back full circle; recollect that Dr. Nayna Patel's first foray into surrogacy was to help a woman deliver her own grandchildren. The discursive shift is however less stark than the policy shift. After all, if it was the logic of furthering the hetero-patriarchal institution of marriage and its primary goal of biological reproduction that permitted the commercialisation of surrogacy in the first place, the very same logic also provided the rationale for limiting the sites for biological reproduction to the domestic familial space where women can now reproduce only for the family (albeit extended family), that too of Indian citizens only, rather than for the market.

Yet there has been considerable criticism of the SRB. While liberals object to depriving single men and women and gay and lesbian couples from resorting to surrogacy, feminists worry about the undue pressure that the familial altruistic model will place on women. The FOGSI (2015) and the Indian Medical Association (IMA) meanwhile registered their protest against the SRB, the IMA calling for 'a liberal and ethical ART law' formulated in consultation with medical bodies and a rethink on the regulation of ART Banks and doctors' liability (Indian Medical Association 2016). All these concerns resonated with the Parliamentary Standing Committee which sought to reverse key proposals of the SRB often harking back to the more measured drafting of the ART Bill 2014 (Ministry of Health and Family Welfare 2014) and proposing yet another policy option—that of the compensated surrogacy model heavily regulated by the state. Whether the government will in fact heed this proposal is yet to be seen. But what is clear is that the law and governance of India's surrogacy sector is far from settled even while the sector itself may be unable to regain its place in the highly dynamic transnational commercial surrogacy market.

Notes

1. Reproductive labour has been defined as 'biological reproduction; unpaid production in the home (both goods and services); social provisioning ([…] voluntary work directed at meeting needs in the community); the reproduction of culture and ideology; and the provision of sexual, emotional and affective services (such as are required to maintain family and intimate relationships)' (Hoskyns and Rai 2007, p. 300).
2. Government of India Ministry of Home Affairs (Foreigners Division). (2015). No. 25022/74/2011-F.I (Vol. III) and Government of India Ministry of Health & Family Welfare (Department of Health Research) (2015). No. V. 25011/119/2015-HR.

3. *B. K. Parthasarthi v. Government of Andhra Pradesh and others*, 2000 (1) ALD 199.
4. *K. S. Puttaswamy v. Union of India* (Writ Petition [Civil]) app no. 494/12, Supreme Court, 24 August 2017.
5. See in *Javed and others v. State of Haryana and others* where electoral laws barring individuals with more than two children from contesting elections to local bodies (*panchayats*) was upheld. *Javed and others v. State of Haryana and others*, AIR 2003 SC 3057.
6. *Jan Balaz v. Anand Municipality and others,* AIR 2016 Guj 21.
7. *Baby Manji Yamada v. Union of India (UOI)*, AIR 2009 SC 84.
8. The postemergency years were when then Prime Minister Indira Gandhi unilaterally imposed a national emergency across the country for twenty-one months between 1975 and 1977. She ruled by decree. Elections were suspended, civil liberties of citizens (particularly those of her political opponents) severely curtailed, press censored and large-scale forced sterilisation of the population undertaken.
9. *G v. G* (Parental order) 2012 (EWHC 1979 [Fam]) [48].
10. *Baby Manji Yamada v. Union of India (UOI) and Anr*, AIR 2009 SC 84.
11. *Union of India and Anr v. Jan Balaz and others*, special leave app. no. 31639/09, Supreme Court of India; *Union of India and Anr v. Jan Balaz & others*, civil app. no. 8714/10.
12. Government of India, Ministry of Home Affairs (Foreigners Division), (2012, 2013). Guidelines regarding foreign nationals intending to visit India for commissioning surrogacy. No. 25022/74/2011/F.I. dated 09.07.2012, 07.03.2013, 16.07.2013 and 14.10.2013.
13. Government of India, Ministry of Home Affairs (Foreigners Division) (2014). No. 25022174/2011-F.1 (Vol. III).
14. *Jayashree Wad v. Union of India and others*, Writ Petition No. 95/15 in Supreme Court.
15. Indian Council of Medical Research (Division of Reproductive Biology and Maternal Health) (2015). Notification, Commissioning of surrogacy—instructions regarding.
16. Government of India, Ministry of Commerce & Industry, Department of Commerce, Directorate General of Foreign Trade (2015). Notification No. 25/2015-2020.
17. Government of India Ministry of Home Affairs (Foreigners Division). (2015). No. 25022/74/2011-F.I (Vol. III).

18. Government of India Ministry of Health and Family Welfare (Department of Health Research). (2015). No. V. 25011/119/2015-HR.

Bibliography

Aravamudan, G. (2014). *Baby makers. The story of Indian surrogacy*. Noida: HarperCollins.

Bailey, A. (2014). Reconceiving surrogacy: Toward a reproductive justice account of Indian surrogacy. *Hypatia, 26*(4), 715–741.

Chang, M. (2009). Womb for rent: India's commercial surrogacy. *Harvard International Review, 31*(1), 11–12.

Deomampo, D. (2016). *Transnational reproduction: Race, kinship, and commercial surrogacy in India*. New York: New York University Press.

Fenton-Glynn, C., Kaan, T., & Scherpe, J. (Eds.). (2018). *Eastern and western perspectives on surrogacy*. Cambridge: Intersentia (forthcoming).

Gupta, J. A. (2012). Reproductive biocrossings: Indian egg donors and surrogates in the globalized fertility market. *International Journal of Feminist Approaches to Bioethics, 5*(1), 25–51.

Hoskyns, C., & Rai, S. M. (2007). Recasting the global political economy: Counting women's unpaid work. *New Political Economy, 12*(3), 297–317.

Indian Council of Medical Research (ICMR). (2005). *National guidelines for accreditation, supervision and regulation of ART clinics in India* [online]. Available at http://icmr.nic.in/art/art_clinics.htm. Accessed 9 May 2017.

Indian Council of Medical Research (ICMR). (2017). *About us* [online]. Available at http://www.icmr.nic.in/About_Us/About_Us.html. Accessed 9 May 2017.

Indian Medical Association (IMA). (2016). *IMA's Viewpoint on the New Draft Surrogacy (Regulation) Bill* [online]. Available at http://emedinews.in/ima/Press_Release/august/26.pdf. Accessed 9 May 2017.

Indian Society for Third Party Assisted Reproduction. (2017). *Highlights* [online]. Available at http://instar.co.in/highlights.html. Accessed 9 May 2017.

Law Commission of India. (2009). *Need for legislation to regulate assisted reproductive technology clinics as well as rights and obligations of parties to a surrogacy* [pdf]. New Delhi: Government of India. Available at http://lawcommissionofindia.nic.in/reports/report228.pdf. Accessed 11 May 2017.

Ministry of Health and Family Welfare. (2008). *The Assisted Reproductive Technology (Regulation) Bill 2008* [pdf]. Available at http://www.prsindia.org/uploads/media/vikas_doc/docs/1241500084~~DraftARTBill.pdf. Accessed 7 Nov 2017.

Ministry of Health and Family Welfare. (2010). *Assisted Reproductive Technologies (Regulation) Bill 2010* [pdf]. New Delhi: Government of India. Available at http://surrogacylawsindia.com/admin/userfiles/ART%20REGULATION%20Draft%20Bill%202010(1).pdf. Accessed 1 Dec 2017.

Ministry of Health and Family Welfare. (2014). *Assisted Reproductive Technology (Regulation) Bill 2014* [pdf]. New Delhi: Government of India. Available at http://www.prsindia.org/uploads/media/draft/Draft%20Assisted%20Reproductive%20Technology%20(Regulation)%20Bill,%202014.pdf. Accessed 7 Nov 2017.

Ministry of Women and Child Development. (2016). *Trafficking of Persons (Prevention, Protection and Rehabilitation) Bill 2016* [pdf]. New Delhi: Government of India. Available at http://www.prsindia.org/uploads/media/draft/Draft%20Trafficking%20Persons%20Prevention%20Protection%20and%20Rehabilitation%20Bill%202016.pdf. Accessed 15 Dec 2017.

Nayak, P. (2014). The three ms of commercial surrogacy in India: Mother, money and medical market. In S. DasGupta & S. D. Dasgupta (Eds.), *Globalization and transnational surrogacy in India* (pp. 1–22). Lanham: Lexington Books.

Pande, A. (2014). *Wombs in labor: Transnational commercial surrogacy in India*. New York: Columbia University Press.

Parliament of India. (2016). *The Surrogacy (Regulation) Bill 2016* [pdf]. Available at http://www.prsindia.org/uploads/media/Surrogacy/Surrogacy%20%28Regulation%29%20Bill,%202016.pdf. Accessed 7 Nov 2017.

Rajya Sabha. (2017). *One Hundred Second Report on the Surrogacy (Regulation) Bill 2016* [pdf]. New Delhi: Parliament of India. Available at http://www.prsindia.org/uploads/media/Surrogacy/SCR-%20Surrogacy%20Bill,%202016.pdf. Accessed 7 Nov 2017.

Rao, M. (2012). Why all non-altruistic surrogacy should be banned. *Economic & Political Weekly, XLVII*(21), 15–17.

Rudrappa, S. (2015). *Discounted life: The price of global surrogacy*. New Delhi: Orient Black Swan.
Sangari, K. (2015). *Solid: Liquid, a (trans)national reproductive formation*. New Delhi: Tulika Books.
The Federation of Obstetric and Gynaecological Societies of India (FOGSI). (2015). *FOGSI policies* [online]. Available at http://www.fogsi.org/fogsi-policies/. Accessed 9 May 2017.

8

Surrogacy in India: Political and Commercial Framings

Sunita Reddy, Tulsi Patel, Malene Tanderup Kristensen and Birgitte Bruun Nielsen

Introduction

Bollywood producer and director Karan Johar, a 44-year-old single man became a father of twins through surrogacy in February 2017 (*The Times of India* 2017). Another film star, Tushar Kapoor, also a bachelor, became a parent with the help of a surrogate mother in June 2016 (Sahadevan 2016). Both stars had always wanted to have children, and they became parents as bachelors and with an 'alternate sexuality' in today's context. These two cases are atypical for India. Lesbian, gay, bisexual, transgender, queer and intersex (LGBTQI) people face legal and social difficulties, and in India, a single man is not eligible to adopt a female child.[1] However, prior to 2012—when transnational gestational surrogacy was prohibited for

S. Reddy (✉)
Centre of Social Medicine and Community Health, School of Social Sciences, Jawaharlal Nehru University, New Delhi, India

T. Patel
Department of Sociology, Delhi School of Economics, University of Delhi, New Delhi, India

S. Mitra et al. (eds.), *Cross-Cultural Comparisons on Surrogacy and Egg Donation*,
https://doi.org/10.1007/978-3-319-78670-4_8

same-sex couples and single parents—India was seen as the 'baby cradle' of the world (Qadeer 2010). Commercial gestational surrogacy was allowed for everyone: singles, gays, lesbians, heterosexuals, Indians and foreigners.

In order to regulate surrogacy, the 'Assisted Reproductive Technologies (Regulation) Bill' was drafted by the government in 2008 and was later amended a number of times in the recent versions of the bill. Even in the absence of any law regulating surorgacy, the practice was however legally controlled and restrcited time and again through executive orders issued by the government, which are briefly discussed in this chapter (Kotiswaran, Chapter 7, for a detailed overview on these bills). A new Bill the 'Surrogacy (Regulation) Bill' (2016) was proposed and approved by the cabinet in 2016 (Press Information Bureau Government of India Cabinet 2016). To date, none of these bills have been passed as an act. In the absence of a definite law, the in vitro fertilisation (IVF) clinics and the intended parents (IPs) of Indian origins are still able to beget children through commercial surrogacy in India.

The aim of the chapter is to unravel the different stages through which surrogacy became a debatable and highly controversial issue in India, creating social, ethical and legal uproar. This chapter is based on a review of the relevant literature and on empirical research. The research was carried out from 2012 till recently in Delhi, the capital of India, and in the southern city of Hyderabad. Twenty-two clinics in Delhi and 4 clinics in Hyderabad were studied. After the new Surrogacy Regulation Bill 2016, the first two authors conducted the recent interviews with the Indian Council of Medical Research (ICMR) official, and the first author conducted further interviews with the chairperson of the drafting committee of the ICMR assisted reproductive technology (ART) guidelines (2005) and the doctors

---

M. T. Kristensen
Gynecology and Obstetrics Department, Aarhus University Hospital, Viborg, Denmark
e-mail: mts@clin.au.dk

B. B. Nielsen
Obstetric Clinic, Rigshospitalet, Copenhagen University Hospital, Copenhagen, Denmark
e-mail: bbn@clin.au.dk

from four clinics in Hyderabad. Our analysis is also based on the interactions and presentations of doctors from IVF clinics at the national conference organised by the Indian Society for the Study of Reproduction and Fertility (ISSRF[2]) and the ICMR on 23–25 January 2017 and previous seminars in 2011, 2012 and 2013. This conference provided us with insights on the views, reactions and strong opposition to banning commercial surrogacy as voiced by the IVF clinics' doctors with regard to the new 'Surrogacy Regulation Bill' (2016). Media reports and debates as well as panel discussions on commercial surrogacy on national television channels like NDTV, CNN-IBN and RSTV[3] are also included in our analysis.

The scholarly debates around surrogacy have been fed by the different perspectives from feminists, academics, women's groups and policymakers that are important for any policy decision. Each of these perspectives comes from a different viewpoint. The findings from empirical research need to be taken into account for providing a local, sociocultural and politico-economic context for any policy recommendation.

Global Debate on Surrogacy

The developments in genetics, genomic research and ART at the end of the twentieth century have raised many new questions and concerns about the meaning of *eugenics* and its ethical and moral status in the modern world. ART offers more choices and more hope of success to those wishing to bear a child, making couples desperate to try again and again (Franklin and Ragoné 1998). It also gives opportunities to create newer forms of families—single parent, gay and lesbian couples (Shanley 2002). All over the world, the debates around ART, surrogacy, Pre-Implantation Genetic Diagnosis (PGD) and designer babies continue. Christine Overall (2012) highlights some significant ethical and social themes embedded in contemporary structures of reproduction: the manipulation of reproductive choices, the denigration of women's bodies, the overvaluing of fertility, genetic links to one's offspring and the commodification of reproduction. Anthropologists and women's groups all over the world have been raising concerns about the obvious health risks ART pose for women. They criticise the regressive perspective towards women's role, their fertility and the medicalisation of their bodies (Inhorn and Van Balen 2002; Sama 2007; Tanderup et al. 2015a, b). Challenging eugenics, genetic research and the New Reproductive Technologies (NRTs), Hartmann (2006) calls for a better understanding in both science and political economy in order to make informed feminist judgments, in regard to what we should be in favour of and what we should not.

Surrogacy is practised differently in different countries. Its application depends on the law of the individual country. In very few countries, surrogacy is legal and commercial, including a few states in the US, Russia, Ukraine, Georgia, Greece and, till recently, Thailand and India. Now, commercial surrogacy is banned both in Thailand and in India. Few other countries, like Canada, Australia, Colombia, UK and the Czech Republic, have provisions for altruistic surrogacy, and in some countries, like Israel, it is highly regulated. In almost all other countries, surrogacy is illegal. Thereby the fact that IPs and commissioning parents (CPs) move to countries like India was a natural development—promoting cross-border medical travel, especially for reproductive services (Mitra, Chapter 5).

Surrogacy in India

Surrogacy has been legal in India since 2002. ICMR came up with the national guidelines for accreditation, supervision and regulation of ART clinics in India in 2005 (Kotiswaran, Chapter 7). The first ART Bill was introduced in 2008; subsequently, various executive orders were passed in 2010, 2012, 2013, 2014 and 2015, as discussed in Table 8.1. IVF centres, including surrogacy centres mushrooming in all the metropolitan Indian cities, have had a lot of business since the early 2000s. It is estimated that there are over 3000 IVF surrogacy clinics in India, of which only 365 are registered with the ICMR so far (ICMR 2010). It is estimated that by 2020, the IVF service market will be worth USD 775 million (Bureau 2017).

Debates and controversies, especially around transnational surrogacy in India, have been extensively studied and analysed. Feminists have raised concerns about the commodification of the female body, the

Table 8.1 Snapshot of the changing policies

Sl. no.	Date	Name of the law/Bill	Important statement
1.	ICMR guidelines (2005) (ICMR) (UPA government)	ART guidelines for IVF clinics	IVF Clinics to register with ICMR. Out of thousands of clinics, only 365 are registered
2.	ART (Regulation) Bill (2008)	The draft ART (Regulation) Bill was passed by the MoHFW	To regulate practice of ART and surrogacy in the semen banks, IVF clinics and research institutions
3.	Law Commission Report 228 (August 2009) (UPA government)	Need for legislation to regulate ART clinics. Rights and obligations of parties to surrogacy or surrogate?	The legal issues related to surrogacy are very complex and need to be addressed by a comprehensive legislation
4.	ART (Regulation) Bill (2010) (UPA government)	ART (Regulation) Bill (2010)	Surrogacy as legal in India, and practice of gestation commercial surrogacy with ICMR guidelines to follow. Surrogacy was open to all irrespective of nationality, gender, sexuality and marital status and also infertility status (most liberal)
5.	ART (Regulation) Bill (2012) modified (UPA government)	ART (Regulation) Bill (2012) amendment	The draft prepared in 2012 has banned surrogacy for foreign gay couples and single people. Only married couples were allowed with at least two years of marriage

(continued)

Table 8.1 (continued)

Sl. no.	Date	Name of the law/Bill	Important statement
6.	ART (Regulation) Bill (2014) (UPA government)	Assisted Reproductive Technologies (Regulation) Bill (2014)	It bars foreign nationals from hiring surrogates in India. However, the draft Bill proposed to allow surrogacy to overseas citizens of India (OCIs), people of Indian origin (PIOs), non-resident Indians (NRIs) and even to foreigners married to an Indian citizen. A foreigner married to an Indian citizen shall produce 'Medical Visa for surrogacy'. Bill shuts the door on homosexuals and people in live-in relationships
7.	Surrogacy (Regulation) Bill (2016) (BJP government)	Surrogacy (Regulation) Bill (2016)	The new Bill gave 12 points to follow, which are discussed later in this chapter

commercialisation and exploitation of poor women in the process of surrogacy in the Indian context (Sama 2009 [including biases against women and gestational surrogates in the 2005 guidelines and 2008 Bill]; Crozier 2010; Qadeer 2010; Qadeer and John 2009; Smerdon 2008; Gupta 2008) as well as their inadequate protection from the perspective of social work (Palattiyil et al. 2010). Further, Pande's (2009) gendered labour analysis of surrogacy as outsourcing of reproductive services provides a perspective on how discourses of materiality are transferred through metaphor and routine, and how enclosures produce factory-like workers: 'perfect mother worker' (Pande 2010, p. 969).

Similar and more complex findings stem from Rudrappa's (2010) study of everyday experiences of women in Bangalore's garment industry, which has become the breeding ground for Bangalore's 'reproductive assembly line'. Further studies on reproductive disruptions (Mitra and Schicktanz 2016) as well as medical and bioethical concerns (Tanderup et al. 2015a) on unethical practices like multiple embryo transplants, foetal reduction, hiring more than one surrogate for a couple and 'C-section only' deliveries are discussed. A number of studies raise questions on the bonding between the surrogate mother and the child during the pregnancy (Pande 2009; Teman 2008) and breastfeeding (Reddy et al. 2015) from the perspective of the rights of the surrogate child.

Changing Policies on Surrogacy in India

The ethical and legal issues emerging from time to time led to changing policies and amendments in the ART Bill and finally, a new 'Surrogacy Bill' emerged. Table 8.1 gives a snapshot of changing bills, their specific features and our observations. All the listed bills are still pending enactment, giving enough scope for the IVF clinics and also IPs to still get what they want.

When executive order banned surrogacy for foreigners in 2014, all the IVF clinics opposed this order.[4] There were several protests by the surrogates themselves. Over 70 women who were pregnant at the time and/or had already acted as surrogates protested in front of the ICMR in Delhi (*The Times of India* 2015b) and around 50 protested in front of the Akanksha Infertility Clinic in Anand, Gujarat (*The Times of India*, 2015a). Several couples of Indian origin also protested against the ban in India. They raised concerns including the fact that a 'large number of embryos are also stuck in India due to the ban' (Sonwalkar 2016).

The new 'Surrogacy (Regulation) Bill' was cleared by the cabinet and tabled in parliament on 25 August 2016 by the current ruling party—the Bharatiya Janata Party who at present form central government (Kotiswaran, Chapter 7). With its restrictive, orthodox and conservative ideology, the Bill bans commercial surrogacy and allows only altruistic surrogacy. This new surrogacy Bill came as a surprise, swinging from one extreme position to another.

The previous central government in India run by the United Progressive Alliance allowed commercial gestational surrogacy, choosing to be liberal with little regulation and no law to monitor the working of the ART clinics. As mentioned earlier, it was also open for all, disregarding sexuality and citizenship. The 2016 'Surrogacy (Regulation) Bill' bans commercial gestational surrogacy for everyone and only allows altruistic surrogacy for heterosexual couples, who have been married for the past five years. This is currently debated across the media with regard to ethics, morality, legality, ideology, reproductive rights and also personal choices.

During the various conferences we organised throughout 2011, 2012 and 2013, some social sciences scholars voiced the opinion to ban the practice of surrogacy itself. They highlighted the emotional pain of surrogate mothers in relinquishing the babies they delivered. They assume that the maternal bond is essentially natural, though anthropological work provides evidence of mixed feelings of maternal bonding with offspring (Scheper-Hughes 1985). On the other hand, a few women's groups, liberal feminists and academicians have, on the basis of their ethnographic and other research, suggested stricter ethical practices, regulations and safety nets for surrogates. The medical and ethical deficiencies in surrogacy practices pointed out by women's study groups (Sama 2009), social-anthropological (Reddy and Patel 2012) as well as interdisciplinary studies (Tanderup et al. 2015a, b), and many media reports were overlooked in passing the 2016 surrogacy Bill. The argument for it was based on the ideology that it is against the 'traditional Indian family' values. Most interestingly, similar resonance was witnessed in the coming together of left and right ideologies in the discussions on population policy at the International Conference on Population and Development (ICPD) with a Cairo Declaration (United Nations Population Information Network 1994). The bigger question therefore is what are the true motivations behind the decision of the current government against commercial surrogacy? While overwhelmingly, the focus is on commercialisation of healthcare and pro-market development. Prohibition on vague moral grounds without a proper assessment of the social impacts and the purposes surrogacy can serve would be problematic.

Under the ART bills (2008) and (2010), IVF clinics thrived, with liberal rules, and no regulation. Some clinics in South India served 50% foreigners as IPs (Reddy and Patel 2015); 3/4 of the IPs in Anand (Gujarat) were foreigners (non-resident Indians) during 2010 (ISSRF 2017). Extensive media coverage of cases gone wrong highlighted concerns, like the *Baby Manjhi* case in 2008, where a commissioning Japanese couple got divorced. In the meantime, the would-be social mother declined to accept the surrogate baby. Eventually, the paternal grandmother took custody of the surrogate baby. In another case, a German couple who received twin sons through surrogacy in 2008 in India were stateless and had to wait for two years to get citizenship, following a long legal battle. A few other cases of abandonment of children by IPs in 2012[5] led to the policy change. There were reported cases of violation of children's rights, abandonment of disabled surrogate babies, issues with visa/citizenship, adoption as well as paedophilia. The media coverage of the death of surrogates and egg donors also impacted the policy shift.

The ART Bill (2014) banned foreigners from the home countries where surrogacy was illegal, due to issues of giving children citizenship. Some stranded in India, waiting to return to their home country. Now, commissioning couples in India are required to be married and also have to produce a certificate saying 'the woman is unable to conceive her own child'. The Bill also proposed to make it mandatory for all couples commissioning surrogacy to accept the custody of the child or children irrespective of any abnormality they may have (this was also in the earlier ART Bill from 2008). Commissioning couples are to submit a certificate indicating that the child/children born in India through surrogacy is/are genetically linked to them and that they will not involve the child/children in any kind of pornography or paedophilia.

The 2016 'Surrogacy (Regulation) Bill' put a complete ban on commercial surrogacy. Only Indian couples who have been married for at least 5 years can opt for altruistic surrogacy, provided at least one of them has been proven to have fertility-related issues. Only close relatives, not necessarily related by blood, will be able to offer altruistic surrogacy to the eligible couples. However, who such relatives are is left to one's own interpretation. A woman can become a surrogate mother only for altruistic purposes and under no circumstances is she to be paid for it, although payment can be made towards her medical expenses. It also bans unmarried people, live-in couples and homosexuals from opting for altruistic surrogacy. Now, foreigners, even Indians from overseas, cannot commission surrogacy. A surrogacy regulation board will be set up at both central and state levels; still, the question remains when and how surrogacy will be regulated. All ART clinics will need to be registered but so far, only a few hundred are registered with the ICMR. The law will be applicable to all of India, except for the state of Jammu and Kashmir. It is not clear why Jammu and Kashmir were left out.

Surrogacy clinics will be allowed to charge for the services rendered in the course of surrogacy, but the surrogate mother herself cannot be paid. The question is: When clinics get profits, why will a surrogate give her services for free? Commercial surrogacy, abandoning the surrogate child, the exploitation of the surrogate mother and the selling/import of human embryos have all been categorised as violations that are punishable by a jail term of at least 10 years and a fine of up to USD 15,555. However, this will be applicable only when the law is passed.

Under the new surrogacy Bill, the clinics will have to maintain records of surrogacy for 25 years. However, empirical research shows that until now only a few clinics have kept such records and many clinics have kept no records. Besides, those which are not registered are not accountable to keep records.

The surrogate child will have the same rights as a biological child. However, if the surrogate mother is a relative, there are a number of possibilities for familial conflicts on property issues as well as emotional turmoil.

The 'Surrogacy (Regulation) Bill' 2016: Experts' Perspectives

After the proposed ban on commercial surrogacy in 2016, interviews were conducted with the members of the drafting committee and doctors of the IVF clinics to get to know their views on the issue. In-depth interviews with two experts (one was the chairman) from the drafting committee of the 2005 ICMR guidelines were conducted; we will call them Dr. B and Dr. C for this chapter. In this chapter, 'the Bill' refers to the ART 2008 and the ICMR guidelines 2005 which were mentioned by each of the clinics we studied in Hyderabad and Delhi. In their opinion, the ban on commercial surrogacy is uncalled for; when the 2005 guidelines were drafted, a large part of the Indian population was consulted in order to include their views. The chairman of the drafting committee said: 'It took us 20 years to draft a bill and reach a level which is satisfactory. To be fair, no bill in the country has been drafted in this manner where the rules and regulations were drafted along with the bill' (sic). Both stressed that several weaknesses in the 2005 guidelines had been modified. This process was not applied before banning commercial surrogacy. The views of these experts were also not sought before proposing the 2016 surrogacy Bill. 'Without the ART bill, it does not make sense as surrogacy includes ARTs'. Secondly, both held '[…] the 2005 guidelines had all points required for ethical practice in surrogacy, only the law had to be enacted. There is no problem with the old Bill (i.e., 2008), the problem is we do not follow the rules' (sic). Dr. B added that it was discussed with experts internationally and that they had received letters from many countries, stating 'Our bill is the best bill'. Thirdly, they are not in support of banning commercial surrogacy, but recommended to further monitor and regulate the ART clinics, 'If surrogacy was done ethically, then it was a win-win situation for everyone'.

Separating surrogacy from ART is only looking at one part of the picture. Proposed in a great hurry, the 2016 Bill is deprived of the wide range of peoples' views obtained for the 2005 ART guidelines. These included public views on surrogacy. The gap between the guidelines and their practice is an implementation problem. For example, thousands of ART clinics in India are functioning without the recommended National ART Registry (NARI). Most ART clinics are also self-regulated in the absence of the enactment of the 2008 Bill.

On the issue of the ban on surrogacy for the LGBTQI community, Dr. B said, 'Since LGBTQI is illegal in India, surrogacy for LGBTQI is automatically illegal. It depends on the country's stand on LGBTQI status and not on surrogacy law. If LGBTQI is legalised tomorrow, surrogacy will become legal' (sic). He held in favour of the legal view on surrogacy for LGBTQI. However, his position does not object to LGBTQI or singles seeking surrogacy before the ban.

Dr. C, who has been instrumental in overseeing all the developments in the ART draft Bill (2008) and ICMR guidelines (2005), urged that the Bill was perfect, only some more safeguards should be added for the surrogate and that it should have been passed as a law. 'Since ART bill 2008 and various amendments are still in the form of bills, thereafter, clinics are not mandated to follow. Out of more than 1,000,000 clinics, only 385 are registered with the ICMR' (sic). He noted that the new 'Surrogacy Bill' (2016) is removed from reality and the social context of India. Having directly collected public opinion on ART and surrogacy for the 2008 Bill, he found people were open to surrogacy. ART was accepted by the pronatalist culture and psyche of Indians. Though such reproductive practice is stratified, it reflects the intense desire for having a child and the IPs pay whatever it costs. ART holds the promise of saving many marriages. Evidence shows that the main burden of a couple's infertility is largely borne by the woman, under constant threat of divorce (Singh et al. 1996). Besides, ethnographic studies have also shown the eagerness of women to be egg donors and surrogates (in casual wage workers and domestic helps in Hyderabad by Reddy and Patel 2012, in Delhi by Tenderup et al. (2015), and the garment workers in Bengaluru by Rudrappa 2010).

The interviews with the IVF clinicians show that they are following the 2016 'Surrogacy Bill', but are unhappy as they have lost a good number of clients. They are also questioning the new law and want to challenge it in a higher court.

A chief embryologist, hereafter referred to as Dr. A, in one of the leading IVF clinics in Hyderabad, questions, 'The rule is sent by ICMR, the problem is we do not know, who set the rules? If it is banned for foreign nationals, we will just abide by the law and not take any more foreign nationals for surrogacy' (sic). Talking about the proportion of CPs who were foreigners, Dr. A said, 'In 2008, we had many foreigners compared to Indians, then slowly it becomes 50:50. Now the foreigners have come down, I mean it is nil now, those who are already registered, they are allowed to finish, after that, we did not allow any more registrations. While foreign clients are not taken up, Indian clients are likely to continue' (sic).

Right to Procreate and the Right of the Surrogate Child

There are mixed reactions among some doctors regarding the welfare of the child. Dr. KC, a chief embryologist from a corporate clinic in Hyderabad, refers to single men becoming fathers through surrogacy, saying that 'I think if we look at the society, it is changing. People are not so conservative. Tushar Kapoor (the film star) is very happy with what he has got, if the law permits everything is different' (sic). He elaborated further 'I think it should be about the welfare of the child. Whatever policy comes, it should come up for the welfare of the child. Anyway if it is trying to look into the welfare of children probably we should encourage that' (sic).

Surrogacy and ART are primarily welcomed when the goal is to have children and to complete an incomplete family. Parenting issues are more significant in India, especially the upbringing of surrogate children. However, excluding single women and lesbians from surrogacy (as done by the 2016 Bill) reflects gender essentialism and a focus on marital union as the gold standard for parenting, on the one hand, and, on the other hand, the rising desire for one's genetic and biological offspring as promised by the advanced genetic science. In India, the magnitude is that there are 17.9 million infertile couples across the nation (Shivaraya and Halemani 2007) and 8.12% (50–80 million) worldwide (Ganguly and Unisa 2010). The consequent voluminous push and pull in opposite directions creates new policy dilemmas accentuated by stratified reproduction in surrogacy. The welfare of the child should be the underlying principle for any policy on surrogacy.

Unethical Practices

The Ban and the Need to Regulate

Earlier studies by the scholars Sama (2007, 2009), Tanderup et al. (2015a, b), Reddy and Patel (2015) and media coverage highlighted the unethical practices of a few IVF clinics. The interviews with doctors raise issues of ethical practice and views against prohibiting commercial surrogacy. Doctors are not against altruistic surrogacy. Many doctors from ART clinics were in favour of encouraging altruistic surrogacy in 2012 (at a seminar organised by the authors). Dr. A mentioned the unethical practice of hiring more than one surrogate, saying that

> It is very clearly mentioned in our ICMR guidelines (2005), that only one surrogate can be hired for one couple at a time. In some cases, the couple hires more than one surrogate and out of the children born, takes back the successful and healthy one and leave the others. That may be the reason why the change in the policy. Out of hundreds of clinics, two, three clinics are not following the rules and then they sacked everything that affects everybody (sic).

Talking about the rights of both the CPs and the surrogates, Dr. A voiced the opinion that 'I think instead of banning for all NRIs, Person of Indian Origin/Overseas Citizenship of India, there should be a proper channel, to regulate for the people who want to have children'. Talking about the right to procreate and to become parents, she said 'Who are we to tell what is right or wrong? We are nobody to tell them to have a child or not. We are misusing our power' (sic). She also raises the issue of business development, economy and the losses to the surrogates. She said,

> We would be missing the development, we are going into a reverse gear. I mean there is no proper channel and proper vigilance. Whatever ban they impose, instead they can probably guide. The poor surrogates who cannot meet their needs, ultimately it is an opportunity lost for them to get money (sic).

Dr. KC suggested more transparent measures. He said,

> At least ten people are waiting for us, and a lot of people whom we have turned down, it's because probably they are NRIs, they have given up their Indian passports, they have been OCIs. Only those, who have Indian passport are entitled to go for surrogacy. Though it is really good, that streamlines very well and may stop exploitation. But more women exploitation is happening by the agents, sometimes it could be the clinic also, or probably it can happen by both. But, rather than stopping all, one has to think that both the parties are getting benefitted, surrogate is getting money and the couples are happy to get their child. So banning surrogacy, I personally feel is not the best option (sic).

He stressed the importance of more transparency and the continuity of surrogacy, 'If they think there is an exploitation of surrogates, there can be another way… probably keep a kind of registry, try to justify their pay and take care of them and their welfare. I mean if there is an organisation that can actually make it transparent' (sic). Talking about surrogacy, which is small in number, yet important, he further states, 'Okay in India, in some big clinics, we don't really do more than ten surrogacies in a month; it is always single digit wherever you probably find it. If we had too many cases, then that would have probably made many changes at a larger level. So it really doesn't matter' (sic). The doctor points out that the ban on surrogacy is not acceptable on the grounds of unethical practices happening in some clinics; these are very few in number. He is in favour of regulating it, as it would benefit not just the couples and the surrogates, but the overall economic growth in India.

Most studies in India (Saravanan 2013; Pande 2009; Rao 2012, and by the present authors) have shown that commercial surrogacy is stratified into better off CPs and into poorer surrogates and egg donors. This is similar in other countries permitting surrogacy. Though social stratification is a societal question, the restraint of success rates and the booming of business through unethical ART practice are anti-Hippocratic and unacceptable (Mitra, Chapter 5). Agents earn more while doing very little compared to egg donors and surrogates, the concern brought out in the ethnographic works by the present authors and also by Rudrappa (2010).

Unethical Practices in Clinics: Keeping the Surrogates Ready

Many doctors and embryologists from fertility clinics mentioned the need to regulate the clinics and to follow ethical practices. Banning commercial surrogacy is not a solution. They counter-questioned: Will altruistic surrogacy eliminate unethical clinical practices automatically?

Dr. V from a tertiary care hospital shares the commercial practice of one of the clinics and thinks it is one of the probable reasons for banning commercial surrogacy, as 'foreign couples demand quick result'.

> I know a lady from the USA who had approached us, was here for the limited length of time. Just in three days' time, they are starting up a surrogacy programme. There is hardly any preparation. What that lady told me was, even before she approached, this particular clinic had four surrogates already worked up and ready. So, they don't have to wait for anything. But that sort of practice needs a lot of business mind that is all I could say. Just like a machine they are working on. That is purely commercial (sic).

Dr. V also shared that in some clinics they implant embryos into the intended couple and at the same time also into the surrogate to have better success rates. If the IPs are successful, then they abort the surrogate. She also mentioned the mixing of gametes in case of failure.

Raising clinical ethics and the need to have a law in place, Dr. KC said,

> Before you do anything we need to have a law. People who are just using IVF clinics, like a money-making machine, are running without all the proper procedure and qualification in place, they are not registered, do not keep all the information; like how many miscarriage happened, how many donors, how many cycles have been done, how many babies born. Without records, it is not proper. It is not mandatory for clinics to register with NARI! (sic)

Dr. N, another senior embryologist, who worked in the Middle East before, spoke about the ban of gay couples,

> There are a lot of problems in making clear guidelines. Some time back, let's say gay marriages were sin. Now gay marriages are acceptable and people are changing their mind. Personally, for me God created a system, gay relations, for me is unnatural; I think we are misusing this technology. Right or wrong I am not really sure but, this is how the society is evolving into (sic).

Similar views were also expressed by a few other practitioners in Delhi and Hyderabad (Tanderup et al. 2015a, b; Reddy and Patel 2015). We found that reservations against LGBTQI parenting rights influence ART practice in some clinics, but not all.

Suggestions for ART Regulation in India

Most of the IVF clinic doctors advocated for ART regulation instead of a blanket ban on all of them. Some clinicians held that, though very rare, the unethical practices by some clinics might have led to the ban.

Dr. V, who moved to India seven years back, said,

> I have worked in the UK for a few years for Human Fertility and Embryology Authority (HFEA), which is the regulatory body in the UK, they do inspections and for every case, the report has to go to them. There is also a law, which says, they can't keep more than two embryos. So, everything is recorded, there are records where you can go and check, and there are witnesses. Any record can be picked up and checked.

> Some honest service is offered there. In India I don't see that kind of culture, we will find all cooked up files. Different stories, which I heard were shocking for me at least for a couple of years, but now I am used to it. So I feel, let me do the good work, that is what I do (sic).

The non-transparent and hierarchical culture of ART clinics in India is not viewed as unethical. Consent and choice are usually clinical decisions made on behalf of lay clientele, more so when the patients are poorer and less educated (Tenderup et al. 2015). Surrogacy processes are no different, except that egg donors and surrogates get better reception in ART clinics than in their other encounters with doctors.

A panel discussion on 'Third Party Reproduction' was organised as part of an international ISSRF conference in Delhi (2017). Critical views were voiced by a few scholars.[6] Four of the panellists were doctors from IVF clinics who were all opposing the 2016 'Surrogacy (Regulation) Bill', arguing for the regulation of IVF clinics and demanding the lift of the ban on commercial surrogacy. They were surprised about the content of the new Bill and protested it, especially the ban on commercial surrogacy that was decided without any discussions and consultation with the stakeholders. They were convinced that if this ban continues, they will all go to the court to reverse this Bill. One doctor said that the regulations should be decided with the help of extensive consultations with all the parties, IPs and CPs who have used ART or who still need them, surrogate mothers, ART clinics, all political parties (both conservative and liberal), social activists and women's rights activists. Overall, it should not be the courts, but government agencies like ICMR that are involved in the process.

Dr. B expressed the view that the goal of regulation should be to provide the benefit of technology to patients in need; it should match the changing social climate and protect the basic human right to reproduce in order to ensure the joy of parenthood. The surrogacy policy should be in line with policies for natural reproduction and other types of parenthood like adoption, i.e., by single parents, OCI/PIO and foreign nationals. The benefits of modern technology should be utilised to their full extent to bring happiness to the thousands of couples or individuals who have no other choice for completing their family except through surrogacy. It also will bring direct and indirect gains to the Indian nation, including the pride that comes from being a favoured destination for advanced technology (ISSRF 2017, pp. 62–63).

Besides many common suggestions discussed earlier, the ethical dilemma of using the available technology and the entitlement of exercising the right to reproduce and to start a family surfaced.

Dr. Nayna Patel of the Akanksha Clinic in Anand expressed the opinion that there are many misconceptions about the exploitation of surrogates. She described them as vulnerable, poor, and outlined how her clinic gives them the opportunity to earn money in a dignified way. Additionally, she has established a foundation that allows them to learn new skills during and after the surrogacy. In her opinion, a complete ban on commercial surrogacy is not the solution, but there is a need for regulation. In the case of altruistic surrogacy, she highlighted various scenarios of exploitation that could happen to the woman familiar with the couple who might be coerced into agreeing to altruistic surrogacy—as it was suggested by the present government. Though Pande (2017) and later on Saravanan (2013) critique the permanent monitoring of surrogates by the Akanksha clinic, they do not (or at least not negatively) comment on the skill training for the surrogates that can benefit their future income. Other measures provided by the clinic are financial guidance, support in buying land, building a house, making fixed deposits with banks and also with other family aspirations and achievements. The restricted view of the 'reproducing' mother certainly forces the surrogate to leave her work within the family to other family members or her spouse to make herself 'bioavailable' (Cohen 2008 on kidney donors in India) for producing babies for others.

Media Coverage and Panel Discussion on Surrogacy

From time to time, the media covers the ethical, social, legal and controversial aspects of surrogacy, as it does with other political matters. Mainstream media channels like CNN-IBN, NDTV, RSTV and Doordarshan have in the past organised a debate whenever there was a change in the Bill. Discussions highlighted the commercial, unethical nature of how the IVF clinics work and also the exploitation of surrogates. Surrogates were also involved in these debates. They talked about how surrogacy helped them improve their lives. As discussed above and as it is shown in empirical studies, IVF clinics do indulge in unethical surrogacy practices. However, a large majority of IVF clinics follow the ICMR guidelines (2005) and the ART Bill (2008).

Many IVF clinic doctors and women's groups are of the opinion that permitting only altruistic surrogacy may not be feasible in India, given the workings of the patriarchal society. Some women in the family will be obliged to be unremunerated 'reproductive gift givers' leading to a stratified reproduction within families. This may create many social, emotional and psychological burdens. As the surrogate mother would remain in close proximity to the child, problems regarding the two mothers might occur. There could be other issues of property ownership. The case of an Indian surrogate who died (R. Lal, 29 years old) was discussed to showcase how altruistic surrogacy can be equally exploitative, only in a different way: Lal was coerced by her husband and in-laws to be a surrogate for her infertile sister-in-law, thereby compensating for the insufficient dowry she had brought into her marriage. Despite her unwillingness and failing health, she became a surrogate mother, only to die of a massive heart attack while giving birth to a surrogate baby boy at a hospital in Ranchi 26 July 2014 (Kusum 2017, p. 88). In the case of altruistic surrogacy, the disclosure of infertility as required as part of the new Bill is a breach of the couples' privacy, dignity and autonomy. All the ART clinicians and those at the 2017 conference were of the opinion that commercial surrogacy should be legalised in order to protect the surrogates and to regulate the clinics more effectively.

It is surprising to see the current political regime, which is aggressively marketing 'Brand India' for medical tourism, looking for economic growth, while at the same time proposing a ban on commercial gestational surrogacy. IVF and surrogacy generate a major share of the income in the medical tourism sector. The question is whether the ideology of preserving the 'traditional values' and a conservative mindset outweighs economic growth.

Conclusion

The conflicting views arising out of the 2016 surrogacy Bill has a social and cultural meaning. Childlessness has serious cultural, demographic, social and health implications. With the changing social values, norms and new forms of families, the desire to have children is gaining justifiability for singles and LGBTQI. However, given the stratified reproduction debate, it becomes a globalised phenomenon and women find their bodies as sites of 'reproductive labour', especially in India (Mitra, Chapter 5). In India, they are often married early and are done with having children of their own by the age of 20–22. They often get sterilised soon after having completed their families, but as they are still in their prime reproductive age, take up surrogacy.

Studies have portrayed surrogates not as voiceless victims but as women actively working with an agency in order to make decisions and to improve their living conditions as a result of becoming a surrogate (Pande 2010, 2017; Deomampo 2013; Tanderup et al. 2015a, b). Given the liberalised economy, insecure livelihoods and the frequent absence of husbands, the burden falls on them to fend for the family. With no other choice, they are willing to take up 'reproductive labour'. It would be an injustice to deny them this opportunity to fend for themselves and their families until secured jobs or sustainable livelihoods are provided to the poor.

Banning commercial surrogacy without consulting the framers of ICMR guidelines or without considering the findings of a full range of empirical studies is less likely to be helpful to any of the interested parties in this business. Altruistic surrogacy will rule out the ability of surrogates to earn money the way they were able to prior to the 2016 surrogacy Bill. Once exposed to the family, the infertile might be further stigmatised, women more than men, as 'deficient'. The emotional complexity that results from the identifiability of surrogates and children born through altruistic surrogacy, including patrimony issues, might yield unanticipated challenges in family relations. Denying LGBTQI entry to the surrogacy market after the August 2017 Supreme Court ruling that gave them the right to freely express their sexual orientation and right to privacy (Ganguly 2013) will likely be challenged.

The important concern is that women should not be exploited either by the IVF clinics, agents/agencies or the IPs (Schicktanz, Chapter 6). Early enactment of ART bills enabling both altruistic surrogacy and commercial surrogacy is suggested along with strict regulation of the IVF clinics. Safeguards for the surrogates need to be ensured in the form of post-delivery care, insurance and legal support in case of any mishap. Before the law is passed, all the voices, especially the surrogates, should be heard. The Bill is currently being discussed in the parliament sessions again. Only time will tell what the surrogacy landscape will look like.

Notes

1. Section (4) of 57 Chapter VIII on Adoption Juvenile Justice (Care and Protection) Act of 2000, amended in 2016.
2. The Indian Society for the Study of Reproduction and Fertility (ISSRF) was established in the year 1988 comprising distinguished scientists, public health executives, programme managers and clinicians from the field of reproductive sciences. The central aim of the society was to provide a platform to the researchers and scientists working in the field of reproductive biology [online]. Available at www.issrf.org. Accessed 6 Dec 2017.
3. The panel discussion can be accessed from the website of Anthropos India Foundation http://www.anthroposindiafoundation.com. Accessed 6 Dec 2017.
4. Government of India. Ministry of Home Affairs (Foreigners Division). *Foreign nationals [including Overseas Citizen of India (OCI) cardholders] intending to visit India for commissioning surrogacy*, no. 25022/74/2011-F.1 (Vol. III)—3 November 2015 [Online]. Available at http://mha1.nic.in/pdfs/surrogacy03112015.pdf. Accessed 6 Dec 2017.
5. A baby boy was abandoned in a clinic in Delhi, as the Australian couple already had boys and wanted a girl child (Carney 2015).
6. First author made a presentation sharing the ethical and social concerns. Sarojini from Sama too presented some critical views apart from some legal concerns shared by law student Sonali Battacharya from NLSIU Bangalore.

Bibliography

Bureau, B. (2017). IVF market in India to reach $775 million by 2022. *Business Standard* [online]. Available at http://www.business-standard.com/content/b2b-pharma/ivf-market-in-india-to-reach-775-million-by-2022-117050200507_1.html. Accessed 13 Aug 2017.

Carney, J. (2015). Baby boy abandoned in India by parents who kept his twin sister but left him behind 'as they already had a boy and couldn't afford twins'. *Mail Online* [online]. Available at http://www.dailymail.co.uk/news/article-3137049/Baby-boy-abandoned-India-surrogate-parents-kept-twin-sister-left-boy-couldn-t-afford-twins.html. Accessed 2 Apr 2017.

Cohen, L. (2008). Operability, bioavailability and exception. In A. Ong & S. J. Collier (Eds.), *Global assemblages: Technology, politics, and ethics as anthropological problems* (pp. 79–90). Hoboken, NJ: Wiley.

Crozier, G. K. D. (2010). Protecting cross-border providers of ova and surrogacy services? *Global Social Policy, 10*(3), 299–303.

Deomampo, D. (2013). Transnational surrogacy in India: Interrogating power and women's agency. *Frontiers: A Journal of Women Studies, 34*(3), 167–188.

Franklin, S. (1997). *Embodied progress: A cultural account of assisted conception.* New York: Routledge.

Franklin, S., & Ragoné, H. (1998). *Reproducing reproduction: Kinship, power, and technological innovation.* Philadelphia: University of Pennsylvania Press.

Ganguly, M. (2013). *India's Supreme Court upholds right to privacy.* Human Rights Watch [online]. Available at https://www.hrw.org/news/2017/08/24/indias-supreme-court-upholds-right-privacy. Accessed 18 Nov 2017.

Ganguly, S., & Unisa, S. (2010). Trends of infertility and childlessness in India: Findings from NFHS data. *Facts, Views & Visions in ObGyn, 2*(2), 131–138.

Ginsburg, F., & Rapp, R. (1991). The politics of reproduction. *Annual Review of Anthropology, 20,* 311–343.

Gupta, A. S. (2008). Medical tourism in India: Winners and losers. *Indian Journal of Medical Ethics, 5*(1), 4–5.

Gupta, J. A., & Richters, A. (2008). Embodied subjects and fragmented objects: Women's bodies, assisted reproduction technologies and the right to self-determination. *Journal of Bioethical Inquiry, 5,* 239–249.

Hartmann, B. (2006). Everyday eugenics. *ZCommunications* [online]. Available at https://zcomm.org/znetarticle/everyday-eugenics-by-betsy-hartmann/. Accessed 19 Oct 2017.

Indian Council of Medical Research (ICMR). (2005). *National guidelines for the supervision and regulation of ART clinics in India* [online]. Available at http://icmr.nic.in/art/art_clinics.htm. Accessed 29 Apr 2017.

Indian Council of Medical Research (ICMR). (2010). *National registry of assisted reproductive technology (ART) clinics and banks in India* [online]. Available at http://www.icmr.nic.in/icmrnews/art/art.htm. Accessed 13 Aug 2017.

Indian Society for the Study of Reproduction and Fertility (ISSRF). (2017, January 23–25). *Souvenir and Abstract Book. International conference on reproductive health with emphasis on strategies for infertility, assisted reproduction and family planning.* Organised by ISSRF, ICMR and Indian fertility society (IFS) and other organisations, New Delhi.

Inhorn, M., & Van Balen, F. (2002). *Infertility around the globe: New thinking on childlessness, gender, and reproductive technologies.* Berkeley: University of California Press.

Kusum, S. (2017). *Souvenir and Abstract Book.* New Delhi: Indian Society for the Study of Reproduction and Fertility (ISSRF).

Markens, S. (2012). The global reproductive health market: U.S. media framings and public discourses about transnational surrogacy. *Social Science and Medicine, 74*(11), 1745–1753.

Ministry of Health and Family Welfare. (2008). *The Assisted Reproductive Technology (Regulation) Bill 2008* [pdf]. Available at http://www.prsindia.org/uploads/media/vikas_doc/docs/1241500084--DraftARTBill.pdf. Accessed 7 Nov 2017.

Ministry of Health & Family Welfare, Government of India, New Delhi. (2010). *The Assisted Reproductive Technologies (Regulation) Bill* [online]. Available at http://icmr.nic.in/guide/ART%20REGULATION%20Draft%20Bill1.pdf Accessed 29 Apr 2017.

Mitra, S., & Schicktanz, S. (2016). Failed surrogate conceptions: Social and ethical aspects of preconception disruptions during commercial surrogacy in India. *Philosophy, Ethics, and Humanities in Medicine* [online]. Available at https://www.ncbi.nlm.nih.gov/pmc/articles/PMC5075174/pdf/13010_2016_Article_40.pdf. Accessed 10 Dec 2017.

Overall, C. (2012). *Ethics and human reproduction: A feminist analysis.* London: Routledge.

Palattiyil, G., Blyth, E., Sidhva, D., & Balakrishnan, G. (2010). Globalization and cross-border reproductive services: Ethical implications of surrogacy in India for social work. *International Social Work, 53*(5), 686–700.

Pande, A. (2009). "It may be her eggs but it's my blood": Surrogates and everyday forms of kinship in India. *Qualitative Sociology, 32*(4), 379–397.
Pande, A. (2010). Commercial surrogacy in India: Manufacturing a perfect mother-worker. *Signs: Journal of Women in Culture and Society, 35*, 969–992.
Pande, A. (2017). Gestational surrogacy in India: New dynamics of reproductive labour. In E. Noronha & P. D'Cruz (Eds.), *Critical perspectives on work and employment in globalizing India* (pp. 267–282). Singapore: Springer.
Parliament of India. (2016). *The Surrogacy (Regulation) Bill 2016* [pdf]. Available at http://www.prsindia.org/uploads/media/Surrogacy/Surrogacy%20%28Regulation%29%20Bill,%202016.pdf. Accessed 7 Nov 2017.
Qadeer, I. (2010). Benefits and threats of international trade in health: A case of surrogacy in India. *Global Social Policy, 10*, 303–305.
Qadeer, I., & John, M. E. (2009). The business and ethics of surrogacy. *Economic and Political Weekly, 44*(2), 10–12.
Rao, M. (2012). Why all non-altruistic surrogacy should be banned. *Economic and Political Weekly, 47*(21), 15–17.
Reddy, S., & Patel, T. (2012). *Reproductive tourism in India: An ethnographic study of surrogacy in Hyderabad*. New Delhi: Ministry of Women and Child Development.
Reddy, S., & Patel, T. (2015). "There are many eggs in my body": Medical markets and commodified bodies in India. *Global Bioethics, 26*(3–4), 218–231.
Reddy, S., Patel, T., Tanderup, M., & Nielsen, B. B. (2015). Breastfeeding and bonding: Issues and dilemmas in surrogacy. In T. Cassidy & A. El-Tom (Eds.), *Ethnographies of breast feeding: Cultural contexts and confrontations* (pp. 157–171). London: Bloomsbury.
Reddy, S., & Qadeer, I. (2010). Medical tourism in India: Progress or predicament. *Economic and Political Weekly, 45*(20), 69–75.
Rudrappa, S. (2010). Making India the "mother destination": Outsourcing labor to Indian surrogates. *Research in the Sociology of Work, 20*, 253–285.
Sama. (2007). Assisted reproductive technologies in India: Implications for women. *Economic and Political Weekly, 42*(23), 2184–2189.
Sama. (2009). Assisted reproductive technologies: For whose benefit? *Economic and Political Weekly, 44*(18), 25–31.
Sarojini, N., Marwah, V., & Shenoi, A. (2011). Globalisation of birth markets: A case study of assisted reproductive technologies in India. *Globalization and Health, 7*(1), 1–9.

Saravanan, S. (2013). An ethnomethodological approach to examine exploitation in the context of capacity, trust and experience of commercial surrogacy in India. *Philosophy, Ethics, and Humanities in Medicine, 8*(1), 10.

Scheper-Hughes, N. (1985). Culture, scarcity, and maternal thinking: Maternal detachment and infant survival in a Brazilian shantytown. *Ethnos, 13*(4), 291–317.

Sahadevan, S. (2016). Tusshar Kapoor becomes father to baby boy using surrogacy. *The Indian Express* [online]. Available at http://indianexpress.com/article/entertainment/bollywood/tusshar-kapoor-becomes-father-to-baby-boy-using-surrogacy-2878996/. Accessed 16 Nov 2017.

Shanley, M. L. (2002). *Making babies, making families: What matters most in an age of reproductive technologies, surrogacy, adoption, and same-sex and unwed parents*. Boston, MA: Beacon Press.

Shivaraya, M., & Halemani, B. (2007). Infertility: Psycho-social consequence of infertility on women in India. *Indian Journal of Social Development, 7*(2), 309–316.

Singh, A., Dhaliwal, L. K., & Kaur, A. (1996). Infertility in a primary health centre of north India: A follow-up study. *Journal of Family Welfare, 42*(1), 51–56.

Smerdon, U. R. (2008). Crossing bodies, crossing borders: International surrogacy between the United States and India. *Cumberland Law Review, 39,* 15.

Sonwalkar, P. (2016). Regulate don't ban surrogacy UK couples urge India. *Hindustan Times* [online]. Available at http://www.hindustantimes.com/world-news/regulate-don-t-ban-surrogacy-uk-couples-urge-india/story-hHxLAjbbvbDF2nqj0gpsaJ.html. Accessed 6 Dec 2017.

Sreenivas, K., & Campo-Engelstein, L. (2010). Domestic and international surrogacy laws: Implications for cancer survivors. *Cancer Treatment and Research, 156,* 135–152.

Tanderup, M., Reddy, S., Patel, T., & Nielsen, B. B. (2015a) Informed consent in medical decision-making in commercial gestational surrogacy: A mixed methods study in New Delhi, India. *Acta Obstetricia et Gynecologica Scandinavica, 94*(5), 465–472.

Tanderup, M., Reddy, S., Patel, T., & Nielsen, B. B. (2015b). Reproductive ethics in commercial surrogacy: Decision-making in IVF clinics in New Delhi, India. *Journal of Bioethical Inquiry, 12*(3), 491–501.

Teman, E. (2008). The social construction of surrogacy research: An anthropological critique of the psychosocial scholarship on surrogate motherhood. *Social Science and Medicine, 67*(7), 1104–1112.

The Times of India. (2015a). *Surrogate mothers stage protest in Anand* [online]. Available at https://timesofindia.indiatimes.com/city/ahmedabad/Surrogate-mothers-stage-protest-in-Anand/articleshow/49589988.cms. Accessed 6 Dec 2017.

The Times of India. (2015b). *Women protest surrogacy ban for foreigners* [online]. Available at https://timesofindia.indiatimes.com/city/delhi/Women-protest-surrogacy-ban-for-foreigners/articleshow/49612904.cms. Accessed 6 Dec 2017.

The Times of India. (2017). *I feel enormously blessed: Karan Johar* [online]. Available at https://timesofindia.indiatimes.com/entertainment/hindi/bollywood/news/i-feel-enormously-blessed-karan-johar/articleshow/57473701.cms. Accessed 16 Nov 2017.

United Nations Population Information Network (POPIN). (1994, September). *International conference on population and development (ICPD)*. Cairo [online]. Available at http://www.un.org/popin/icpd2.htm. Accessed 2 Apr 2017.

Vora, K. (2013). Potential, risk, and return in transnational Indian gestational surrogacy. *Current Anthropology, 54*(S7), 97–106.

Whittaker, A. (2011). Cross-border assisted reproduction care in Asia: Implications for access, equity and regulations. *Reproductive Health Matters, 19*(37), 107–116.

9

Indian Surrogates: Their Psychological Well-Being and Experiences

Nishtha Lamba and Vasanti Jadva

Introduction

In 2013, India's surrogacy industry was estimated to be worth Rs. 25,000 crores (approximately 2 billion GBP) with approximately 200,000 clinics offering fertility treatments (Malhotra and Malhotra 2013). India became a popular global destination for intended parents (IPs) due to its minimal legal regulations, access to advanced medical technology, the prospect for IPs' to be named on the birth certificate, and the cheaper costs involved (Smerdon 2008; Karandikar et al. 2014). Estimates suggest that up until 2015, 25,000 surrogacy children were born in India (Söderström-Anttila et al. 2016). However, during this time, India was also being criticised globally for enabling rich westerners

N. Lamba (✉) · V. Jadva
Centre for Family Research, University of Cambridge, Cambridge, UK

V. Jadva
e-mail: vj227@cam.ac.uk

S. Mitra et al. (eds.), *Cross-Cultural Comparisons on Surrogacy and Egg Donation*,
https://doi.org/10.1007/978-3-319-78670-4_9

to exploit poor and vulnerable Indian women (DasGupta and Dasgupta 2014) with growing media stories on babies being left stateless and parentless.

The psychological impact of commercial surrogacy on Indian surrogates has not so far been examined. Research has shown that (non-surrogate) women from low-income populations in developing countries tend to struggle with emotional problems, primarily due to economic difficulties (Pereira et al. 2007). Given that an overwhelming majority of surrogates come from this group of women, it is vital to evaluate their psychological well-being. Furthermore, concerns over the treatment of surrogates in low-income countries, and the way in which surrogacy is practiced and legislated in these countries, raise additional questions about the possible psychological harm experienced by surrogates.

This chapter reports from a larger research study carried out in India, from February 2014 to November 2016, which examined the psychological well-being and experiences of surrogates hosting pregnancies for international IPs. The surrogates were recruited from a surrogacy clinic in Mumbai, India. This clinic, established in 2010, provides a number of infertility services, including egg and sperm donation, embryo donation, in vitro fertilisation (IVF) and surrogacy. At the time of the study, the clinic was accommodating around fifty surrogates in a surrogacy house situated near the clinic. Fifty surrogates were interviewed during four to nine months of pregnancy, and 45 of these were followed up for a second interview four to six months after the birth of the surrogacy child.

A comparison group of pregnant women was included in order to compare the questionnaire measures of psychological well-being and experiences of prenatal bonding. This would allow us to see if levels of anxiety, depression and stress in surrogates were different to that experienced by women who were pregnant with their own child. The group of expectant mothers came from public hospitals in Mumbai and Delhi and were closely matched to the surrogates on key factors such as age, socio-economic status, educational background and marital status. Sixty-nine expectant mothers were interviewed during four to nine months of pregnancy, and 49 of these were followed up for a second interview four to six months after the birth of the child. All interviews were conducted in Hindi in a private room at the clinic or hospital.

The psychological measures used in this study assessed the separate yet related constructs of anxiety, depression and stress which are common indicators of psychological health (Lovibond and Lovibond 1995). Anxiety is defined as an emotional state characterised by feelings of excessive restlessness, tiredness, increased activity in the autonomic nervous system as well as a general lack of attention (Spielberger and Rickman 1990). Symptoms of depression include continuous sadness and a general loss of interest in daily activities. It negatively affects one's perceptions, judgements, memories and overall quality of life and further increases the risk of self-harming behaviours including suicide (World Health Organisation 2017). Stress is experienced when an individual feels unable to cope with the demands and strains from their environment (Lazarus 1966). The symptoms of anxiety, depression and stress have been found to overlap, as they share a common aetiology (Glover 2014). Together, higher levels of anxiety, depression and stress may give rise to feelings of helplessness and loneliness (Green 2005).

This chapter explores how factors associated with Indian surrogacy may contribute to surrogates' psychological well-being. We firstly compare surrogates psychological well-being and her bonding to the foetus to that of women expecting their own child. We then look at how different aspects of a surrogacy arrangement may impact on surrogate's psychological health. Specifically, whether the surrogate sees or meets the newborn and IPs, the secrecy and social stigma associated with surrogacy, the availability of social support from family and other surrogates and the role of financial compensation. We end by evaluating the relevance of these findings in relation to the new policy of altruistic surrogacy being introduced in India. This research has strong implications for policy by highlighting the features of surrogacy that can affect the well-being of those involved.

Psychological Well-Being of Surrogates

There are a multitude of factors involved in surrogacy. Psychological, sociological and biological processes are all important for a successful surrogacy arrangement (Baslington 2002). Whilst there has been in-depth research on other processes of surrogacy, there hasn't been any research on the psychological well-being of surrogates in low-income countries (Söderström-Anttila et al. 2016).

It has been suggested that factors associated with surrogacy, such as relinquishing the baby, lack of contact with the IPs, hormonal changes and having a strong bond with the foetus, may lead to emotional difficulties for a surrogate (Ciccarelli 1997; Braverman et al. 2012; Jadva et al. 2014). Ciccarelli (1997) undertook a longitudinal study in the US assessing the psychological well-being of 14 surrogates. The study found that six of the surrogates were experiencing emotional discomfort due to their role as a surrogate; however, only one attributed this to the relinquishment of the baby. A study of 34 surrogates in the UK found that 35% of surrogates faced minor psychological difficulties a few weeks after the birth of the child reducing to 6% one year later (Jadva et al. 2003). Twenty of these surrogates were assessed ten years later and were not found to be showing any signs of depression (Jadva et al. 2014). Overall, in terms of surrogates' experiences of relinquishment, research in the UK and US has repeatedly shown that surrogates are generally able to hand over the baby they carry (Jadva et al. 2003, 2012) and that this does not appear to lead to psychological difficulties for the surrogate in the long-term (Jadva et al. 2014; Imrie and Jadva 2014).

Compensated surrogacy in developing countries encompasses additional features that may act as psychological stressors for surrogates. It has been reported that as surrogates in India are predominately illiterate, they are highly susceptible to neo-colonial exploitation, such that risks, impacts and basic information regarding pregnancy and surrogacy are (often deliberately) not communicated (Deonandan et al. 2012). Furthermore, the majority of surrogates enter surrogacy arrangements due to economic desperation (Karandikar et al. 2014), most do not see or meet the newborn or IPs (Lamba et al. 2018), nor do they receive professional counselling (Karandikar et al. 2014). In summary, such circumstances, which are unique to cross-border surrogacy arrangements in India, may leave a person highly vulnerable to psychological problems.

In the present study, participants were administered the anxiety, depression and stress scale (ADSS) (Bhatnagar et al. 2011) during pregnancy and a few months after the birth. The ADSS was developed and standardised on participants that included those from illiterate and marginalised groups in India. Participants responded "yes" or "no" to questions such as "I feel more nervous and anxious than usual", "often I want to be alone" and "I find myself getting agitated by everything".

Comparisons between surrogates and expectant mothers revealed that there were no differences between them in either anxiety or stress during pregnancy or a few months after the birth of the child suggesting that women carrying a surrogate pregnancy did not differ in terms of her anxiety or stress from women pregnant with their own child. However, surrogates were found to be suffering from significantly greater depression during pregnancy and after the birth, in comparison with expectant mothers (Lamba et al. 2018). Excerpts from the interview transcripts provide insight into surrogates' feelings. For example, one of the surrogates explained her circumstances by saying: "All I get to sit and talk to others and get to share our grief. All come here because they have problems, no one comes happily. Each woman has her problems and they come here because of that. If they have a good home and stuff, why would they do this?"

It is possible that the two groups of women may have differed in their levels of depression before pregnancy; however, the present study did not assess the women before pregnancy was achieved.

Bonding with the Unborn Child

It has been argued that the surrogates' level of bonding to the foetus may affect their psychological well-being after relinquishment (Ragoné 1994; Teman 2010; Jadva et al. 2014). Only a few studies have examined the prenatal bonding experienced by surrogates. Fischer and Gillman (1991) found that American surrogates experienced less bonding to the foetus than non-surrogates. In addition, van den Akker (2007) found that, in comparison with intended mothers, British surrogates were less positive and less concerned about the well-being and health of the foetus. For Indian surrogates, Pande (2010) revealed how they are "trained" to stay detached from the unborn child/ren, by for example, being told not to look at the baby after relinquishment.

In the present study, an adaptation of the Maternal–Fetal Attachment Scale (MFAS) (Cranley 1981), a widely used measure of assessing the extent to which pregnant women bond with the unborn baby, found that surrogates affiliated and interacted less with the foetus than expectant mothers and that surrogates displayed greater care and attention towards the foetus' needs than expectant mothers (Lamba et al. 2018). This finding is perhaps culturally rooted and could be explained using Pande's (2010) work on surrogate's facing a "worker-mother" duality. As a "worker", she is expected to be attentive to the needs of the foetus; however, as a "mother", she is expected to affiliate and interact less with the foetus, thus staying detached (Lamba et al. 2018).

The interview transcripts provide further insights into this finding by showing how most surrogates developed an emotional boundary between themselves and the foetus. For example, one commented: "I already know that I have to give this baby to them. That is why there is no benefit by feeling close".

When asked about feeling connected to the child, many women replied with a simple "no". If probed, they would often add a brief explanation like the following: "It is not ours, this is others' belonging" (Teman, Chapter 4).

The following quote illustrates how surrogates felt a duty of care towards the unborn children: "I didn't take care much in my kids as I do now for these babies, because this is something based on medicines and tablets, I am scared [for the babies]".

In the present study, contrary to concerns raised in the past, no significant association was found between feelings towards the unborn child/ren and psychological well-being (Lamba et al. 2018) suggesting that relinquishing the newborn may not cause psychological harm to the surrogate.

Feelings Towards the Surrogacy Baby

In India, as the surrogate rarely sees or meets the newborn (Pande 2010), the surrogate's bond with the unborn child is broken abruptly after the birth of the child, which has the potential to cause psychological distress

(British Medical Association 1996). The satisfaction experienced with seeing, holding and meeting the baby is an important aspect of a successful surrogacy arrangement (Hohman and Hagan 2001). No research has up until now discussed the psychological impact of not being able to see or meet the surrogacy child as it is unheard of in the West. In the UK, the US and Israel, surrogates often develop a direct relationship with the IPs and child that often continues as the child grows up (Jadva et al. 2014; Imrie and Jadva 2014; Teman 2010).

In the present study, all the surrogates had expressed a desire to see the newborn; however, none had met the newborn on the day of delivery and two-thirds had not seen the baby after the birth (Lamba et al. 2018).

Although many of the surrogates had not met the baby at the time of delivery, most said they had caught a glimpse during birth. For example: "After delivery they were taking him to clean. At that moment I turned my head and looked at him. They don't allow seeing, I have heard".

Surrogates, who did not manage to catch even a glimpse of the baby, spoke about hearing the baby cry: "I heard her crying. They just had given me oxygen and gave me some medicines on my stomach and I did not realise when they had cut my stomach and removed the baby. I just heard the baby cry and my eyes closed and I became unconscious".

Many surrogates spoke about wanting to keep a photograph of the baby, although only one-fourth of surrogates received one. Some mentioned wanting to show the baby's picture to their relatives. For most surrogates, having a photograph was seen as important to provide them with a *momento*. One surrogate expressed: "They should have come to meet us and ask us how we were doing at least… and this is a caesarean and the marks would remain. So just for that sake we should at least get the photographs of the baby". Another stated: "I will meet them again if they come. I will be happy, I will take photo with baby and the family and keep in my house so that I can remember". One of the surrogates expressed anger and disappointment about not being shown or given a photograph of the newborn:

> I met the parents only once and they had not even shown me the picture of the baby. They did not do at least that much for me. I had given them a baby after 18 years… They must be thinking that I will run away with

> the baby. If I have to run, I would have run when the baby was in my tummy only. It is ok if they do not want to show the baby but at least they can give me the photo.

A few surrogates who had seen the baby expressed feeling content with the surrogacy arrangement and the handover experience: "Yes, I could deal with the situation easily after seeing them... Otherwise I would have always thought how they look and how they are. I am peaceful now. After all I have kept the babies in my womb for nine months, I deserve to see their faces at least". Interestingly, this surrogate spoke about not feeling a sense of closure without having seen the baby.

Lastly, it is important to note here that none of the surrogates reported a desire to establish a direct relationship with the surrogacy child. However, given that their wish to meet the newborn was not met, it is perhaps unsurprising that they did not seek or expect a relationship with the child. In relation to their thoughts on whether the child should be told about their birth using surrogacy, approximately half of the surrogates said they do not feel the parents should disclose. For example: "I felt that the parents had waited for long for this day and so I should give them. They took my photo and said they will tell the child that I gave birth to him, but I asked them not tell him that, otherwise... They will think how his mother can feel the pain when she did not give her birth".

To summarise, whilst most of the surrogates wished to meet the baby, in reality, few did so which led to feelings of disappointment. For those who did see the baby, this may have provided them with a sense of closure.

Feelings Towards the IPs

In India, it is usual for the IPs and surrogates to not meet during pregnancy or after the birth and an overwhelming majority do not stay in touch (or establish a direct relationship) with each other (Lamba et al. 2018; Mitra and Schicktanz 2016; Pande 2011). Cultural differences can be observed between Indian surrogates and Western surrogates, as those from the West are more likely to consider a long-lasting relationship with the IPs and the child as an essential part of the surrogacy arrangement (Braverman et al. 2012). In fact, some researchers suggest that the relationship developed between the IPs and surrogate is vital to a surrogacy arrangement (Teman 2010). In contrast, in India, the existence of a relationship between the surrogate and the IPs is left at the mercy of IPs and clinic heads, leaving the surrogates wishes largely ignored. In fact, clinics have been reported to discourage IPs to develop a direct relationship with the surrogates (Mitra and Schicktanz 2016).

Usually, surrogates in the West are able to develop a harmonious relationship with the IPs. Most of the surrogates (but not all) stay in touch with the IPs during pregnancy although the frequency and type of contact maintained can vary (Jadva et al. 2003, 2012; Braverman et al. 2012). Research has shown that surrogates show ease in separation when they know where the child is going, suggesting that the relationship may develop to the couple rather than with the child (Baslington 2002). It has been suggested that Indian surrogates hope for a lasting bond with IPs and wish for a reciprocal relationship and generosity from IPs in the form of "gift giving" (Pande 2011; Vora 2014).

In the present study, the majority of surrogates expressed a desire of wanting to stay in touch with IPs. Some surrogates also talked about either expecting gifts from the IPs as an act of reciprocity during pregnancy or about their disappointment when they did not receive any gifts following the birth. This disappointment was usually stronger when their close friends at the surrogacy hostel had received gifts from their respective IPs. The gifts included cash, gold jewelery or chocolates for the surrogate's children.

In our sample, just over half of the surrogates had not had contact with the IPs following the birth. It was common for most of the surrogates to not know anything about the IPs, including their age or nationality. Similar to their feelings regarding not seeing the child, most of the surrogates expressed their disappointment about not meeting the IPs, especially after the birth. For example: "I could not meet the baby and the parents. I really wanted to meet them. I wanted to see who are there in their family. I wanted to ask these but it could not happen. It is ok… Let's move on".

The lack of a standard protocol at the clinic about the level of contact surrogates should have with the IPs, led to feelings of uncertainty for most surrogates. One of the surrogates expressed: "In the surrogate house, the staff and the surrogate mothers say that some parents show the babies to the surrogate mothers and some don't… Some surrogates have met the parents… It all depends on one's luck".

So it was common for surrogates to blame their fate for not being able to meet the newborn or IPs. Since these surrogates lived together, it was common for them to share stories with each other. Consequently, surrogates often reported their feelings towards meeting (or not meeting) the IPs relative to whether their friends at the surrogacy house had met (or not met) the IPs. For example, one said:

> Actually one of my friends could not even meet the baby. She would feel good if she meet her at least once. They used to complain to me that they could not even meet the baby. I was also thinking the same that they would not meet me, but they did. It felt good that they came and asked about my health. I had worked for them and in turn they too had thought about me. It feels good.

Another said: "Yes, as every surrogate says that their party has come and met them and brought gifts for them… The other lady's baby was not healthy and there was a lot of blood loss for her. We used to think that we had given our client such a healthy baby and they did not meet us at least once".

Surrogates in the present study were often aware of the clinics role as mediator, for example one of the surrogates said: "They were talking in English and madam translated their conversation. I had asked for the contact number but the clinic doesn't allow to do that. The parents had told me that they would send the photos of the babies to me… I don't know whether they have already send it to the clinic or not".

For those surrogates who had met the IPs after birth, their meetings were usually brief, ranging from five to twenty minutes. As mentioned before, these meetings were supervised by a member of staff from the clinic, who translated their conversation. Most of the surrogates who had met the IPs expressed relief during the second interview, for example one said: "I felt happy to meet them. At least I know where the child is going. Who are the parents? I was relaxed".

In fact, surrogates' happiness with the handover was positively correlated with meeting the IPs after the delivery, reinforcing the assumption that surrogates felt a sense of satisfaction when they met IPs. It is important to gain more information on the thoughts, motivations and decision-making process of IPs who travel to developing nations for fertility treatments to understand their reasons for not meeting the surrogate and what impact this has on them and the resultant child (König, Chapter 13).

Secrecy and Social Stigma Associated with Surrogacy

Surrogacy is stigmatised in Indian society and is kept a secret, buried under societal pressure as the surrogates' family considers it immoral (Pande 2010, 2011). Research has shown that surrogates face social humiliation, criticism and might be shunned by neighbours or family members (Deonandan et al. 2012; Karandikar et al. 2014). This emotionally draining experience of keeping surrogacy a secret could contribute to poorer psychological outcomes for surrogates, as according to inhibition theory, secrecy can affect the physical and subjective well-being of the secret keeper (Pennebaker 1985; Kelly 2002).

In our sample, all of the surrogates were hiding their involvement in surrogacy to some extent. A third of the surrogates were hiding surrogacy from everyone, and approximately two-thirds of the surrogates were hiding surrogacy from most people. Surrogates created stories and used different lies to conceal their surrogacy from family, neighbours and the wider community. For instance, one said: "Yes, neighbours ask me about it and I repeat the same, that I have a disease". Another expressed: "I am not hiding my pregnancy. I just tell people that it is my baby. After my delivery I will say that I have given my baby to my sister or my brother".

Moreover, in order to explain their absence to relatives and neighbours, they often said they had taken up a job as a domestic helper in a nearby town. For example: “My mother keeps making excuses that I am at work or I have been to my in-laws place or something else, whatever she feels like saying. We are telling them 100 lies in order to hide a single truth”. Another, who could not hide her surrogacy successfully and faced criticism, said: “My neighbours say that it is ‘paap’ (sin) in their opinion because we get money by giving a child”.

In addition, people’s lack of education and awareness about what surrogacy involved further confounded and complicated whether to tell others. A few surrogates expressed their fears that people would not understand how a surrogacy pregnancy is achieved and might associate it with “dirty sex work”. This was a bigger concern for surrogates who were separated, divorced or widowed. One of the surrogates said that: “Surrogacy is a good thing as I am the reason for someone’s happiness and I am able to relieve them from a burden. It is the problem of people if they do not understand it but I have to hide it as I don’t have a husband with me and for what reason then I am pregnant”.

Accordingly, some surrogates said that they might tell their family about surrogacy once the surrogacy arrangement is completed. For example, one commented: “My parents and in laws, they don’t know what surrogacy is and it is difficult to make them understand so I am hiding from them now and will tell them later”.

A few surrogates mentioned that they did not conceal their involvement in surrogacy, for instance one said: “No I did not hide from anyone. I thought when I go home and they will see my stomach and later ask me where is the baby? So I told them all and they are ok with it”. Another surrogate spoke about how the support she received from her husband enabled her to be more open with others: “If my husband is with me why should I be scared? I didn’t hide from anyone, this is nothing bad and I am not afraid”.

A few surrogates challenged society’s “moral” attitude towards surrogacy whilst explaining reasons for not keeping it a secret. For instance, one said: “The baby has not been formed due to wrong methods, right. Whoever asks me, I tell them about it and slowly all the people are coming to know about it”.

In these cases, surrogates used the words "good" and "not wrong" to explain why they did not hide surrogacy. Another surrogate emphasised how it was her choice: "Why to hide? It is your body and you have taken the risk then why to hide from others".

Since most of the surrogates were hiding surrogacy from people, it was unlikely for them to receive criticism. The majority of surrogates instead used anticipation or fear of criticism, as a reason to hide their surrogacy. For example, one surrogate said: "I don't want neighbors or anyone else to know about it or my husband and my kids will be affected the most as their names will be spoilt".

In the present study, it was found that hiding surrogacy and facing criticism were associated with feeling depressed, a few months after the birth of the child but not during pregnancy. It is possible that during the pregnancy, criticism and social disapproval was less of an issue as living in the surrogacy hostel protected surrogates from this. However, after the birth of the surrogacy child, surrogates often found it difficult to readjust to life at home where family members (except husbands) were usually unaware of their surrogacy pregnancy. Thus, they had to continue this secrecy even after the surrogacy arrangement had ended. Interview transcripts offered additional information on how most surrogates were constantly negotiating whether and to what extent they should conceal surrogacy from their family, friends, neighbours and society. Overall, the creation of false stories and keeping up with their lies may cause emotional distress to the surrogates, during and after the surrogacy arrangement.

The Role of Social Support

Research in the West has shown that surrogates often receive less social support during pregnancy in comparison with general expectant mothers (Fischer and Gillman 1991) or intended mothers (van den Akker 2007). However, in Western countries, it is common for surrogates to receive psychological counselling which can provide them with a source of support (Söderström-Anttila et al. 2016). In India, however, surrogates may be more vulnerable to mental health problems as these services are not usually available to them (Karandikar 2014). In the present study, two-thirds of the surrogates felt that they had received sufficient support during pregnancy with some feeling unsupported. The two main sources of social support for an Indian surrogate were from their family and from other surrogates.

Support from the Family

In the current sample, most surrogates discussed surrogacy with prominent female figures in their lives, e.g. their mothers and sisters, which provided them with a sense of support, especially during the pregnancy. The support came in different forms, for example regular phone calls to ask after their health, taking care of their children whilst they lived in the surrogacy house or visiting them at the surrogacy house. Surrogates rarely disclosed their surrogacy to their children, as they were mostly very young. For example, one said: "They ask me, mummy what has happened to your stomach. I say it is gastric problem".

It was common for husbands to initially reject the idea of surrogacy, calling it "dirty work" usually because they could not understand the medical procedure involved in gestational surrogacy. For example, one of the surrogates said:

> He was scared but later he understood. I explained to my mother-in-law and husband that one does not have to make physical relationship. My husband would not hear that. He used to think something different. That is why he would not send me. Later on I made him understand how it is done. A baby is given birth by injections. That is why he sent me.

Surrogates also reported that their husbands struggled with the idea of how they would explain the pregnancy to others. However, it was common for surrogates to report that they had persisted in convincing their husbands to allow them to do surrogacy, with some surrogates arranging for their husband to meet the agent or staff at the fertility clinic to learn more about what surrogacy involves.

Support at the Surrogacy House

In India, it is common for surrogates to live in surrogacy houses during their pregnancy, which are usually located near fertility clinics. Although the Western media has been negative in their reports about the existence of surrogacy houses, the majority of surrogates enjoyed living in them and developed strong friendships with each other. As has been reported by others, the surrogacy house provided them with a feeling of kinship and sisterhood (Pande 2011; Vora 2014), which translated into feeling supported. For example, in the present study, one of the surrogates expressed: "It was easy living in the hostel. At first I felt strange, but then the other ladies said, 'Don't worry, we will take care of you.' I felt happy and good being there".

Social support during pregnancy was also found to have a negative relationship with depression in surrogates, a few months after delivery (Lamba et al. 2018). That is, surrogates who had felt less supported during pregnancy were more likely to feel depressed, after the birth. This could be attributed to having to move back home where they might have found themselves unable to share their feelings and experiences of surrogacy and handover with others. For example, one of the surrogates said: "I am fine now. I was happy here. After going home, I had to listen to everybody criticising me and no one used to do that here".

Moreover, most of the surrogates reported how they missed living in the surrogate house with their friends. For instance, one said: "I miss being there. All are loved ones". Another said, "Nowadays I just think of being at the surrogate house because there only I could sleep peacefully".

In summary, living at the surrogacy house provided surrogates with support by enabling them to share their thoughts and feelings with other surrogates, and removing the burden of maintaining a secret and hiding surrogacy from those around them. The surrogacy house helped in keeping the surrogacy pregnancy hidden, further providing a protective environment and an escape from social disapproval.

The Role of Financial Compensation

In commercial surrogacy, the surrogate is compensated for gestating, giving birth and relinquishing parental rights to the intending parents (Wilkinson 2003). Indian surrogates have been found to be primarily motivated by financial gain with the money earned being spent on children's education, renting/buying a house and paying off debts (Pande 2011; Karandikar et al. 2014; DasGupta and Dasgupta 2014). Therefore, satisfaction with the money received from surrogacy is likely to affect the surrogates' experience of and attitude towards surrogacy.

In the present study, all the surrogates reported a financial incentive as their motivation for becoming a surrogate. Surrogates received approximately £3000 for carrying one baby and £3250 for carrying twins. Approximately £50 was paid as a monthly income for living in the surrogacy house which was included in the total amount. In the narratives about their life following surrogacy, the surrogates often spoke about their satisfaction or dissatisfaction with the money received. For example: "Everything has become good, we don't have any tensions now as we had earlier, now we started a small business and life is going on well". For this surrogate, the money had been used to buy a business and improve her family's lifestyle, releasing her from the stresses she faced prior to being a surrogate.

Surrogates built different narratives around payment in commercial surrogacy. One of them explained how payment puts both the IPs and surrogates on an equal footing. She said: "To gain something you have to lose something right so, they need a child and we need money. Nobody will purposely become a surrogate. Give blood, take tablet what do they all mean? We are also going through something and they are giving money for that not for free".

Another surrogate explained how payment helps maintain an emotional distance from the unborn child, she said: "I was alright… The agreement is there that I will never have any right over the baby… The baby belongs to the parents and I have been paid for that". This narrative is in line with previous research which found that being paid for surrogacy helps the surrogate to remain detached from the foetus (Baslington 2002).

During the follow-up interview, the majority of the surrogates said that they had no complaints regarding payment and most of them expressed satisfaction with the payment they had received. Surrogates who were married also reported that they, rather than their husbands, took the decisions on how to spend the money earned via surrogacy.

Conclusion

The present study suggests that sociocultural factors, such as lack of social support and secrecy and stigma associated with surrogacy, may contribute to higher levels of depression in Indian surrogates. Furthermore, other features of surrogacy unique to the Indian context, such as satisfaction with payment, meeting the IPs and the newborn and feeling supported at the surrogacy house, may be associated with positive experiences of surrogacy. Contrary to expectations, the surrogates' bond to the foetus does not lead to significant emotional difficulties in relinquishing the baby.

Whilst this study is the only study to examine the psychological impact of surrogacy for surrogates in low-income countries, it is important to note that the present study was carried out prior to the change in the law on surrogacy in India. The new surrogacy bill allows "altruistic surrogacy" to only Indian infertile couples, who have been married for a minimum of five years and who have not had a child previously. The surrogate should be a close family relative who can only receive a payment that covers her basic expenses (Kotiswaran, Chapter 7).

Some of the findings from the present study raise concerns regarding the new restrictive form of altruistic surrogacy in India. Previous research has shown that payment in compensated surrogacy enables surrogates to feel less connected to the foetus. Within the new altruistic model, it would be interesting to observe whether lack of payment would change a surrogate's feelings towards the unborn child/ren during pregnancy and/or newborn after birth.

Within the commercial context, the present study interestingly found that satisfaction with payment was associated with lower levels of stress during pregnancy. However, it remains to be seen how the motives of surrogates change within an altruistic surrogacy arrangement and how that may impact on her psychological health. In fact, surrogates have been found to report both extrinsic (e.g. payment) and intrinsic (e.g. helping others) motives for becoming a surrogate (Jadva et al. 2003), leading to moral theorists to challenge a polarised image of a surrogate's motives, which are viewed as either purely for money or as an act of kindness (Stuhmcke 1995). Furthermore, it has been argued that in terms of the actual money obtained by surrogates, there is little difference between the amounts a surrogate receives within an altruistic arrangement such as the UK and a compensated one such as the US (Jadva 2016).

An additional concern relates to the requirement for the surrogate to be a family member as this could lead to women being coerced into acting as a surrogate, particularly if she is a subordinate member of the family. During pregnancy, difficult decisions related to aborting the foetus or unpleasant events, such as miscarriages, can lead to blame and guilt within a close family environment. Once born, the likelihood of close proximity of the surrogate in the child's life may undermine the mother's parental role.

Moreover, the stigma associated with surrogacy in India may mean that the child is not told about their birth using surrogacy and may not realise that a member of their family is also their surrogate. Given that secrecy can burden relationships and can take a great deal of cognitive effort (Pennebaker 1985; Kelly 2002), the psychological impact of this on surrogates in the Global South needs to be evaluated.

Bibliography

Baslington, H. (2002). The social organization of surrogacy: Relinquishing a baby and the role of payment in the psychological detachment process. *Journal of Health Psychology, 7*(1), 57–71.

Bhatnagar, P., Singh, M., Pandey, M., & Amitabh, S. (2011). *Manual for anxiety, depression and stress scale*. Agra, UP: National Psychological Corporation.

Braverman, A., Casey, P., & Jadva, V. (2012). Reproduction through surrogacy: The UK and USA experience. In M. Richards, G. Pennings, & J. B. Appleby (Eds.), *Reproductive donation: Practice, policy and bioethics* (pp. 289–307). Cambridge: Cambridge University Press.

British Medical Association. (1996). *Changing conceptions of motherhood. The practice of surrogacy in Britain*. London: Greener Books.

Ciccarelli, J. C. (1997). The surrogate mother: A post-birth follow-up study. *Dissertation Abstracts International, 58*(3-B), 1522.

Cranley, M. S. (1981). Development of a tool for the measurement of maternal attachment during pregnancy. *Nursing Research, 30*(5), 281–284.

DasGupta, S., & Dasgupta, S. D. (Eds.). (2014). *Globalization and transnational surrogacy in India: Outsourcing life*. Lanham, MA: Lexington Books.

Deonandan, R., Green, S., & Van Beinum, A. (2012). Ethical concerns for maternal surrogacy and reproductive tourism. *Journal of Medical Ethics, 38*(12), 742–745.

Fischer, S., & Gillman, I. (1991). Surrogate motherhood: Attachment, attitudes and social support. *Psychiatry, 54*(1), 13–20.

Glover, V. (2014). Maternal depression, anxiety and stress during pregnancy and child outcome; what needs to be done. *Best Practice & Research Clinical Obstetrics & Gynaecology, 28*(1), 25–35.

Green, C. (2005). *Conquering stress* [e-book]. http://Lulu.com. Available at https://itunes.apple.com/gb/book/conquering-stress/id443345946?mt=11. Accessed 12 Nov 2017.

Hohman, M. M., & Hagan, C. B. (2001). Satisfaction with surrogate mothering: A relational model. *Journal of Human Behavior in the Social Environment, 4*(1), 61–84.

Imrie, S., & Jadva, V. (2014). The long-term experiences of surrogates: Relationships and contact with surrogacy families in genetic and gestational surrogacy arrangements. *Reproductive Biomedicine Online, 29*(4), 424–435.

Jadva, V. (2016). Surrogacy: Issues, concerns, and complexities. In S. Golombok, R. Scott, J. B. Appleby, M. Richards, & S. Wilkinson (Eds.), *Regulating reproductive donation* (pp. 126–139). Cambridge: Cambridge University Press.

Jadva, V., Blake, L., Casey, P., & Golombok, S. (2012). Surrogacy families 10 years on: Relationship with the surrogate, decisions over disclosure and children's understanding of their surrogacy origins. *Human Reproduction, 27*(10), 3008–3014.

Jadva, V., Imrie, S., & Golombok, S. (2014). Surrogate mothers 10 years on: A longitudinal study of psychological well-being and relationships with the parents and child. *Human Reproduction, 30*(2), 373–379.

Jadva, V., Murray, C., Lycett, E., MacCallum, F., & Golombok, S. (2003). Surrogacy: The experiences of surrogate mothers. *Human Reproduction, 18*(10), 2196–2204.

Karandikar, S., Gezinski, L. B., Carter, J. R., & Kaloga, M. (2014). Economic necessity or noble cause? A qualitative study exploring motivations for gestational surrogacy in Gujarat, India. *Affilia, 29*(2), 224–236.

Kelly, A. E. (2002). *The psychology of secrets*. New York: Springer.
Lamba, N., Jadva, V., Kadam, K., & Golombok, S. (2018). The psychological well-being and prenatal bonding of gestational surrogates. *Human Reproduction, 33*(4), 646–653.
Lazarus, R. S. (1966). *Psychological stress and the coping process*. Mc-Graw Hill: University of Michigan.
Lovibond, P. F., & Lovibond, S. H. (1995). The structure of negative emotional states: Comparison of the depression anxiety stress scales (DASS) with the beck depression and anxiety inventories. *Behaviour Research and Therapy, 33*(3), 335–343.
Malhotra, A., & Malhotra, R. (2013). *Surrogacy in India: A law in the making India*. New Delhi: Universal law Publishers.
Mitra, S., & Schicktanz, S. (2016). Failed surrogate conceptions: Social and ethical aspects of preconception disruptions during commercial surrogacy in India. *Philosophy, Ethics, and Humanities in Medicine, 11*(1), 9.
Pande, A. (2010). Commercial surrogacy in India: Manufacturing a perfect mother-worker. *Signs: Journal of Women in Culture and Society, 35*(4), 969–992.
Pande, A. (2011). Transnational commercial surrogacy in India: Gifts for global sisters? *Reproductive Biomedicine Online, 23*(5), 618–625.
Pennebaker, J. W. (1985). Traumatic experience and psychosomatic disease: Exploring the roles of behavioural inhibition, obsession, and confiding. *Canadian Psychology/Psychologie Canadienne, 26*(2), 82.
Pereira, B., Andrew, G., Pednekar, S., Pai, R., Pelto, P., & Patel, V. (2007). The explanatory models of depression in low income countries: Listening to women in India. *Journal of Affective Disorders, 102*(1), 209–218.
Ragoné, H. (1994). *Surrogate motherhood. Conception in the Heart*. Boulder, CO: Westview.
Smerdon, U. (2008). Crossing bodies, crossing borders: International surrogacy between the United States and India. *Cumberland Law Review, 39*, 15–85.
Söderström-Anttila, V., Wennerholm, U. B., Loft, A., Pinborg, A., Aittomäki, K., Romundstad, L. B., et al. (2016). Surrogacy: Outcomes for surrogate mothers, children and the resulting families—A systematic review. *Human Reproduction Update, 22*(2), 260–276.
Spielberger, C. D., & Rickman, R. L. (1990). Assessment of state and trait anxiety. In N. Sartorius, V. Andreoli, G. Cassano, L. Eisenberg, P. Kielholz, P. Pancheri, & G. Racagni (Eds.), *Anxiety. Psychobiological and clinical perspectives* (pp. 69–83). New York: Hemisphere Publishing.

Stuhmcke, A. (1995). For love or money: The legal regulation of surrogate motherhood. *Murdoch University Electronic Journal of Law, 2*(3), E4.

Teman, E. (2010). *Birthing a mother: The surrogate body and the pregnant self.* Berkeley, CA: University of California Press.

van den Akker, O. B. (2007). Psychosocial aspects of surrogate motherhood. *Human Reproduction Update, 13*(1), 53–62.

Vora, K. (2014). Experimental sociality and gestational surrogacy in the Indian ART clinic. *Ethnos, 79*(1), 63–83.

Wilkinson, S. (2003). The exploitation argument against commercial surrogacy. *Bioethics, 17*(2), 169–187.

World Health Organisation. (2017). *Depression: 'let's talk' says WHO, as depression tops list of causes of ill health* [online]. Available at http://www.who.int/mediacentre/news/releases/2017/world-health-day/en/. Accessed 25 Mar 2017.

10

Surrogacy and Social Movements in India: Towards a Collective Conversation

Vrinda Marwah and Sarojini Nadimpally

Introduction

Surrogacy is located within a matrix of complex contemporary social issues. The surrogacy industry in India grew rapidly in recent years—raising several questions around appropriate regulation, medical protocol and human rights—before the government announced the Surrogacy (Regulation) Bill 2016 (Ministry of Health and Family Welfare 2016). This latest bill proposes to ban commercial surrogacy in India. It will bar foreigners, gay couples, people in live-in relationships and single individuals from accessing surrogacy. Only childless, heterosexual Indian couples, married for a minimum of five years, will be

---

The authors would like to dedicate this chapter to Preet Rustagi, a friend and comrade, who passed away unexpectedly in 2017.

V. Marwah (✉)
The University of Texas, Austin, TX, USA
e-mail: vrinda.marwah@utmail.utexas.edu

S. Nadimpally
Sama-Resource Group for Women and Health, New Delhi, India

S. Mitra et al. (eds.), *Cross-Cultural Comparisons on Surrogacy and Egg Donation*,
https://doi.org/10.1007/978-3-319-78670-4_10

eligible for surrogacy. The surrogate must be married and a 'close relative' of the couple, who has birthed a healthy child prior to the surrogacy. She is permitted to be a surrogate only once in her lifetime. She cannot receive any money for her 'selfless' act (Kotiswaran, Chapter 7).

Despite the government's introduction of this proposed bill, there are several disagreements about what a law on surrogacy should delineate. Activists are divided about whether there should even be a ban, or if different types of surrogacy should be regulated differently (commercial versus altruistic surrogacy, genetic versus gestational surrogacy). What should the rights and duties of different parties in the surrogacy arrangement be? These questions are rife with contestation. And this contestation is perhaps symptomatic of a larger conundrum; given the complexity of surrogacy and its intersection with several other issues—such as child rights, sexual rights, disability—there is little agreement on what are and are not, or rather should and should not be seen as, the problematic aspects of the industry.

This chapter is an attempt by Sama, a New Delhi-based resource group for women and health,[1] to move towards more collective clarity on the question of surrogacy. It reflects thematically on interviews[2] with activists and experts from key social movements in India on which surrogacy has a bearing. We shortlisted fourteen interviewees based on two criteria: (1) relevance of area of work to surrogacy, and (2) knowledge of ongoing debates in surrogacy, and conducted individual interviews lasting one to three hours in duration. Consent was taken from the interviewees to use their names.

This paper is divided into three sections, thematised according to the issues that were raised by respondents. The first section 'Connecting the micro and macro of surrogacy: body, agency, and exploitation' considers broad political questions around how to understand surrogacy. It discusses commercialisation, parallels with sex work, and the larger labour market context in which surrogacy is taking place. The second section 'Constituencies for consideration in surrogacy' examines the implications of surrogacy for LGBTQI rights, people with disabilities, people living with HIV, and child rights. The third section 'Ethical and regulatory issues in surrogacy' highlights a public health and bioethics perspective on surrogacy, ethical dilemmas like genetic determinism and medical indication, and regulatory frameworks.

Given that social issues like disability, sexuality, and labour rights are broad-ranging and complex, this chapter does not claim to represent a monolithic perspective from any one movement. Each social movement included in this chapter is only partially represented by the actor interviewed; the movements themselves are undoubtedly dynamic, diverse and multiple; as such, readers must allow for plural, even conflicting, perspectives on surrogacy from within each social movement. Moreover, the nature of the relationship between surrogacy and social movements cannot be understood in a simplistic or linear fashion. Public health scholar and activist Imrana Qadeer suggests approaching the question of surrogacy through social movements, rather than *vice versa*,

> If you believe people need better wages to survive, better health care, housing and water, then you will organise people to get these things. You will join political movements. Just because one is researching surrogacy, one cannot turn the question on its head and ask, what are movements doing about surrogacy? It is for us to connect with movements and then from there understand why women are doing surrogacy. This is much more than a medical issue. It is a larger socio-economic and political issue.

This chapter is an attempt to understand how key actors from different social movements view and construct the issues in surrogacy from their own unique locations, thus enabling a deeper understanding of surrogacy and its many themes and contests.

Connecting the Micro and the Macro of Surrogacy: Body, Agency and Exploitation

Understanding Surrogacy: What Approach Do We Take?

In recent decades, the assisted reproductive technologies (ARTs) industry has expanded rapidly, and with it, so have surrogacy services. However, the arrangement of surrogacy has been mired in much more controversy than ARTs themselves; often surrogacy is banned in countries where ARTs otherwise flourish. Surrogacy has received attention from media persons, policymakers, scholars and activists. In particular, feminist scholars and activists have tried to understand the why and how of surrogacy, with the purpose of crafting a response to it through policy or movements or other channels. Most of our interviewees, rather than exceptionalising surrogacy, tried to locate it within larger trends in women's employment, health, etc., and suggested continuities with older concerns regarding women's control over their bodies, the devaluation of their labour, etc. Prabha Nagaraja, a sexual and reproductive health and rights activist working with Delhi-based NGO TARSHI, said,

> Even before there was technology for assisted reproduction, there was a form of surrogacy; what we today call altruistic genetic surrogacy has always been around in families. However, now that the technology has become an industry, the potential for exploitation in this exchange has grown exponentially.

Women's studies and disabilities studies scholar Renu Addlakha concurs with this view. She calls the rise of the surrogacy industry 'the murky world of a reproductive subterranean', one that marshals both innovative new technologies and regressive old ideas. Addlakha explains that it is after all the strong social desire to continue family lineage that makes business-minded clinics, run by otherwise conservative gynaecologists, conduct profitable but secretive surrogacy arrangements. At the same time, Addlakha also highlights the challenge that surrogacy poses to marriage. She points out, 'Normatively, marriage is supposed to lead to a child between two people, but in surrogacy a third party is brought in to achieve this function, thus exposing its naturalised character'. Nagaraja agrees with this reading. For Nagaraja, surrogacy has the subversive potential to challenge mainstream notions of motherhood. When considered theoretically, apart from the practical challenges of regulation, Nagaraja argues that surrogacy does indeed broaden the choices in people's lives, by introducing the possibility of the separation of reproduction from sexuality, and even from marriage. She goes on to give the example of TARSHI's trainings on gender and sexuality. In discussions on the nature of womanhood, these trainings use the issue of surrogacy to argue that motherhood is not a romantic eternal bond; women can and do rent out their wombs. And so, just as reproductive rights include the right not to have children, they also include the right to decide how to have children, including through assisted means.

As we consider the terrain of surrogacy and its many concomitant challenges, Nagaraja emphasises the need to be non-judgmental about people's choices.

> In the sexual and reproductive health and rights movement, we talk about being non-judgmental, respecting diversity, allowing for difference, etc. Every person makes decisions based on factors in their lives that we cannot know from the outside. There may be exploitation, and the parties involved need to be aware of it, but it is still their choice. Just because we think something is not right we cannot stop others from taking that decision. This principle holds, whether it is surrogacy or abortion or sex work.

Nagaraja's point here is an important one to consider: someone's reproductive choices may make others uncomfortable, but on principle, the alternative should not be to prescribe to women what they should do with their bodies.

The Exploitation in Surrogacy: Is Commercialisation the Problem?

The notion that the commercialisation of surrogacy is exploitative is a key pivot around which the acceptability of surrogacy has been debated (Sándor, Chapter 3). Are we okay with the commercialisation of surrogacy into a multi-million dollar industry? Is it not exploitative and extractive that poor women have to resort to surrogacy at considerable risk simply to educate their children or build their homes?

Amar Jesani, bioethicist and founder-editor of the Indian Journal of Medical Ethics (IJME), sees the issue of exploitation as a difficult one to address. He points out that the word exploitation itself has both positive and negative connotations; it can mean making the most of a situation, or taking unfair advantage of another. As inequities rise, India's growing middle class can afford to pay for the services of a surrogate as a quick and easy panacea for their fertility troubles. Nonetheless, Jesani asks, in the absence of a larger project of social change in the foreseeable future, do we have the paternalistic right to stop poor women from earning using their bodies? Can we take away one of the few options available for their survival when they themselves might want to use that option? He compares this to clinical trial participants who earn anywhere between Rs. 5000–15,000 in bioavailability and bioequivalence studies.[3] While we may find this exploitative, Jesani argues that because the participants belong to a class that is already employed in low paying and high-risk jobs, they consider the sum of payment attractive and the risks acceptable. Therefore, he concludes that it is not easy to define exploitation in India, especially at the micro-level.

Speaking of the macro level, he adds,

> There are philosophical traditions like Marxism that are attentive to the coercion of the capitalist system, which leaves the wage labourer with no choice but to enter into a contract with the owners of the means of production. However, fields like clinical medicine and bioethics do not recognise systemic coercion. The violation of autonomy in bioethics has to be a physical and visible violation. If one has not been forced into a contract, then one is considered to have freely entered it. It is only now that health system research is illuminating systemic ethical challenges, which is a welcome development.

Commenting on particular practices, Jesani points out that unlike organ trade—about which there is near consensus that a market should not be allowed to proliferate—practices like surrogacy and sex work relate to regenerative bodily capacities, which makes the latter more debatable. However, he is clear about his own position; Jesani is opposed to the rampant marketisation of surrogacy. He stresses the difference between a medically infertile couple seeking a third party arrangement, and rich couples who hire surrogates only because they can. He says categorically, 'For the latter, it is not pathology but purchasing power that is the determinant'. This question of purchasing power is one that comes up again and again. Child rights activists Bharti Ali and Enakshi Ganguly point to perhaps the most famous cases in India—Bollywood actors Aamir Khan and Shahrukh Khan, who had children through surrogacy. In a tweet following the birth, Aamir Khan expressed his gratitude towards everyone involved in arrangement except the surrogate mother. Ali and Ganguly see this as an instance of the invisibilisation of the surrogate. As Ali puts it,

> Look at the difference between those who are opting to hire surrogates and those who are becoming surrogate mothers. It is after all a completely class stratified industry. The poor woman who is becoming a surrogate may not have done so if she had other ways of sustaining herself and her family. Where is the choice then? It is simply laissez-faire economics.

Public health and ethics scholar Mala Ramanathan echoes this discomfort and wonders if the most autonomous woman imaginable—educated, upper class—would do commercial surrogacy. A different perspective on this question is offered by Prabha Nagaraja and queer feminist activist Pramada Menon. For Pramada, whether a woman would willingly opt for surrogacy or sex work if she had other options is impossible to say. However, the question itself seems to come from a moralistic, classist place—as though the decision to have sex or give birth for money can only be viewed with suspicion. This idea that money corrupts is empirically false and is mostly mobilised against already marginalised sections, particularly poor women (Zelizer 1994) in care work. Pramada holds that whatever the reason, we should not stand in judgment on these decisions. She argues that we should instead invest our energy in ensuring that women receive all necessary protections. And given that so much of life is already commodified, Nagaraja wonders how we can turn the clock back on commercialisation. From a rights-based perspective, she suggests the best we can do is to raise awareness, so that women opting for surrogacy have information about the issues involved and can make their peace with their decisions rather than feeling guilty or victimised.

Pramada also points out that any decision regarding the body is complicated to conclusively analyse because it involves varied preferences. She illustrates this with an example from her trainings on sexuality and sexual rights,

> We do an exercise with participants to understand both the individual and collective nature of choice. We tell the story of a married woman who goes to meet her lover everyday across a river. One day in filmy style it begins to pour on her way back. Now, she has three choices. First, a boatman can take her safely across the river if she has sex with him. Second, she can walk across a bridge, but on the other side of the bridge

lives a man who has been accused of rape many times but has never been convicted. The third option is for her to jump into the river and kill herself because she can't swim. What should she do?

Over the years, fewer and fewer people pick the third option, but every now and then someone will say 'she is a sinner so she may as well die'. The real difficulty lies in choosing between the boatman and the rapist on the bridge. And it is so interesting how so many women prefer the rapist on the bridge! When you ask why they want to risk confronting a potential rapist by going across the bridge, they say maybe we will be able to defeat him, maybe he will not want to rape me, maybe I will be able to negotiate with him, maybe today is his off day, maybe he is not really a rapist and on and on. All these arguments come up in favour of the potential rapist, but what about taking the simple and straightforward option of having sex with the boatman? What about negotiating with him? But because that has to do with your body, being agentic is not really an option. The casual transactional sexual act is somehow not a real choice. The real choice is in risking rape. See with the bridge option, you may face violence but you can hold on to the notion of your body being sacrosanct. But with the option of the boatman, somehow you only see violence, you don't see any choice. Why? It is because of how deeply ingrained our notions of the purity and passivity of the body are.

Feminist debates in surrogacy and sex work

Comparisons are often drawn between sex work and surrogacy, given that both commercialise sexual and reproductive labour otherwise considered intimate and familial. Unlike sex work though, surrogacy is seen to be 'for a noble cause'.

The movement for sex workers' rights has shifted the understanding of sex work from one of violence to one of work. This movement defines sex work as the adult consensual provision of sexual services for money and argues that women who make this choice not only pick the best labour market option available to them, but in doing so, subvert the dominant social norm of passive female sexual conditioning. It firmly opposes the vocabulary of prostitution, as well as the notion that the experience of sex work is homogenous, inherently violent and non-consensual, and deserving of a ban.

Sex workers' rights advocate Meena Seshu and Aarthi Pai point out that sex and reproduction are glorified when offered for love, but demonised when offered for money. They argue that there is an inherent infantilisation of women in debates about women's bodies, including within feminist movements. For decades, the women's movement in India has engaged with issues related to the body. It has, for instance, opposed the conflation of sexuality and reproduction, and the reduction of women to their wombs and vaginas. However, Seshu and Pai argue that even among these feminist activists, there is a quiet squeamishness about sex work as a legitimate livelihood option. Seshu and Pai wonder if we are seeing the same squeamishness with reproductive rights advocates about surrogacy. They caution that these moral double standards, and the unwillingness to listen to the voices of sex workers and surrogates themselves, hinder broader alliance-building. Women's right to self-determination has always occupied a central place in feminism; sex work and surrogacy offer the litmus test of precisely this right. Seshu and Pai insist that in order to pass this test—in order to truly advance a non-protectionist and rights-based approach to sex work and surrogacy—we need to develop a sharper understanding of consent. Seshu says

> Interference from families, states, and civil society in women's lives is premised on a discourse of victimhood and victimhood alone; in actuality this serves to legitimise the control of these institutions over women's bodies. It was in response to this that the sex workers movement came up with the slogan 'my body is my business', pun totally intended. The key question is this: does a woman have the ability to consent to whatever kind of labour she wants to provide in whatever kind of situation? The problem is that we do not really believe that women in constrained economic circumstances have the ability to consent. We need to make this leap. Women can and do take decisions about their lives and their work by weighing the pros and cons, in any situation. We need to accept this.

Seshu and Pai add that the potential for exploitation is present in every exchange where there is a power difference between the parties involved. The answer to this, they maintain, cannot be to ban every such exchange. Instead, we need to regulate these exchanges, like we would do with any other trade. Both sex work and surrogacy need regulation.

Under the Immoral Traffic Prevention Act 1956, sex work in India is rendered practically criminal; therefore, the regulatory model that sex workers seek from the state is decriminalisation. A criminalised environment creates unsafe working conditions for sex workers, such as exploitative practices by third parties and lack of access to non-discriminatory health services. There is a lesson here for women's groups that seek regulation of the surrogacy industry and the many links in its large chain. Seshu and Pai warn that if commercial surrogacy is criminalised, we would be making the same mistakes that we are now trying to correct with sex work in India, i.e., the further victimisation of women in the name of their protection.

Feminist scholar and activist Nivedita Menon notes that feminist lessons from sex work are tested in surrogacy because surrogacy is like no other form of labour. Nivedita argues that, without being biologically essentialist, we must remember that pregnancy is a unique condition where a woman has another life inside her body. She draws parallels with feminist debates around abortion. Feminists believe that there should be safe and free access to abortion under whatever conditions or for whatever reasons a woman chooses to abort. Here, she points out that feminists are in effect saying there is only one individual in a pregnancy, i.e. the woman. The foetus is only part of her body. However, this position is tested when it comes to sex- or disability-selective abortion. She argues that in saying no to sex or disability selection, we are prioritising an abstract category of 'women' or 'the disabled' on the basis that they have the right to be born, while making completely irrelevant the actual body of the woman who is bearing the female or disabled foetus. According to Menon, this is troubling. Instead of making it so that a woman has to bear whatever foetus she is carrying, we need to change the circumstances in which women seek sex- or disability-selective abortions. It is these feminist debates that are extended and tested in surrogacy. First of all, how do we understand the human being who is the surrogate—is it one or is it two? And does the foetus have priority or does the woman bearing the foetus have priority? Further, Nivedita points out that a commercial surrogate performs labour, but she is also the 'raw material' of the industry. By this she means, it is not just the work the surrogate's body does, it is also her body *itself* that is the resource in surrogacy. This situation where the body is both labour and raw material is unique. How does one make the conditions of work fair in such a situation? What kind of contract, if at all, can make this fair?

Sociological scholarship by Melinda Cooper and Catherine Waldby (2014), as well as by Sharmila Rudrappa (2015), understands surrogacy as a unique form of labour, produced as much by gendered emotions, networks and subjectivities, as it is by the organisation of markets and developments in technology. Other respondents in our sample also grappled with these themes. Imrana Qadeer emphasised that surrogates should be paid better on the basis that their work requires biological input, which cannot be at par with other resources. The product of surrogacy is another sentient being, and an unwanted baby cannot be returned like a defective product in any another industry. Qadeer suggests that the surrogate be compensated for much more than the nine months of pregnancy and delivery, since the process and impact of surrogacy begins before and continues after this time. Addlakha agrees with this and adds that we should apply a broader understanding of the word labour. Labour in surrogacy is everything from having the baby, to keeping the baby a secret from your family and community.

Articulating a Position: Surrogacy as Labour

Even before but especially since the government's 2016 Surrogacy Regulation Bill, the conversation on surrogacy has been polarised around whether the arrangement should be banned or regulated. Citing the example of practices like child labour, Ali and Ganguly argue that bans do not work in a country like India. Jesani stated his categorical support for altruistic surrogacy, but considered commercial surrogacy problematic. On the other hand, Qadeer called the arguments for a ban on commercial surrogacy ethically unconvincing:

> I no more argue for a ban. I no more argue for altruistic surrogacy. Why should I? As it is the woman is dying and you want her to be altruistic? There is no ethics in fighting for these middle-class values. Let those women survive the way they think they can survive and the issue is how we can help them where they are today... the only ethical position is to

> push for some kind of responsibility on the part of institutions that are doing surrogacy.

By zooming out on the arrangement of surrogacy and considering it within the larger context of women's labour in India, labour scholar Preet Rustagi emphasises the need to work out the modalities of a safe and ethical surrogacy rather than debating its acceptability. Rustagi stresses that the country is at a vulnerable juncture with a mass of people who are so poor that putting their bodies at risk does not discourage them from taking up certain kinds of work. Rustagi argues that surrogacy is similar to other livelihood options for this class of women, that is, options that are characterised by a lack of information and rights, illiteracy, no standard pay and insecurity, while also providing a much-needed economic opportunity.

Nonetheless, because surrogacy transports motherhood from the domain of family to that of contract, it demystifies reproduction by treating it like any other form of unpaid work done by women in families. According to Rustagi, a lot of the polarisation around surrogacy comes from the 'emotional baggage' of seeing motherhood in a market context. She contends that it is this 'emotional baggage' that gives people a misplaced confidence in the institution of the family, and enables their preference for altruistic rather than commercial surrogacy. Rustagi is deeply aware of the strong and almost-interchangeable association of womanhood with motherhood, because it persists in other contexts too. She gives the example of crèches; it is assumed that women who are child care workers do not need to be trained for their job because as women they have a natural affinity towards children. Rustagi stresses

> There are more and less pertinent questions to ask of surrogacy. It is valid to be interested in what class and caste these women are drawn from, but it is not valid to assume that all mothers feel a natural affinity for all their children, no matter what. Women are different and have different approaches to the same situation. To assume only one possible relationship of a woman to her motherhood is a fallacy that betrays our inability to think of a woman as an entity in herself rather than as deriving her status and meaning from others in her life.

Both Rustagi and Qadeer stress the need for surrogates to be organised. Rustagi cautions that collectivisation will not be easy—women do not think of the work they do in farms and homes as unpaid labour, egg donors, and surrogates often act in secret for fear of stigma, and labour unions are under attack in the country. However, she also holds that collectivisation is possible and necessary, as domestic workers and sex workers have demonstrated. Additionally, Nivedita Menon stresses that the ethical questions in surrogacy will be closer to resolution when surrogates themselves organise. Nivedita notes that when sex workers collectivised and articulated their perspectives, the feminist movement learnt things that it had not previously considered. She suggests this might happen with surrogates too, when 'we' no longer speak on 'their' behalf.

Constituencies for Consideration in Surrogacy

Lesbian, Gay, Bisexual, Transgender, Queer and Intersex (LGBTQI)

In general, respondents took a favourable view of the ability of ARTs and surrogacy to enable queer and single people to have biological families. For instance, Mala Ramanathan argues for the regulation of surrogacy because she believes a ban would be disadvantageous to the long-term interests of LGBTQIs. She also points out that commercial surrogacy is important to these constituencies, given that non-conforming individuals are often alienated from their family and kin networks, i.e. the contexts within which altruistic surrogacy occurs.

Pramada Menon discusses the complex nature of the relationship between surrogacy and LGBTQI rights. Within the field of sexual and gender justice, she insists the separation of reproduction from sex is indeed a paradigm shift, a 'light bulb moment' in how we understand and perform gender. And just as many heterosexual women want to be mothers—and the more traditional ones want to be mothers within marriage—many queer people want to have biological children. While

Pramada holds that the emotional desire to be a parent need not be glorified, she points out that it cannot be dismissed either. Queer activists have always promoted alternative family structures; with ARTs and surrogacy, LGBTQIs can theoretically and in many countries practically create a new kind of families, as long as they have the money to do so.

Nevertheless, Pramada is also quick to note that new developments like queer marriage and parenthood remain mired in older structures of power. She gives the example of the matrimonial newspaper advertisement of a prominent gay activist that specified a preference for matches from his Brahmin caste. With ARTs and surrogacy, Pramada wonders to what extent similar caste preferences can be stated and babies 'designed'.

People with disabilities

In 2014, the Baby Gammy case made international headlines when an Australian couple abandoned one of the twin children—a baby with Down syndrome—delivered to them by a Thai surrogate. Referencing this case, Renu Addlakha stresses that commissioning parents cannot be allowed to abandon a child in the event of a disability. However, she is also aware that making it legally binding for commissioning parents to accept a disabled child will not ensure that they will raise and care for the child like they should. According to Addlakha, the disability rights movement needs to contend with these and other unexplored implications of ARTs and surrogacy for disability:

> We need to engage as vociferously with the medical issues in surrogacy as we do with the social issues. One cannot be ignorant. You have to become a mini embryologist to talk with embryologists.

For Addlakha, disabled women commissioning surrogacies is a much more likely scenario than disabled women becoming surrogates. She discusses some cases that she is aware of where disabled women have availed of ARTs and surrogacy services. She explains, even when there are no functional problems with a disabled woman's reproductive system, her health may be compromised in other ways because of her disability. For example, polio can affect other organs of the body in addition to the limbs. It is also made very difficult for single disabled people to adopt, even though it is technically legal; this is just one instance of how disabled women are not regarded as—and often do not often regard themselves as—sexual and reproductive beings. As such, Addlakha understands surrogacy by situating it within the full spectrum of social and medical issues in the reproductive lives of disabled women.

People Living with HIV (PLHIV)

While the HIV movement in India has not advocated for access to ARTs and surrogacy services, some issues raised by PLHIV desirous of using these services were pointed out by advocates Veena Johari and Kajal Bharadwaj.

When PLHIV want to have biological children, there is fear of parent to child transmission of HIV. For serodiscordant couples where the man is HIV-positive and the woman is HIV-negative, ARTs like sperm-washing and Intracytoplasmic sperm injection (ICSI) allow the couple to have biological children without putting the mother or child at risk of HIV transmission. In the case of HIV-positive women who want to use ARTs, proper antiretroviral therapy and prevention of parent to child transmission services render the risk of HIV transmission through pregnancy negligible.

However, neither the ICMR guidelines nor the successive drafts of ART bill allow PLHIV to access ARTs and surrogacy. Additionally, there continues to be anxiety about whether PLHIV will live long enough to care for their children. Adoption laws are not HIV friendly either. PLHIV who want to adopt or can afford ARTs are rejected after their medical screening, and may even find their confidentiality compromised and HIV status revealed.

Given that HIV is now considered a chronic disease, Johari stresses that it is time the laws on adoption and ARTs remove all provisions that discriminate against PLHIV and queer people. And yet, when it comes to the sexual and reproductive rights of PLHIV, the battle for fair and non-discriminatory laws promises to be a long one. As Johari notes, this is evidenced by the upholding of India's anti-sodomy law, Section 377 of the Indian Penal, by the Supreme Court in 2013.

Child rights

Ali and Ganguly point out possible violations of children's rights within the surrogacy industry. For instance, there is an absence of necessary checks on commissioning parents; an Israeli man who had a child through surrogacy in India was later reported to be a convicted paedophile. Similarly, the nationality of a child born through surrogacy can be contested by governments, leaving the child potentially stateless such as in the Baby Manji case. Some of these concerns were addressed by the government; a 2012 Ministry of Home Affairs notification mandated that foreign couples coming to India furnish documentation that their countries would recognise the children born of surrogacy.

Ali and Ganguly note that other issues also need attention. They argue that as with adoption, children born through surrogacy should have the right to do a 'root search' to trace their gestational or genetic parents. While biological parents should also retain their right to refuse to reconnect, these children should be able to explore their roots with the knowledge that the search may or may not end fruitfully.

Significantly, Ali and Ganguly raise questions about our collective understanding of women's agency, and its conflicts with child rights. Ganguly said,

> Look at how conversation on juvenile justice emerged after the Nirbhaya case. Many women's groups backed the move to try children as adults in case of serious crimes. This is nothing but further victimisation of the country's poorest and most vulnerable children. What was the logic behind making the minimum sentence for sexual assault seven years under the Criminal Law Amendment Act? Don't we know that many of the cases of alleged rape are in fact booked by unhappy families when young couples elope? Look at the issue of child marriage. On the one hand, women's groups say motherhood and marriage should not be glorified, but when it comes to child marriage they are concerned that the girl's life will be ruined if her child marriage is annulled. What kind of confused messaging is this? Is a woman who has a child marriage an eternal hapless victim? During the formulation of the Prohibition of Child Marriage Act 2006, there were many who did not want child marriages to be declared illegal and automatically annulled. So now we have this

> mishmash law wherein child marriage is prohibited, but if the marriage takes place, it is voidable, not void. I don't know what we achieved by being neither here nor there. Where is the seriousness on the question of children's rights? These issues may seem unrelated to surrogacy but they are not. Surrogacy is once again generating debate between the rights of the mothers and the rights of the children born to them.

Ali and Ganguly emphasised the need for more dialogue between women's rights and child rights groups to build common ground on surrogacy and other issues.

Ethical and Regulatory Issues in Surrogacy

Public Health Perspective

What is, or should be, a public health perspective on surrogacy? According to public health activist and scholar Imrana Qadeer, surrogacy ranks low on the list of public health priorities, after the fulfilment of agendas such as primary level care, including maternal care, for all. While the prevention and cure of infertility is important, Qadeer holds that surrogacy should ideally happen in tertiary care, where a person who requires it will look for another person who is willing to participate in such an exchange (Qadeer and John 2009; Reddy and Qadeer 2010).

However, she also notes that the health context within which surrogacy is currently being practised leaves much to be desired. On the one hand, health markets and medical tourism are expanding, and on the other, public health is becoming narrow and technology-oriented. She highlights the relationship between the decline of public health and the rise of private health care. Qadeer points out that the state is unwilling to invest even 1% of its resources in public health, and of the little that is invested, there is now a move to prioritise insurance and public–private partnerships, which effectively redirect public money to the private sector. This deteriorating public health scenario is throwing health open to the private sector, but without the necessary regulatory mechanisms to ensure quality (Qadeer 1998).

Qadeer calls this a distortion of public health, part of 'a bigger disease'. She stresses the importance of returning to a model of public health which is about creating healthy societies. In this model, medicine has an important but limited role to play. The true aim of public health is to raise the level of clinical, preventive and promotive services by making them accessible to the maximum number of people, while taking into consideration factors such as total population coverage, costs, side effects, as well as nutrition, water and so on. Qadeer stresses that it is in fact the failure of the public health system to provide basic gynaecological and obstetric services that then leads to infertility in a significant proportion of women. This should be the key focus for infertility assistance by the government. Instead, she points out that surrogacy has become a money-spinner for private providers, with no monitoring of ART clinics and their protocols.

In India today women opt to be surrogates because they need the money. According to Qadeer, this is an economic problem and not a medical one, because the country is not able to create jobs, give minimum wages, provide free education for children or provide adequate housing. She insists that in the final analysis, we must be clear about the direction in which we wish to move,

> What is our vision for an ethical society? What will the nature of reproductive autonomy be for women in the future? The issue is much more than just biological; it is also social, emotional, and personal. Without taking moral positions, we must be very explicit about the exploitation inherent in surrogacy—it is an exploitation that makes women go against their integral nature and lead a fragmented existence.

Bioethics Perspective

Jesani discusses the dilemmas in surrogacy from a bioethics perspective and in doing so highlights the limits and potential of bioethics as a field. In bioethics, analysis is done at the macro-level (systems), meso-level (institutions) and micro-level (individuals), including connections and contradictions across these levels. According to Jesani, paradigms like Marxism that do systemic analyses do not offer much direction for people who are not 'professional revolutionaries' but want to 'do right by others in their everyday lives'. Bioethics as a field emerged from below to offer direction for the day-to-day challenges in medical practice. He gives the example of a doctor who has harvested one kidney and is faced with two patients in need of transplants; irrespective of whether she has a Marxist understanding of systemic inequality, the answer to her predicament will come from bioethics.

According to Jesani, the role of bioethics in surrogacy is to determine whether the arrangement infringes on women's autonomy; whether the benefits of surrogacy outweigh the health risks; and more mundane questions like whether a woman is a good candidate medically to be a surrogate. He also believes that along with providing answers for practitioners at the micro-level, bioethics should go hand-in-hand with public health and clinical medicine perspectives, especially when policy formulation is a goal. He gives the example of Western countries like the UK banning commercial surrogacy within their borders while allowing their citizens to travel to India for the same (Gamble 2009). For Jesani, such a policy smacks of ethical double standards.

Genetic Determinism vs. Adoption

Several respondents took issue with the genetic determinism in ART and surrogacy, which privileges the option of having some genetic continuity with one's child over the option of adoption. For Jesani, the idea that one's child must have something of one's genes is the most disturbing feature of these technologies. According to him, giving primacy to genetic parenthood over other biological contributions or over social parenting is regressive; it is in fact reminiscent of the eugenic perspective in fascist ideologies that propagates one's own race and blood as supreme. For Ramanathan, this is unsurprising; she points out that genetic determinism goes hand-in-hand with the *brahmin-ized* ethos of India's growing middle class.

Nivedita Menon fears that with the proliferation of surrogacy and ARTs, adoption rates will fall. And yet, she is aware that even though India has thousands of abandoned and orphaned children in need of homes, it is extremely difficult to arrive at a policy prescription on this issue. The only logical position would be to disallow surrogacy because adoption is the more ethical option, but it is not feasible or desirable to do so. However, Pramada Menon points out that all family formations come with their own set of complications and restrictions. While activists might want to promote adoption, individual perspectives widely differ and many do not consider adoption an easy option, for reasons that range from logistical to emotional.

Medical Indication

As with previous drafts, the latest surrogacy bill (Ministry of Health and Family Welfare 2016) permits surrogacy only when it is medically indicated for eligible couples, i.e. when the commissioning woman cannot carry a foetus safely to term (Kotiswaran, Chapter 7). Jesani agrees with this proviso, stating that surrogacy cannot be disconnected from the pathology for which it is intended. For him, turning surrogacy into a pure service is classist, eugenicist and unjust; it will effectively mean that the poor must live with their infertility whereas the rich can outsource their reproduction for a price.

Addlakha agrees that surrogacy cannot be allowed to become a legitimate option for rich people who want genetically related children but do not want to birth them. She argues that rather than giving free rein to the consumerist desires of individuals, we must pause to consider the kind of society we want to build. In addition, Seshu and Pai wonder about what options are available for poor women who might also want to enjoy motherhood without the pain, or for trans people who want to be parents. They highlight the class-segregated character of the surrogacy industry. Seshu and Pai nonetheless argue that our collective focus should be on surrogates receiving the highest possible safety and wages (Reddy et al., Chapter 8).

However, in these discussions a significant question remains unanswered—by insisting that commercial surrogacy be available only in cases of medical indication, might we be drawing a line in the sand out of moralism around motherhood? If we can ensure that the surrogacy is happening under fair conditions of work agreeable to all parties, does it matter what the motive for commissioning the surrogacy is?

Regulation-Related Concerns

Respondents welcomed the government's move to regulate the surrogacy industry, but took issue with (a) the process followed for the drafting of the various iterations of the ART/surrogacy bill, and (b) the content of these bills (Kotiswaran, Chapter 7). Even though the issue of surrogacy, embedded as it is in complex social problems, cannot be satisfactorily resolved in a single law with a narrow focus, Nagaraja points out that any draft is an important first step. The government cannot abandon its responsibility towards the minimisation of the negative and exploitative aspects of choices available to people today, she says. Cases like Yuma Sherpa's (*The Indian Express* 2014), a migrant to Delhi from Darjeeling who died during an egg extraction procedure, serve as a grim reminder of the urgent need to check medical negligence and regulate harmful side effects like ovarian hyper-stimulation.

Many respondents expressed disappointment regarding the contents of the various drafts of the ART bill. Pramada Menon notes that the provisioning of ART and surrogacy—meant to enable new family formations—is being planned in very patriarchal ways. She notes, for instance, that there is complete insistence on donor secrecy. As a result, the only known woman whose eggs can be used in IVF is a man's wife; in case a donor is needed, it must be an unknown woman and not a friend, acquaintance or relative. Similarly, a woman needs to be married and have her husband's consent to be a surrogate. For Ali and Ganguly, not only is this outrageous, but also contradictory. On the one hand, the government legitimises the surrogacy industry on the basis of what it sees as women's agency in acting as surrogates. At the same time, it distrusts and delegitimises that very agency by requiring that husbands act as gatekeepers for women who want to become surrogates. Ali and Ganguly argue that single women should have the choice to be surrogates too.

Addlakha stresses that the government should do more to ensure the availability of good quality counselling for commissioning parents, surrogates and surrogates' families (Lamba and Jadva, Chapter 9, for vulnerabilities of surrogates in India to psychological problems). Nivedita Menon wonders how the law will ensure the surrogate's right over her body—will the decision to abort the surrogate pregnancy for whatever reason, or to reduce the number of foetuses in the event of a multiple pregnancy, be the commissioning couple's or the surrogate's?

For Qadeer, any proposed legislation is 'anti-surrogate' if it allows only gestational surrogacy and not genetic. She points out that even if the surrogate's eggs are viable, intrauterine or artificial insemination is not allowed in recent drafts. Instead, in vitro fertilisation is to be done with eggs other than the surrogate's and the resultant embryos are to be transferred into the surrogate. Qadeer points out that while this preserves the genetic separation between the surrogate and the foetus, this separation is notional and mandates a procedure (IVF) that is much more invasive (than IUI). Further, upon delivery the baby must be immediately handed over to the commissioning parents, and breast-feeding by the surrogate is not allowed.

Others expressed concern about the exclusionary nature of the drafting process. Rustagi stresses the need to involve more voices through consultation, while Addlakha wonders if civil society voices carry any weight for political dispensations, which irrespective of the party in power tend to be conservative on issues of gender and sexuality. Most respondents echo this concern; the proposed surrogacy bill propounds a skewed notion of the family by considering only a heteronormative family legitimate and suitable for surrogacy. By disallowing commercial surrogacy, it also risks pushing surrogacy underground, thus compounding the vulnerability of surrogates.

From the case of surrogacy, the larger issue of the shortcomings of feminist organising becomes salient. Pramada Menon emphasises the need for a feminist think tank that will monitor upcoming legislation and alert relevant groups and activists. This is necessary because NGOs and social movements tend to focus most of their energies on their specific areas of work. While this is understandable, Pramada notes that it leads to many missed opportunities for dialogue and collaboration on common concerns.

Conclusion

Many respondents felt that it was important to have conversations within their movements on surrogacy, as well as to bring newer movements and perspectives into ongoing conversations. Pramada Menon suggests engaging newer constituencies such as young people, trade unionists and feminist economists. She also suggests newer strategies like the creation of a national union of surrogates.

Nivedita Menon emphasises that a lot more needs to be done to humanise the women who become surrogates, and to highlight the conditions under which they work. Nagaraja adds that we need to be constructive in how we talk about surrogacy,

> As activists we are usually on the back foot, catching up. It is difficult to think and plan ahead of developments. The challenge with surrogacy is to counter the fan-fare without sounding like you want a ban. When you position something as 'anti' it either fizzles out or you get backlash. But if you offer a positive alternative, the right thing to do, then people run with it. So let's not talk about ban versus regulation, or commercial versus altruistic, but ask what an ethical surrogacy might look like. That seems to me more constructive than just saying, this is a problem.

While Sama believes in the importance of a law that centres the women who act as surrogates, we also remain cognisant of the natural limits of the law in confronting a multifaceted social issue like surrogacy. We believe deliberation within social movements is an important way in which we can move forward on complex questions. This chapter represents a cross-movement conversation that is a step towards enriching our collective perspective on surrogacy. In their reflections, activists and scholars highlight concerns regarding surrogacy from the perspective of their respective social movements. In discussing these concerns however, they also illuminate points of consonance and dissonance between movements, such as between the child rights and women's movements. Some responses reveal conceptual debates around women's agency within exploitative circumstances that, though done in the context of surrogacy here, are in fact much older debates and have analogies in issues other than surrogacy, such as in sex work. Still other responses contain perspectives, such as on 'women's essential nature' which would be imminently debatable. However, almost all responses bring to the fore social problems that the Surrogacy (Regulation) Bill 2016 addresses inadequately, ignores, or even compounds. This chapter is not a comprehensive assessment of surrogacy. It thematises and represents only those issues that came up during the course of the interviews. Without doubt, there are other issues and other perspectives worthy of engagement and representation. We hope to see these taken up in multiple forums and across multiple movements in the future.

Notes

1. Sama is a resource group based in Delhi, working on issues related to women and health. Sama was initiated in 1999 by a group of feminist activists who were involved in the autonomous women's movement which views health from a broader perspective and finds linkages of women's well-being with various determinants of health. Website available at http://www.samawomenshealth.in/.
2. Interviewees and Affiliations- Nivedita Menon, Women's movement/ Jawaharlal Nehru University; Enakshi Ganguly, Child rights/ HAQ: Centre for Child Rights; Bharti Ali, Child rights/ HAQ: Centre for Child Rights; Pramada Menon, Queer movement and women's movement/ Independent consultant; Preet Rustagi, Labour rights/ Institute for Human Development; Veena Johari, Rights of People Living with HIV/AIDS/ Human rights lawyer; Kajal Bharadwaj, Rights of People Living with HIV/AIDS/ Human rights lawyer; Imrana Qadeer, Public health scholar and activist/ Council for Social Development; Prabha Nagaraja, Sexual and Reproductive Health and Rights/ TARSHI-Talking About Sexual and Reproductive Health Issues; Amar Jesani, Bioethicist/ Indian Journal of Medical Ethics; Mala Ramanathan, Public health and ethics scholar/Achutha Menon Centre for Health Science Studies (AMCHSS) and Sree Chitra Tirunal Institute for Medical Sciences and Technology (SCTIMST); Meena Seshu, Sex worker rights activist/SANGRAM- Sampada Gramin Mahila Sanstha; Aarthi Pai, Sex worker rights activist/SANGRAM- Sampada Gramin Mahila Sanstha; Renu Addlakha, Disability rights movement/Centre for Women's Development Studies.

3. Bioavailability refers to the extent and rate at which the active moiety (drug or metabolite) enters systemic circulation, thereby accessing the site of action. Bioequivalence indicates that the drug products, when given to the same patient in the same dosage regimen, result in equivalent concentrations of drug in plasma and tissues (Lee 2017).

Bibliography

Cooper, M., & Waldby, C. (2014). *Clinical Labor*. Duke University Press.

Gamble, N. (2009). Crossing the line: The legal and ethical problems of foreign surrogacy. *Reproductive Biomedicine Online, 19*(2), 151–152.

Lee, J. (2017). Drug bioavailability. *MSD Manuals* [online]. Available at http://www.msdmanuals.com/professional/clinical-pharmacology/pharmacokinetics/drug-bioavailability. Accessed 28 Aug 2017.

Ministry of Health and Family Welfare (2016). *Surrogacy (Regulation) Bill 2016 [pdf]*. New Delhi: Government of India. Available at http://www.prsindia.org/uploads/media/Surrogacy/Surrogacy%20(Regulation)%20Bill,%202016.pdf. Accessed 28 Aug 2017.

Qadeer, I. (1998). Reproductive health: A public health perspective. *Economic and Political Weekly, 33*(41), 2675–2684.

Qadeer, I., & John, M. E. (2009). The business and ethics of surrogacy. *Economic and Political Weekly, 44*, 10–12.

Reddy, S., & Qadeer, I. (2010). Medical tourism in India: Progress or predicament? *Economic and Political Weekly, 45*, 69–75.

Rudrappa, S. (2015). *Discounted life: The price of global surrogacy in India*. New York: NYU Press.

The Indian Express. (2014). *Egg donor's death: Internal bleeding, ovaries severely enlarged, says report* [online]. Available at http://indianexpress.com/article/india/india-others/egg-donors-death-internal-bleeding-ovaries-severely-enlarged-says-report/. Accessed 30 Aug 2017.

Zelizer, V. A. R. (1994). *The social meaning of money: Pin money, paychecks, poor relief and other currencies*. Princeton: Princeton University Press.

Part III

A Restrictive System: The Case of Germany

11

German Law on Surrogacy and Egg Donation: The Legal Logic of Restrictions

Sabrina Dücker and Tatjana Hörnle

Introduction

German law is restrictive with regard to assisted reproductive technologies (ART).[1] More precisely, according to German law, it is illegal to tamper with 'natural motherhood', while the donation of sperm is legal.[2] Prohibitions with the threat of criminal punishment aim to deter medical staff in Germany from performing reproductive techniques that involve egg donation and surrogacy. In this chapter, we will sketch the legal prohibitions and the ways in which intended parents (IPs) circumvent them by travelling abroad to obtain surrogacy and egg donation (König, Chapter 13; Teschlade, Chapter 14, for a detailed, empirical account). The current practices of cross-border egg donation and cross-border surrogacy raise a number of legal questions. From the

S. Dücker (✉) · T. Hörnle
Humboldt-Universität zu Berlin, Berlin, Germany

T. Hörnle
e-mail: Tatjana.Hoernle@rewi.hu-berlin.de

S. Mitra et al. (eds.), *Cross-Cultural Comparisons on Surrogacy and Egg Donation*,
https://doi.org/10.1007/978-3-319-78670-4_11

perspective of German criminal law, one important consideration is whether people who give out information about the details of surrogacy and egg donation in other countries could be prosecuted as accomplices to a criminal offence. For IPs, the main issue is whether and how children born abroad can be registered as their children in Germany. We describe the complicated legal framework regarding the status of children born as a result of cross-border surrogacy as well as important court decisions in this area. At the end of our chapter, we will come to a critical conclusion regarding the German law. In a globalised world, national prohibitions cannot prevent citizens from seeking surrogacy and egg donation abroad and from finding ways to obtain the desired legal status for the infants as their children. Although the law claims to treat citizens equally, it does in fact give an advantage to affluent and well-informed people. Rather than sustaining the absolute prohibitions in German law, it would be preferable to allow and, at the same time, survey and regulate practices.

Prohibitions in German Law

The prohibitions debated here are embedded within the Embryo Protection Act (*Embryonenschutzgesetz*, ESchG) from 1990. This law addresses several issues. Parts of it target the use of embryos in research. Other sections aim to suppress reproductive techniques such as egg donation and surrogacy. According to the ESchG, it is a criminal offence to transfer an unfertilised egg that was produced by another woman (Sect. 1 I no. 1 ESchG), to perform artificial insemination (AI) for any other purpose than bringing about a pregnancy in the woman from whom the egg originated (Sect. 1 I no. 2, II ESchG) and to remove an embryo before completion of implantation in the uterus in order to transfer it to another woman (Sect. 1 I no. 6 ESchG). In addition to techniques that involve egg donation or embryo donation, the ESchG also criminalises surrogacy without egg donation. It is an offence to carry out an AI if the woman is prepared to give up her child permanently (Sect. 1 I no. 7 ESchG). German law does not discriminate between commercial and altruistic forms of egg donations and surrogacy and does not make exceptions for the latter. Sanctions are relatively moderate (fines or imprisonment for up to three years).[3] Women who donate or receive an egg and surrogate mothers as well as IPs are not criminally liable.[4]

Another measure against surrogate motherhood is the prohibition to act as a broker between IPs and women who are willing to become surrogate mothers (Sect. 13 c Law on the Mediation of Adoptions, *Adoptionsvermittlungsgesetz*, AdVermiG). Sanctions are criminal fines or imprisonment for up to one year, or two years or three years if the broker obtains a fee or runs a professional business (Sect. 14 b I, II AdVermiG). Again, IPs and potential surrogate mothers are not criminally liable (Sect. 14 b III AdVermiG). Nonetheless, there are also prohibitions concerning IPs and surrogate mothers directly, with the threat of a non-criminal fine (Sect. 14 I no. 2 AdVermiG): It is against the law to publicise one's demand for surrogacy services or to announce them publicly, in newspaper articles or advertisements or other forms of public communication (Sect. 13d AdVermiG).

In the legislative deliberations, the main argument against egg donation was that 'split motherhood' must be prevented as it might have negative consequences for the child's development. Fears were that splitting motherhood into a genetic and a social mother might create tensions within the child's sense of identity, and that it could cause conflicts if the genetic mother wants to stay in contact with the child.[5] Defenders of the prohibition also point to health risks due to the hormonal treatment for the stimulation of ovaries (Hüppe 2015, p. 126) and argue that the donor's human dignity was violated, at least in the case of commercialised egg donation (Lehmann 2007, p. 176). With regard to surrogacy, the claim of potential harm for children as a consequence of 'split motherhood' reappears. In addition, concerns were that either the surrogate mother would not develop a psychological relationship with the foetus during pregnancy and thus harm its intrauterine development,[6] or that the disruption of an existing familiarity after birth might be detrimental to the children's well-being.[7] Possible shifts of intentions and affections were cited in support of prohibitions, too, with the argument that the surrogate mother might not be willing to give up the child after birth, or that the IPs might not develop a relationship with the child, particularly if it is handicapped.[8] Again, human dignity arguments were used (Beier, Chapter 12).[9]

Egg Donation and Surrogacy in Practice

German couples (or single individuals) might in rare cases manage to obtain a longed-for baby through surrogacy and/or egg donation without going abroad. This would be possible should German reproductive clinics ignore the restrictions of the ESchG, or if IPs choose the traditional, non-criminal version of surrogacy, that is, without medically ART. The first alternative is probably not of great practical importance. There are no indications that physicians in Germany ignore the criminal prohibitions on a larger scale. Court statistics show hardly any convictions for violations of the ESchG.[10] Searching the Internet for press reports uncovers one criminal trial against a German reproductive physician who was convicted for using donated eggs in his practice.[11] Of course, one has to take into account that conviction statistics do not mirror reality (cases remain undetected or end with discretionary decisions before they come to trial). Nonetheless, it is unlikely that egg donation or surrogacy are practiced in noteworthy numbers in Germany. Medical staff would usually not only be deterred by the threat of criminal sanctions, but even more so by the risk of losing their professional standing and their licences.

German law does not criminalise traditional surrogacy (without ART and thus beyond the scope of the ESchG), but there are considerable practical obstacles. IPs would need to find a woman who is willing to get pregnant in an awkward procedure, outside of neutralising clinical settings, and to give the baby, which is genetically her own, away after birth. As the AdVermiG prohibits assistance by an agency as well as a public search for a surrogate mother, it would be necessary to inquire among close friends and relatives. If the circle of potential surrogate mothers is limited to friends and family, it is not very likely to find a woman who is willing to go through the psychological, physiological and social challenges involved.

The most promising solution for infertile couples, gay couples or individuals is to go abroad and to make use of the global market for egg donation and commercial surrogacy (Werner-Felmayer, Chapter 2; Sándor, Chapter 3). Advertisements of reproductive clinics on the Internet indicate a well-established practice and stable demand from potential parents from Germany. If one types the German word for egg donation (*Eizellenspende*) into a search engine, advertisements of centres for reproductive medicine in European countries other than Germany pop up—in German and with detailed information. The restrictions in German law regarding the public communication about demands and offers are about as effective as the German law's attempts to keep the Internet free from pornography. In the age of globalised communication, it is a pointless attempt to try to suppress information. There are no statistics on how many people make use of cross-border egg donation and/or surrogacy, but researchers report substantial numbers of Germans going to Eastern Europe (for Poland see Wilson 2017), Spain and the US, for instance (Knecht 2017, for the differences between the states in the US see Conklin 2014). We must therefore assume that the restrictive German approach is ineffective.

The fact that parents have the option of travelling to a different country in order to get access to egg donation and surrogacy does, however, not dissolve all legal issues (Mitra, Chapter 5). The use of cross-border reproductive services still carries legal risks for German physicians and counsellors who can become subject to criminal investigations. For IPs, the main challenge is to obtain legal parentage for the child after their return to Germany (König, Chapter 13).

Criminal Liability for Preparatory Acts

It is legal to treat and counsel patients after the use of reproductive techniques abroad, however, assistance in the preparatory stages is somewhat risky even if the law abroad does not prohibit egg donation or surrogacy. Prosecutors occasionally make attempts to defend the German stance against egg donation and surrogacy by investigating physicians and brokers who disseminate information on the topic. As a general rule, the German Criminal Code (*Strafgesetzbuch*, StGB) does not apply to acts committed outside of the state's territory,[12] but for accomplices, there are rather strict provisions. Section 9 II no. 2 StGB stipulates that an accomplice who acts within Germany will be punished according to German criminal laws, even if the principals (in our case: the staff in a clinic for reproductive medicine abroad) act in full accordance with their national law. Therefore, whoever 'causes' another person's decision for cross-border egg donation or surrogacy can be punished for incitement (Sect. 26 StGB) to criminal offences under the ESchG. Besides incitement, aiding is also an offence (Sect. 27 StGB).

The *Kammergericht* (the appellate court for Berlin) has pointed out that a physician who assists future egg donation by preparing his patient in Germany would be criminally liable.[13] In the same judgment, the Court argued that one could also become an accomplice to a criminal offence by providing patients with the names and addresses of physicians abroad.[14] The *Kammergericht* made these remarks in a judgment concerning competition law. The plaintiff (acting for a reproductive clinic in Berlin) had demanded that the defendant discontinues events in German cities that informed about Czech clinics' successes with egg donation. This civil suit was ultimately unsuccessful,[15] but the statements about criminal liability should not be ignored. In Bavaria, prosecution offices have opened investigations against doctors, infertility counselling services and brokers (Spiewak 2013). In at least one instance, the case was dismissed, other investigations were terminated after the payment of fines.[16] There is no reliable data regarding the law enforcement strategies. They probably vary across Germany (it seems unlikely that punishable behaviour is restricted to the region of Bavaria).

The Child's Legal Status

The main strand of legal discussions concerns matters of family law and private international law that arise once a baby has been born through surrogacy and/or egg donation. It depends on these legal rules whether IPs can live in Germany with the child as a legally recognised family.

The Child's Legal Status After Egg Donation

If a woman got pregnant after an in vitro fertilisation (IVF) that involved donated eggs and returned to Germany afterwards, the child would be registered as hers. There is no need to mention the circumstances of fertilisation or genetic descent for obtaining a birth certificate or in any other legal contexts. German family law follows the old rule *mater semper certa est.* The German Civil Code (*Bürgerliches Gesetzbuch*, BGB) states: 'The mother of a child is the woman who gave birth to it' (Sect. 1591). Whether and how the mother's husband or partner can secure legal parentage depends on the same general rules as in other cases of childbirth. If a man is married to the gestational mother, he automatically has the status of fatherhood (Sect. 1592 no. 1 BGB);[17] if not, he can acknowledge paternity (Sect. 1592 no. 2 BGB) without explanations concerning genetic descent. He can also adopt the child (Sect. 1741 II no. 1 BGB); this is also an option if the woman has a female registered partner (Sect. 9 VII *Lebenspartnerschaftsgesetz* (Life Partnership Law)) or is married to a female partner.[18]

The Child's Legal Status After Traditional Surrogacy Within Germany

As mentioned above, it is conceivable (and might happen occasionally) that a woman in Germany agrees to use sperm provided by the intended father and to leave the baby to the IPs. As long as there is no use of ART the law does not prohibit this. After the birth of a child, cooperating parties can settle the legal issues quietly. The woman who has given birth is the legal mother, however, IPs have options to gain parental status. At first, the intended father will, with the approval of the woman who gave birth (Sect. 1595 I BGB), acknowledge paternity (Sect. 1592 I no. 2 BGB).[19] After that, if he is living with a spouse or a registered male partner, the spouse or partner can become a co-parent by adopting the child (Sect. 1741 II no. 3 BGB).[20]

The Child's Legal Status After Cross-Border Surrogacy

In the case of cross-border surrogacy, IPs who wish to live in Germany with the child have to overcome legal obstacles. The first obstacle is often to obtain documents for travelling with an infant that was born in another country (Engel 2014, pp. 206–207). German citizens can approach embassies and administrative courts in order to obtain a passport for the baby. Such attempts will only be successful if they can prove that the child is a German citizen, and to clarify this point, they have to expect inquiries concerning the circumstances of the birth. If the IPs manage to arrive in Germany with the baby, they encounter the next difficulty: the birth and their parentage need to be registered. In daily life, a couple or single adult would face a lot of problems if a child, who de facto lives in the same household, is not formally registered as their child. Whether surrogacy is a viable option, therefore depends on whether IPs are able to obtain formal recognition of legal parenthood in Germany. They have to expect some resistance in their encounters with registrar's offices and courts—after all, German law prohibits surrogacy. Unlike egg donation or a confidential traditional surrogacy arrangement amongst friends or family, cross-border surrogacy is easier to detect. By now, German authorities are familiar with the practices and popular destinations.[21] The question then is: How do officials at the registrar's offices and German courts decide about the children's legal status?

German courts have consistently refuted two conceivable, easy solutions. First, it will not suffice to submit a contract concerning surrogacy, arguing that this contract should be sufficient to be registered as parents. The BGB declares contracts that are incompatible with public morals (Sect. 138 BGB) invalid, and this clause is applied to surrogacy contracts.[22] Secondly, biological fathers should not set their hope on DNA evidence. In the case of a baby born in India, the male plaintiff presented evidence that he was the biological father. Nonetheless, the couple did not obtain the desired passport for the child because under German law, genetic relation does not establish legal fatherhood.[23]

Which National Law?

In order to decide about passports, registrations, citizenship, etc., legal descent needs to be established. In cases with cross-border context, the very first question is, which national law is applicable—German family law or another national law? Officials at German registrar's offices and judges will begin their legal analysis with Article 19 Introductory Act to the German Civil Code (*Einführungsgesetz*, EGBGB). The first rule in Article 19 EGBGB says: 'The descent of a child is governed by the law of the place where the child has his or her habitual residence'. If IPs returned to Germany with a child with the intention of staying here, this will be the child's 'habitual residence', and therefore German law is applicable (Duden 2015a, p. 97). German courts have not accepted Ukraine or India, for instance, as the 'habitual residence' of a baby if the child is several weeks or a few months old and is waiting to be relocated to Germany.[24] Only if the child has stayed in the country of birth for a longer period of time, might IPs profit from the baby's 'habitual residence' in the other country and thus from a legal system's greater permissiveness with regard to surrogacy. However, when the child finally arrives in Germany, the 'habitual residence' changes, and it is debated whether this signifies a change of the legal status as well.[25] The second rule in Article 19 EGBGB is: 'In relation to each parent, the descent can also be determined by the law of the country of this parent's nationality'. Again, this rule will in most cases lead to the application of German family law.[26]

German Family Law

There are no provisions in the BGB that *explicitly* address legal paternal status after surrogacy and egg donation. This does, however, not preclude that IPs might be successful in gaining legal status as parents according to the general rules. These rules are gender-specific (for the German law on maternity and paternity see Dethloff and Ramser 2007, pp. 179–184; Gössl 2013, pp. 135–137; Coester-Waltjen 2016, pp. 209–213).

The legislature explained the introduction of Section 1591 BGB (*mater semper certa est*) in 1998 with the goal to prevent surrogacy.[27] Accordingly, courts insist: Whatever the circumstances, the initial legal status of the mother is reserved *exclusively* for the gestational mother, independent of genetic relations.[28] Male IPs are in a somewhat better position: If the surrogate mother is not married, they can obtain parental status through a declaration that acknowledges paternity.[29]

If a man has gained the legal status of the father for himself, co-parental status and full custody rights for his spouse (or a male registered partner) can be secured with an adoption (Sect. 1741 II no. 3 BGB).[30] This second step on the way to becoming a family in the legal sense requires more efforts and more waiting time, and there is the risk of failure. German judges who are aware of the surrogacy background have in the past applied strict standards for deciding whether or not the intended mother can adopt the child under German Law (for legal arguments regarding the applicable norm see Dethloff 2014, p. 930).[31] If a lesbian married couple or a lesbian couple in a registered partnership chooses a surrogacy arrangement abroad, they will be in an even more difficult position under German family law than couples including at least one male[32]: A male person can secure legal parenthood qua acknowledgement, but females only have the possibility to adopt—with all the risks involved, and, in the meantime, none of them have custody rights.

Obtaining a Court Decision Abroad

If IPs carefully plan how to obtain parental status in Germany before they embark on a surrogacy arrangement, a key element is to obtain a court decision in a more liberal legal system outside of Germany. Then, German courts can simply recognise the foreign court's decision (Act on Proceedings in Family Matters and in Matters of Non-Contentious Jurisdiction, FamFG, Sect. 108). The precondition is that a 'court' has specifically addressed the matter of 'legal parentage'. Neither an entry into a public registrar will suffice[33] nor a ruling that is limited to biological fatherhood.[34] If IPs can present a court decision that assigns them legal parental status, they have a chance to get around German family law—unless the foreign court decision is considered incompatible with the *ordre public*. This notion of *ordre public* is expressed as: 'would lead to a result which is manifestly incompatible with the fundamental principles of German law, particularly if recognition would be incompatible with civil rights' (Sect. 109 I No. 4 FamFG).

Since the end of 2014, chances are good that a foreign court decision on legal parenthood will be recognised. In a landmark decision on 10 December 2014, the German Federal Court of Justice (*Bundesgerichtshof*, BGH, highest appellate court in civil and criminal matters) emphasised, that the *ordre public* clause does not stand in the way of such rulings. The two plaintiffs were men in a registered partnership who had found a surrogate mother in California; one of the plaintiffs had donated sperm. Shortly before the child was born, they had obtained a ruling by the Superior Court of California that declared both men parents of the child. After the baby was born, they returned to Berlin with the child. The registrar's office in Berlin at first refused to register them as parents, but their lawsuit was eventually successful: The BGH ruled that the decision by the Californian court must be recognised.[35]

This outcome was not self-evident. The judges at the BGH could have argued that surrogacy is manifestly incompatible with the fundamental principles of German law. In the same case, the lower German court had expressed the view that surrogacy violates human dignity and pointed out that human dignity is a fundamental value and core of German law.[36] The French Court of Cassation in a case called Mennesson also found that surrogacy agreements are null and void on grounds of public policy, and that it would be contrary to the essential principle of French law to give effect to such agreements.[37] The legislative explanations to the ESchG and the AdVermiG stated that surrogacy is incompatible with children's well-being. However, the BGH did not pick up these arguments.

One explanation for this liberal turn of the BGH can be found in prior decisions by the European Court of Human Rights (ECtHR) (Sándor, Chapter 3). The above-mentioned French case (*Mennesson vs. France*) went to the ECtHR.[38] The European judges found that France had violated the children's rights protected by Article 8 I European Convention on Human Rights (right to privacy). In this case and in the similar case of *Labassee vs. France*,[39] the ECtHR pointed out that domestic law must recognise the children's relationship with their biological father when determining their legal status. Important features of the German case were similar to those in the French cases (surrogacy in the US, the intended father was the biological father, a prior ruling by the Superior Court of California). The BGH followed the path taken by the ECtHR.

The decision by the BGH was most helpful for the second male plaintiff, who is not biologically related to the child, as well as for gay couples, in general. The biological father had formally acknowledged paternity and the surrogate mother was not married—under these circumstances, he could have acquired the desired legal status in Germany anyhow, even if the court had not accepted the Californian ruling. For his male partner the cross-border surrogacy in combination with the recognition of the Californian judgment led to an outcome that otherwise would not have been available under German family law: to be registered with co-parental status, without the need for an adoption procedure.[40]

To sum up the current case law in Germany: Since the BGH's landmark decision in 2014, IPs can, under certain conditions, obtain parental status after cross-border surrogacy. These conditions are: first, a foreign court decision that declares or confirms legal parentage, second, biological descent from one of the persons who claim legal parentage. The BGH left it open whether or not a lack of biological relation makes a difference, but it would be logical to assume this. After all, the ECtHR had argued in favour of the child's rights emphasising the importance of biological parentage as a component of identity.[41] Also, in a newer decision, the ECtHR accepted the Italian authorities' decision to take a small child away from the IPs.[42] The crucial difference to the French cases *Mennesson and Labassee* was that in the Italian case the child was not biologically related to the complainants. Third, the BGH made it a requirement that the surrogate mother has freely agreed to the surrogacy arrangement and has handed the baby over voluntarily. To confirm the required voluntariness, the Court demands documentation regarding the surrogate mother's identity and the content of agreements.[43]

Still, IPs cannot be sure that all German courts will decide in this uniform and predictable way. There is some heterogeneity in outcomes despite the fact that the highest German appellate court has developed guidelines. It seems that some judges perceive it as a severe affront that the BGH and the ECtHR have opened the door for well-organised and affluent citizens to circumvent the German prohibitions against surrogacy. In a decision in 2017, the appellate court in Braunschweig complained about 'reproductive tourism' and refused to grant legal parenthood despite the fact that the IPs met the requirements the BGH had asked for.[44] For the Braunschweig judges' disapproval it may have played a role that the IPs, husband and wife, were 67 and 64 years old and were obviously very wealthy (the contract with the surrogate mother in Colorado, US, granted her the payment of more than USD 500,000 for carrying their twins to term). At the time of writing this chapter, the Braunschweig decision is not final, but whatever the final outcome will be: it sheds a light on tensions within the judiciary. Not all German judges are willing to simply accept foreign courts' rulings that contradict the stance against surrogacy taken by the German law.

Conclusion

The current legal situation in Germany can be described as complicated, ineffective, unpredictable with regard to outcomes, as well as unjust. The prohibitions in the ESchG from 1990 and the AdVermiG do not keep infertile couples and individuals from seeking egg donation and surrogacy. These services can be obtained abroad. Medical staff in centres for reproductive medicine in Germany will be confronted with their patients' requests to be advised and prepared for such procedures, but if they respond, they might be treated as an accomplice to acts the German law classifies as criminal offence. The risk of an actual criminal conviction is probably small, but this is not much of a comfort since enforcement practices are unpredictable. For medical staff, it is an unfortunate situation to be torn between patients' desire to have a local doctor's support and the demands of the law. From the perspective of IPs, it is hard to understand that their German physicians are not even allowed to inform them about chances and risks of techniques that are common medical practice in other countries.

The situation is unjust because loopholes out of the otherwise restrictive German law are accessible to some, but not all infertile individuals and same-sex couples. The procedures require a lot of resources: good legal advice on where to go, the ability to communicate with foreign authorities and foreign courts, time, energy and first and foremost: money. In June 2017, the BGH had to decide about a civil suit against a health insurance company.[45] In 2011, the plaintiff had spent 11.000 Euro for her treatment in a Czech clinic that involved egg donation (a successful treatment: she has given birth to a child). Afterwards, she tried to get reimbursed by her health insurance, however, her lawsuit was unsuccessful. Wealthy and well-organised IPs, however, have good chances of fulfilling their desires. Additionally, inequality is not only a matter of financial resources, knowledge and competence: the ECtHR' strong focus on biological fatherhood and genetic descent places lesbian couples at a disadvantage. Male same-sex couples have better chances of obtaining legal parental status after surrogacy because sperm donation is not very complicated.

Another problem is that IPs cannot be sure that all registrar's offices and courts in Germany are willing to follow the lead taken by the BGH and the ECtHR. The highest courts now take a pragmatic approach towards surrogacy that to a certain degree accepts the normative power of the factual. The underlying logic is that once children are born, a forward-looking perspective, oriented towards their well-being, is more important than backward-looking frowns about the circumvention of German law. However, not all state officials and judges at the lower courts are comfortable with this development. Those who feel obliged to defend the anti-surrogacy decisions of the German legislature will search for reasons to refuse registrations, passports, the recognition of foreign court rulings and adoptions. Therefore, the outcome of procedures to obtain legal parenthood remains not entirely predictable.

What should be done? Over the last years, observers of the situation in Germany have increasingly demanded legislative action. One possible solution focuses on the children's legal status after surrogacy, with the recommendation that German law should be simplified by granting legal parentage according to the law of the country where the child was born (Helms 2016, p. F101; Dethloff 2015, p. 681). The more far-reaching demand is to abolish existing prohibitions in the ESchG and to enact modern legislation on reproductive medicine (Kreß 2013; Müller-Terpitz 2016, pp. 53–54; Gassner et al. 2013; Valerius 2017, pp. 26–27). This is, in our view, the recommendable approach. The ECtHR has emphasised legislatures' duty to monitor developments in reproductive medicine and to consider amendments of the law.[46] For the German legislature, it is time to overhaul the ESchG and to be less restrictive about egg donation and surrogacy in Germany.

To support this conclusion, it is not sufficient to point out that prohibitions are not effective because people have discovered ways to circumvent them. This diagnosis would, *as such*, not necessarily lead to an abolitionist position. For the vast majority of criminal laws, it is possible to observe gaps between law and social reality, but without the legal prohibition, more harmful conduct would occur. If 'changed social realities in globalised societies' were the only line of argument, we would face some difficult follow-up questions, such as: Does it undermine the legal order if citizens have unequal access to institutions that serve vital interests (such as the interest to have children), favouring the well-educated and wealthy who are able to pursue a globalised approach in fulfilling their interests? To what extent can legal orders uphold their legitimacy in the face of unsystematic, unequal law enforcement? Fortunately, for our topic, these general questions can be left aside. The crucial argument to demand changes of the restrictive German law is more straightforward. There were never good reasons to categorically prohibit egg donation and surrogacy, but only reasons to regulate and monitor practices.

Justifying the prohibition of egg donation and surrogacy by pointing to human dignity has not been convincing to start with. In these contexts, human dignity arguments tend to embellish an underlying anti-modernist sentiment against 'tampering with nature' (Hörnle 2013, p. 748; for a critical view on the German obsession with human dignity in the context of reproductive medicine, see Birnbacher 1996). Now, the BGH points out that surrogacy per se does not clash with the surrogate mother's human dignity.[47] Arguments that deserve more attention concern the children's well-being and the autonomy and well-being of the women who donate eggs or become surrogate mothers. Neither of these aspects can, however, justify an absolute prohibition. The German legislature's focus on the dangers of 'split motherhood' for the development of children has exaggerated risks. Fears that egg donation and surrogacy are in general detrimental to the children's well-being are not corroborated by evidence. The level of IPs' care for the children seems to be above average (Mayer Lewis 2017; Wiesemann 2017) and during the first years, the level of warmth and interactions with the child is higher than with mothers that conceived naturally—this effect seems to vanish over time, but follow-up studies still conclude that, 'overall, the findings indicate that these families continue to function well in the early school years' (Golombok et al. 2011). Regarding the psychological well-being of surrogate mothers after birth, there are no indications that surrogacy necessarily has a negative impact (Jadva et al. 2014; the authors do, however, mention that more research concerning cross-border surrogacy is needed, p. 378).

While an absolute prohibition is not defensible, there are reasons to carefully consider conditions for legalised egg donation and surrogacy. Contracts and their enforcement should be screened with an eye to voluntariness, health issues and best practices concerning the long-term well-being of all persons involved. Another point in need of regulation is the children's fate if either the surrogate mother or the IPs change their mind at some point. Procedural details (documentation that grants transparency for all persons involved if they later wish to track their biological relatives, counselling, visiting rights, etc.) deserve attention, too. The preferable alternatives to absolute prohibitions are laws that allow egg donation and surrogacy, but regulate the modalities and provide rules in case of complications.

From this perspective, the current German legal situation is absurd. Insistence on absolute prohibitions drives IPs out of the country, which means that there is no information about the medical and social conditions of reproductive practices and the treatment of women involved. If the surrogate mothers and egg donors remain unknown figures in some foreign country, it is impossible to ascertain whether they have made autonomous decisions (Dethloff 2016, p. 209). There might be exceptional situations when the surrogate mother's human dignity is indeed violated (if she is forced to conceive a child or if her refusal to give up the child after birth is ignored), but they cannot be detected if surrogacy is displaced to countries with insufficient supervision (for the exceptional situations that amount to violations of human dignity, see Duden 2015a, pp. 201–203). Rather than insisting on the purity of the law in the books and turning a blind eye to realities, the German legislature should develop rules for ART with the aim to protect the vulnerable (Resnik 2001; Duden 2015b, p. 660).

It is, of course, not realistic to expect that allowing surrogacy and egg donation in Germany would put an end to cross-border activities. People make this choice not only because of the current prohibitive German law but also because the costs are lower. The rules of health insurance companies therefore also need to be reconsidered. Some IPs might still continue to seek ART abroad to circumvent legal procedures and restrictions (Engel 2014, p. 205). However, it would still be an improvement if large parts of the current practice could be scrutinised and regulated.

Notes

1. See for a comparison of countries the ECtHR, *Labassee vs. France*, app. no. 65941/11, at no. 31–32, judgment of June 26, 2014: of 35 member states of the Convention on Human Rights, 14 explicitly prohibit surrogacy and 7 allow non-commercial surrogacy (Albania, Georgia, Greece, Russia, Ukraine, Netherlands, UK). In other states, the legal situation is unclear or surrogacy is tolerated informally. A comprehensive study on the law in European countries: Brunet et al. (2013).
2. There are restrictions for gender selection and an absolute prohibition to use the sperm of a man after his death (Sect. 3 and Sect. 4 I no. 3 ESchG). Also, anonymity is increasingly contested: court decisions emphasise that children must be enabled to track the identity of their genetic father, see Oberlandesgericht Hamm, app. no. I–14 U 7/12, judgment of February 6, 2013.
3. Readers might think that three years imprisonment would be rather serious, but one has to keep in mind that German criminal laws provide a wide sentence range while the courts typically do not choose sentences in the vicinity of the maximum allowed by law.

4. Section 1 III ESchG: "The woman from whom the egg cell or embryo was collected, and likewise the woman to whom the egg cell or embryo will be transferred, and in the cases of subsection 1, number 7, the surrogate mother and likewise the person who wishes to permanently take care of the child, shall not be liable to punishment".
5. Bundestags-Drucksache (BT-Drs.) 11/5460, p. 7.
6. BT-Drs. 11/4154, p. 6.
7. BT-Drs. 11/5460, p. 8.
8. BT-Drs. 11/4154, p. 1, 7; BT-Drs. 11/5460, p. 8.
9. BT-Drs. 11/4154, p. 7.
10. In some years, there are no convictions at all on the basis of the ESchG (which includes more prohibitions than the ones discussed here). In 2009, one person was convicted, the same in 2010 (Strafverfolgungsstatistik, published yearly by the Statistisches Bundesamt).
11. The Landgericht Hof sentenced him to five years imprisonment for a number of different offences, including large-scale tax evasion. Die Welt, 2015, October 20. Kinderwunsch-Arzt zu fünf Jahren Haft verurteilt. https://www.welt.de/regionales/bayern/article147815610/Kinderwunsch-Arzt-zu-fuenf-Jahren-Haft-verurteilt.html. Accessed June 27, 2017.
12. Section 3 StGB: "German criminal law shall apply to acts committed on German territory". German courts have jurisdiction for some crimes committed abroad, for instance, genocide, piracy, human trafficking, kidnapping, but not for acts falling under the ESchG (Sect. 5, 6 StGB).
13. Kammergericht, app. no. 5 U 143/11, at no. 53–56, judgment of November 8, 2013.
14. Ibid., at no. 57. If it cannot be proven that a specific person acted upon the advice, the giving of the advice is not punishable because German law requires the actual commission of the principal offence, see Section 27 StGB. If a patient does, however, make use of the address given to her, it does neither matter that she had already planned to seek medical help abroad before the consultation nor that the same address could have easily been obtained otherwise, for instance via Internet, see Oberlandesgericht Oldenburg, app. no. 1 Ss 185/12, judgment of February 18, 2013 (in this case, the doctor had given his patient the address of a Dutch clinic for an abortion).
15. BGH, app. no. I ZR 225/13, judgment of October 8, 2015. The BGH argued that the Act Against Unfair Competition cannot be applied because the prohibitions in the ESchG do not serve the purpose to protect German clinics against economic competitors.

16. See for the dismissed case Focus, June 27, 2014. Im Visier der Justiz: Ermittlungen gegen Kinderwunsch-Ärzte laufen. http://www.focus.de/familie/kuenstliche_befruchtung/im-visier-der-justiz-ermittlungen-gegen-kinderwunsch-aerzte-laufen_id_3952202.html. Accessed June 27, 2017. See reports about proceedings before the Amtsgericht München: Redaktion Beck aktuell, becklink 2001563; Rost (2014), Münchner Merkur, 2014, June 27. Ermittlungen gegen Kinderwunsch-Ärzte. https://www.merkur.de/bayern/ermittlungen-gegen-kinderwunsch-aerzte-3659476.html. Accessed June 27, 2017.
17. Section 1592 no. 1 BGB: "The father of a child is the man who is married to the mother of the child at the date of the birth".
18. Same sex marriage is possible in Germany since 1 October 2017 (Bundesgesetzblatt (BGBl.) 2017, I, p. 2787).
19. If the mother is married, the legal steps are more complicated (the husband's paternity must be contested as a first step, Sect. 1600 BGB), but not impossible.
20. If the intended father has not yet acknowledged paternity, a married couple adopts the child together, Section 1741 II no. 2 BGB. The up-to-then mother who loses this status qua adoption must declare approval (Sect. 1747 I BGB, eight weeks must have passed since birth, Sect. 1747 II no. 1 BGB).
21. See for an unsuccessful attempt to claim having born a baby in India Verwaltungsgericht Berlin, app. no. 23 L 79/11, judgment of April 15, 2011.
22. Kammergericht Berlin, app. no. 1 W 413/12, judgment of August 1, 2013; Verwaltungsgericht Köln, app. no. 10 K 6710/11, judgment of February 20, 2013.
23. Verwaltungsgericht Berlin, app. no. 23 L 79/11, judgment of April 15, 2011.
24. Verwaltungsgericht Berlin, app. no. 23 L 283.12, judgment of September 5, 2012; Verwaltungsgericht Berlin, app. no. 23 L 79/11, judgment of April 15, 2011; Oberlandesgericht Celle, app. no. 17 W 48/10, judgment of March 3, 2011.
25. Oberlandesgericht Celle, app. no. 17 W 48/10, judgment of March 3, 2011.
26. 'Parent' in the sense of Article 19 EGBGB is the person who claims to be parent (Duden 2015a, pp. 68–69). If one of the IPs has another citizenship, assessing the legal situation will be even more demanding, and to make things worse: there is also an option relating to the marital legal status of the mother, Article 19 I, no. 3 EGBGB.

27. BT-Drs. 13/4899, p. 82.
28. Oberlandesgericht Stuttgart, app. no. 8 W 46/12, judgment of February 7, 2012; Verwaltungsgericht Berlin, app. no. 23 L 283.12, judgment of September 5, 2012.
29. Verwaltungsgericht Köln, app. no. 10 K 6710/11, judgment of February 20, 2013.
30. Parentage via successive adoption (either of the biological child or of a child already adopted by one partner) is possible for same-sex couples if they live in a legally registered partnership (Lebenspartnerschaft), see Section 9 VII Civil Partnership Act or if they are married.
31. The Amtsgericht Hamm, app. no. XVI 192/08, denied the wife of a man who had gotten parental status via acknowledgment the desired adoption, judgment of February 22, 2011; but see also Landgericht Frankfurt, app. no. 2/09 T 50/1, judgment of August 3, 2012.
32. However, if they are citizens of a country in which both women in a gay marriage are legal parents of a child, they can use this as the basis to be both registered in Germany as parents, too, without adoption procedures, BGH, app. no. XII ZB 15/15, judgment of April 20, 2016.
33. Oberverwaltungsgericht Münster, app. no. 19 A 2/14, judgment of July 14, 2016.
34. Verwaltungsgericht Köln, app. no. 10 K 2043/12, judgment of November 13, 2013.
35. BGH, app. no. XII ZB 463/13, judgment of December 10, 2014.
36. BGH, app. no. XII ZB 463/13, at no. 13–14, judgment of December 10, 2014.
37. Cited by the ECtHR, *Mennesson vs. France*, app. no. 65192/11, no. 82, judgment of June 26, 2014.
38. ECtHR, *Mennesson vs. France*, app. no. 65192/11, judgment of June 26, 2014.
39. ECtHR, *Labassee vs. France*, app no. 65941/11, judgment of June 26, 2014.
40. Both German highest courts, that is the Federal Constitutional Court (Bundesverfassungsgericht) and the BGH now explicitly declare that it makes no difference for the well-being of children if they grow up with a male and female or with two parents of the same sex: Bundesverfassungsgericht, app. no. 1 BvL 1/11, 1 BvR 3247/09, at no. 80–81, judgment of February 19, 2013; BGH, app. no. XII ZB 463/13, at no. 44, judgment of December 10, 2014.

41. ECtHR, *Mennesson vs. France*, app. no. 65192/11, at no. 100, judgment of June 26, 2014.
42. ECtHR, *Paradiso and Campanelli vs. Italy*, app. no. 25358/12, judgment of January 24, 2017.
43. BGH, app. no. XII ZB 463/13, at no. 51, judgment of December 10, 2014.
44. Oberlandesgericht Braunschweig, app. no. 1 UF 83/13, judgment of April 12, 2017. The Court uses far-fetched arguments, for instance, challenging on purely formal grounds DNA-expertise that established the biological fatherhood of the male plaintiff, and arguing that the District Court in Boulder, Colorado, had decided about the legal parenthood six weeks before birth (but in the case decided in 2014 by the BGH, this was the case, too).
45. BGH, app. no. IV ZR 141/16, judgment of June 14, 2017.
46. ECtHR (Grand Chamber), app. no. 57813/00, at no. 117–118., *S.H. and others vs. Austria*, judgment of November 3, 2011.
47. BGH, app. no. XII ZB 463/13, at no. 49, judgment of December 10, 2014.

Bibliography

Birnbacher, D. (1996). Ambiguities in the concept of Menschenwürde. In K. Bayertz (Ed.), *Sanctity of life and human dignity* (pp. 107–121). Heidelberg: Springer.

Brunet, L., Carruthers, J., Davaki, K., King, D., Marzo, C., & Marzo, C. (2013). *A comparative study on the regime of surrogacy in EU member states*. Available at http://www.europarl.europa.eu/thinktank/en/document.html?reference=IPOL-JURI_ET%282013%29474403. Accessed 27 June 2017.

Coester-Waltjen, D. (2016). Family forms and parenthood in Germany. In A. Büchler & H. Keller (Eds.), *Family forms and parenthood. Theory and practice of article 8 ECHR in Europe* (pp. 201–236). Cambridge: Intersentia.

Conklin, C. (2014). Simply inconsistent: Surrogacy laws in the United States and the pressing need for regulation. *Women's Rights Law Reporter, 35*, 67.

Dethloff, N. (2014). Leihmütter, Wunscheltern und ihre Kinder. *Juristenzeitung, 69*, 922.

Dethloff, N. (2015). Changing family forms: Challenges for German law. *Victoria University of Wellington Law Review, 46*, 671.

Dethloff, N. (2016). Anmerkung. *Juristenzeitung, 71,* 207.

Dethloff, N., & Ramser, C. (2007). Tensions between biological and social conceptions of parentage in German Law. In I. Schwenzer (Ed.), *Tensions between biological and social conceptions of parentage* (pp. 177–210). Cambridge: Intersentia.

Duden, K. (2015a). *Leihmutterschaft im Internationalen Privat-und Verfahrensrecht.* Tübingen: Mohr Siebeck.

Duden, K. (2015b). International surrogate motherhood: Shifting the focus to the child. *Zeitschrift für Europäisches Privatrecht, 23,* 637–660.

Engel, M. (2014). Cross-border surrogacy: Time for a convention. In K. Boele-Woelki, N. Dethloff, & W. Gephart (Eds.), *Family law and culture in Europe. Developments, challenges and opportunities* (pp. 199–216). Cambridge et al.: Intersentia.

Gassner, U. et al. (2013). *Fortpflanzungsmedizingesetz, Augsburg-Münchner Entwurf.* Tübingen: Mohr Siebeck.

Golombok, G., Readings, J., Blake, L., Casey, P., Marks, A., & Jadva, V. (2011). Families created through surrogacy: Mother-child relationships and children's psychological adjustment at age 7. *Developmental Psychology, 47,* 1579–1588.

Gössl, S. (2013). Country-report: Germany. In K. Trimmings & P. Beaumont (Eds.), *International surrogacy arrangements. Legal regulation at the international level* (pp. 131–142). Oxford and Portland: Hart.

Helms, T. (2016). *Rechtliche, biologische und soziale Elternschaft: Herausforderungen durch neue Familienformen. Gutachten F zum 71. Deutschen Juristentag.* München: C.H. Beck.

Hörnle, T. (2013). Menschenwürde und Ersatzmutterschaft. In J. Joerden, E. Hilgendorf, & F. Thiele (Eds.), *Menschenwürde und Medizin. Ein interdisziplinäres Handbuch* (pp. 743–754). Berlin: Duncker & Humblot.

Hüppe, H. (2015). Legalisierung der Eizellspende? *Zeitschrift für Rechtspolitik, 48,* 126.

Jadva, J., Imrie, S., & Golombok, S. (2014). Surrogate mothers 10 years on: A longitudinal study of psychological well-being and relationships with the parents and child. *Human Reproduction, 30,* 373.

Knecht, M. (2017). *Reproduktives Reisen und die Herstellung von Verwandtschaft.* Available at http://www.ethikrat.org/dateien/pdf/fb-22-03-2017-simultan-mitschrift.pdf. Accessed 13 Aug 2017.

Kreß, H. (2013). Samenspende und Leihmutterschaft: Problemstand, Rechtsunsicherheiten, Regelungsansätze. *Familie, Partnerschaft, Recht,* 240–243.

Lehmann, M. (2007). *Die In-vitro-Fertilisation und ihre Folgen*. Frankfurt am Main: Peter Lang.

Mayer Lewis, B. (2017). *Lebensweltlicher Ansatz, individuelle Sicht der Problematik* [pdf]. Available at http://www.ethikrat.org/dateien/pdf/fb-22-03-2017-simultanmitschrift.pdf. Accessed 13 Aug 2017.

Müller-Terpitz, R. (2016). 'ESchG 2.0': Plädoyer für eine partielle Reform des Embryonen-schutzgesetzes. *Zeitschrift für Rechtspolitik, 49,* 51.

Resnik, D. B. (2001). Regulating the market for human eggs. *Bioethics, 15,* 1–25.

Rost, C. (2014). Kinderwunsch in der Grauzone. *Süddeutsche Zeitung* [online]. Available at http://www.sueddeutsche.de/muenchen/prozess-in-muenchen-kinderwunsch-in-der-grauzone-1.2049308. Accessed 27 June 2017.

Spiewak, M. (2013). Strafsache Kinderwunsch, *DIE ZEIT* [online]. Available at http://www.zeit.de/2013/34/kinderwunsch-justiz-embryonenschutzgesetz/komplettansicht. Accessed 27 June 2017.

Valerius, B. (2017). Kinderwunschbehandlungen im Ausland. *Zeitschrift für Medizinstrafrecht, 3,* 20–27.

Wiesemann, C. (2017). *Streitgespräch: Der ethisch angemessene Umgang mit den bestehenden Problemen*. Available at http://www.ethikrat.org/dateien/pdf/fb-22-03-2017-simultanmitschrift.pdf. Accessed 13 Aug 2017.

Wilson, T. (2017). *Unravelling orders in a borderless Europe? Cross-border reproductive care and the paradoxes of assisted reproductive technology policy in Germany and Poland*. Available at http://dx.doi.org/10.1016/j.rbms.2017.02.002. Accessed 27 June 2017.

12

Ethical Objections About Surrogacy in German Debates: A Critical Analysis

Katharina Beier

Introduction

German law defines surrogacy as practice in which a woman is willing to permanently give up her child after birth. Since 1991, surrogacy is banned by the Embryo Protection Act (*Embryonenschutzgesetz*, ESchG) (for the legal regulation of surrogacy in Germany, see Dücker and Hörnle, Chapter 11).

Debates on surrogacy began well before the enactment of these legal provisions. Following surrogacy's practice in the US since the late 1970s, the first children were expected to be born from surrogacy also in Germany (Coester-Waltjen 1982). Societal and legal debates were fueled by attempts to establish agencies for matching potential surrogates and intended parents (IPs) in Germany as well as by three failed surrogacy arrangements that resulted in legal disputes (Bernhard 2014).

K. Beier (✉)
Department of Medical Ethics and History of Medicine,
University Medical Center Göttingen, Göttingen, Germany
e-mail: Katharina.Beier@medizin.uni-goettingen.de

S. Mitra et al. (eds.), *Cross-Cultural Comparisons on Surrogacy and Egg Donation*,
https://doi.org/10.1007/978-3-319-78670-4_12

Although the enactment of the ESchG and the adaptation of the Adoption Law did close the door for surrogacy within Germany quite effectively, the participation of German citizens in transnational surrogacy is not covered by these provisions (Mitra, Chapter 5). This leads to the paradoxical situation that, despite the existence of a legal ban, Germany is confronted with the ethical and legal repercussions of surrogacy.

While the legal framework for surrogacy is analysed elsewhere in this volume (Dücker and Hörnle, Chapter 11), this chapter will focus on the German ethical debate. Specifically, it aims to provide a better understanding of the ethical objections that feed Germany's rather restrictive stance towards surrogacy. To this end, after presenting some information on surrogacy and its current practice in Germany, the chapter will not only describe ethical objections being explicitly or implicitly raised in the academic discourse from the late 1980s until today, but, as a second step, also examine their soundness by highlighting their underlying premises and confronting them with insights from international analyses. Given that concerns about children's right to get to know their origin and surrogate mother's instrumentalization remain as most cogent objections from this analysis, I will outline some conditions for surrogacy that would have to be in place in order to mitigate these concerns.

Surrogacy in Germany: Some Background Information

Given Germany's restrictive position on surrogacy, transnational surrogacy is currently the only way for German citizens to get access to this practice. While the use of reproductive services across borders is a widely known phenomenon, information on numbers, reasons and type of reproductive care remain largely in the dark. A survey from 2010 revealed that German citizens represent a significant share among those who use reproductive services abroad, including in vitro fertilisation (IVF) with or without intracytoplasmic sperm injection (ICSI), intrauterine inseminations (IUI), preimplantation genetic diagnosis (PGD), gametes and embryo donation (Shenfield et al. 2010). On the basis of this it has been extrapolated that each year between 3300 and 4200 treatment cycles are performed on German patients who are mostly driven by the wish to evade legal restrictions (Thorn and Wischmann 2010).

However, valid figures documenting the frequency of surrogacy are generally sparse. Due to inconsistencies in reporting this even applies to countries where surrogacy is allowed. For countries with legal bans in place only estimates can be found. For Germany it has been assumed that between 50 and 100 couples make use of transnational surrogacy every year (Konigorski 2013).[1] In fact, a recent judgement by the German Federal Court of Justice suggests that more couples will feel motivated to engage in cross-border surrogacy arrangements (Dücker and Hörnle, Chapter 11). While India (Reddy et al., Chapter 8) Thailand and Mexico recently introduced restrictions for the provision of surrogacy to foreigners, Russia, Cambodia, Guatemala and the Ukraine are increasingly taking the lead in transnational surrogacy. Also some states of the US make surrogacy available to foreigners. It can thus be assumed that these countries are preferred destinations for German IPs, too (Mitra, Chapter 5; König, Chapter 13; Teschlade, Chapter 14).

However, there is not only a lack of reliable quantitative but also qualitative data on surrogacy in Germany. Families who owe their existence to an illegal practice mostly avoid publicity[2] and even if they speak about their experiences it is difficult to obtain an authentic picture from these self-descriptions. In particular, those who make use of surrogacy are caught between glossing over the difficulties of this practice in order to overcome the societal stigmatisation of their way of family-building or being open about their experiences, inclusively negative ones, with the danger of (further) delegitimising surrogacy in the public (Meyer-Spendler 2015; König, Chapter 13).

The Ethical Debate About Surrogacy in Germany

That surrogacy is currently the most controversial and polarising practice of assisted reproductive treatment (ART) is also reflected by the German discourse. Since surrogacy became publicly known, critical assessments dominate the public and scientific debate. This becomes not least apparent through polemic descriptions. For example, surrogacy has been referred to as "reproductive prostitution" (Felberbaum 2009, p. 625).[3] Scepticism is also reflected by a non-representative survey conducted by the German Ethics Council in 2014. Asked, whether surrogacy should be legalised in Germany, 43% of the respondents were opposed to this, whereas 39% spoke out in favour of it and 18% remained undecided. For comparison, egg donation, which is also banned by German law, is supported by 63% of respondents, with only 27% rejecting it and 10% being undecided.[4]

With regard to academic publications on surrogacy two phases can be distinguished. The first one dates back to the beginning of the 1980s and lasts until the beginning of the 1990s. Since surrogacy was unregulated at that time in Germany, legal issues, for example with regard to family law, parentage and legal status (Coester-Waltjen 1982; Dietrich 1989) have dominated the debate, whereas only few works systematically have addressed the ethical and psychosocial consequences of surrogacy (Hass et al. 1988). The second phase gained momentum in the second decade of the twenty-first century and lasts until today. Due to the rise of transnational surrogacy, conflicts of law and comparative legal analyses are at the centre of current debates (Diehl 2014; Duden 2015) (Dücker and Hörnle, Chapter 11). At the same time, there is an increasing recognition of surrogacy's ethical challenges (Schleissing 2014; Bleisch 2013). However, compared to the large body of legal works, ethical analyses of surrogacy are still in the minority.[5]

The following examination is based on the premise that ethical arguments about surrogacy are not only invoked by ethicists but by a broad range of scholars from different disciplines, including, philosophy, law, social sciences, medicine and theology. The perusal of German publications revealed two major clusters of objections to surrogacy. The first cluster takes up concerns about children's welfare with regard to their potential objectification and instrumentalization, challenges to responsible parenthood and impediments to children's development of an own identity. The second involves caveats in terms of surrogates' autonomy and welfare, particularly with regard to their potential exploitation and instrumentalization.

While these objections are also present in international debates on surrogacy, there is at least one peculiarity with regard to their normative framing in the German discourse (also Schicktanz, Chapter 6). In particular, criticism about children's and surrogate mothers' instrumentalization and commodification tend to culminate in claims about human dignity violations (Schliesser 2016).[6] Given that human dignity is defined as core value by the German constitution, the rejection of surrogacy in these terms indicates fundamental concerns about its practice. The latter are most prevalent in the context of commercial surrogacy. On the one hand, this can be explained by the fact that commercial arrangements are currently the only accessible option for German IPs. On the other hand, criticism on commercial surrogacy is spurred by a deontological Kantian understanding of autonomy that stresses the priceless value of every human being. However, given that dignity is a rather controversial concept in bioethics claims about surrogacy's instrumentalizing and potentially even dignity-violating effects require further analysis and explanation.

The Child's Welfare

While it has been argued that children's welfare is a neglected aspect in the debate on surrogacy (Schliesser 2016), this observation is at least not backed by the academic discourse. Concerns about children's welfare in the context of surrogacy have been raised early on (Dietrich 1989; Hass et al. 1988). In the debates preceding the ESchG, it was assumed, for example, that the surrogate mother's suppression of a psychosocial relationship with the foetus during pregnancy and the child's relinquishment after birth would negatively affect its development (Dietrich 1989, Jamba and Ladva, Chapter 9).[7]

Although claims about negative effects on children are increasingly challenged by empirical studies on surrogacy (Golombok et al. 2013) these findings prompt different reactions in the German debate. While some take these as confirmation of rather small risks that can be contained by regulation (Günther et al. 2014), others are more cautious, by pointing to the limited value of these studies insofar most of them have been conducted by just one research group in the UK and the children included were still quite young (Diehl 2014).

Arguments that support the prohibition of surrogacy in terms of children's welfare can be found with regard to three aspects: (A) children's insecure moral and legal status, (B) children's objectification and instrumentalization and (C) changes to the concept of responsible parenthood and resulting impediments to children's development of an own identity.

Moral and Legal Insecurity

One threat to children's welfare is expected to result from their ambivalent moral situation during a surrogacy pregnancy. For example, it is feared that "nobody needs to feel bound to the child until its birth" (van den Daele 2002, p. 60). However, from a moral perspective this can be doubted for both, the surrogate mother and the IPs. First, even though the surrogate may be determined to not raise the child herself, this does not absolve her from moral responsibilities for its well-being. In fact, given that she is the only one who can take care of the foetus during pregnancy, she "cannot choose not to be morally responsible for the fetus" (van Zyl 2002, p. 170). Second, the lack of a gestational relationship does not per se preclude bonding between the IPs and the foetus (Teman 2009). Insofar the child would not have been conceived without the IPs' intense wish for it, it is also doubtful to generally suppose that they will be ambivalent of the child or hesitant to take on their parental responsibilities.

However, by arguing that parental responsibility is not only constituted by an act of will but rather emanates from a common history with the child in the mother's womb, this assessment has been questioned. Although bodily mediated experiences are not the only source of parental responsibility, they are still considered such an important involvement that their abandonment is said to undermine parental responsibility and consequently the child's welfare (Schleissing 2014). Strictly speaking, however, this is not a concern about parental responsibility in general but rather about *maternal* responsibility. In fact, given that social fatherhood is widely acknowledged in Germany, the absence of a shared prenatal history is rarely perceived as token of irresponsible fatherhood. Consequently, surrogacy's assumed threat to responsible parenthood arises against the background of a specific understanding of the maternal role which is expected to involve the experience of gestation.

Given that German citizens can only use surrogacy abroad, the unclear 'legal' situation of children arriving in Germany after such arrangement has been pointed out as another welfare-compromising aspect (ibid.). Strictly speaking, however, this threat is not inherent to reproduction via surrogacy but an effect of the current legal framework. For this reason, the ethical question about children's welfare in surrogacy needs to be decoupled from this issue.

Children's Objectification and Instrumentalization

A second complex of welfare-related objections refers to the potential objectification and instrumentalization of children, particularly in the context of commercial surrogacy. Since arguments of children's objectification and instrumentalization are employed with regard to different aspects, their meaning requires further analysis.

A first aspect regards the involvement of payment in surrogacy. The child is said to have a price at which it can be sold (van den Daele 2002) and in this way becomes a commodity. While this appears as common argument to support the claim that surrogacy violates the dignity of children (Diehl 2014), its premises can be challenged.[8] Strictly speaking it is not the child that is exchanged for money, but the effort and burden of the woman who carries it to term. Proponents of the child-trafficking argument typically counter this argument by pointing to the circumstances of payment (Thomale 2015). For example, there is evidence from India that surrogates receive the full amount of money only after the child's handing-over (Saravanan 2013). Although this is clearly a problematic aspect in the treatment of surrogate mothers, it is not easy to see how it turns children into objects or commodities. There is no evidence that payments in surrogacy prompt the IPs or the surrogate mother to think of the child as object, let alone mere object. Insofar it has been noted that not every form of instrumentalization is morally problematic (Herrmann 2006), the claim may need further specification, for example, by asking to what extent payments in surrogacy would do harm to the child's integrity as person. Admittedly, the answer to this question is still limited by the lack of long-term empirical studies on children's well-being after surrogacy. However, other commercialised reproductive contexts, such as IVF, do at least not suggest the occurrence of any harmful effects. In fact, if children's integrity would be violated by the mere fact that money has been involved in their creation, we would also need to be concerned about costly infertility treatments.

Moreover, although surrogacy involves payments from the side of the IPs, they do not "own" the child, but rather obtain parental responsibilities to act in the child's best interest (Arneson 1992, p. 149). One might still argue that instrumentalization occurs in case that the IPs reject the child after birth. Although such cases recently received high attention by the media, the rejection of children, for example due to disability, is not unique to the practice of surrogacy but may occur in any reproductive setting. In order to substantiate the claim that rejection constitutes an instance of children's instrumentalization one would thus need to show that parents in surrogacy are particularly prone to this reaction. However, this can be doubted. For example, a study revealed that the rate of women who decided for prenatal testing after IVF is "roughly half that reported in the general prenatal testing literature" (Caleshue et al. 2010, p. 579). In light of their prior infertility experiences, the women included in this study frequently stressed their readiness to "accept any baby" (ibid., p. 577) as reason for their decline of prenatal testing.

Another concern implied by the objection of children's objectification and instrumentalization is that they would not be seen or treated as ends in themselves due to their assisted 'production' within a contractual framework for the need satisfaction of childless couples (Schliesser 2016; Dietrich 1989). In support of the claim that children are not "born" in the first line but "produced" it is argued, for example, that surrogacy not only involves "a targeted correction of the natural birth process" but places the birth of children into "a completely new, mediated context" (Thomale 2015, pp. 6f.). Obviously, this argument associates surrogacy with reprehensible reproductive interventions by comparing it with "similarly disquieting appropriations of reproduction for external purposes" in German history (ibid.). While this allusion to eugenic practices in the past raises questions of its own, the following discussion will challenge surrogacy's frequent association with 'production' and the exchange of tangible goods in the German discourse.

First, it is important to note that the claim of children being no longer born but produced in surrogacy implies a rejection of interfering with the 'natural' process of reproduction, for example, by 'correcting' infertility with the help of a third person's contribution. It is interesting to note that particularly in the context of new reproductive technologies, reference to nature is frequently made with a normative intention.[9] However, normative recourses to nature have been criticised for their epistemic limitations (Birnbacher 2006). Second, the argument of children's production is also empirically contestable. Specifically, although it is true that surrogacy expands the circle of persons who are biologically involved in a child's procreation, it is still directed at the reproduction of a 'particular child' that is born to 'predetermined parents' rather than an arbitrary (mass) production of offspring that is detached from any relationship. There is even no reason to assume that children would no longer be born 'for their own sake'. In fact, children are born for a variety of reasons, which are more or less focused on the child itself, e.g. completing one's life by a child, ensuring the continuation of family, providing subsistence to older family members. All these motives, however, do not turn children into objects, nor are these immoral reasons as long as it is ensured that they are cared for by responsible parents (Wiesemann 2006).

Insofar objections in terms of children's objectification and instrumentalization seem difficult to maintain,[10] I will take up another concern about surrogacy that engages with the notion of parental responsibility and children's right to get to know their biological origin.

Challenges to Responsible Parenthood and Children's Development of an Own Identity

A fundamental concern that underpins the whole argumentative complex about children's welfare in the German discourse is related to changes of the very concept of parenthood in the context of surrogacy.

While divisions between social and biological mother- and fatherhood are not unusual, for example in the context of blended families or adoption, surrogacy, at least its gestational form, leads to a new type of divided parenthood, i.e. split motherhood. The latter implies that two women do have a biological relationship with the child—one by providing her eggs and the other one by taking on the gestational part. Since egg donation involves a similar constellation, objections in terms of split motherhood are mostly raised in this context but in principle also extend to surrogacy.

A crucial concern in this regard stresses impediments to children's development of an own identity[11] as result of split motherhood (Hass et al. 1988).[12] Particularly the duplication of the mother's role is expected to disturb the child. This perception must be seen against the background that sperm donation occurred as a rather hidden practice at this time. In light of the prevalent notion of family as a biologically defined relationship, parents were advised to keep sperm donation a secret, not least to the child. Interestingly, the issue of (non-) anonymity was raised in the context of surrogacy early on. For example, by contrasting surrogacy with anonymous sperm donation the practice has been defended for the reason that the surrogate does not use her reproductive capacities anonymously outside any human communication and without caring about the emergence of the new human being (Coester-Waltjen 1982). Although this argument rightly stresses the intimate relationship between surrogate and the child-to-be during pregnancy, there is no guarantee that the surrogate's identity will be disclosed to the child later on.

In light of the current discourse on third-party ART this touches on a key point. Given that a child's right to get know her biological origin is increasingly acknowledged in Germany, most recently in the context of sperm[13] and embryo donation (German Ethics Council 2016), surrogacy might also need to be brought under the scope of this regulation (Arbeitskreis Abstammungsrecht 2017). Since many donor-conceived children are interested in their donors and even potential half-siblings (Freeman et al. 2014) it can be assumed that children who were born by a surrogate mother might also regard her as part of their origin and thus feel an urge to learn about her identity. This would exert particular pressure on transnational arrangements insofar these are prone to prevent disclosure by imposing not only regulative but also spatial, linguistic and cultural barriers.

In a governmental report prior to the ESchG, for example, surrogacy is also criticised for contradicting the principle "that parenthood cannot be disposed of at will".[14] According to this perception, parenthood is not only a biological given, but genetic lineage is also identified with lasting, non-disposable relationships. Therefore, split parenthood is seen as threat to the latter. However, it is important to note that fatherhood—irrespective of the existence of a biological relationship with the child—is already at couples' and particularly women's disposal insofar, for example, the mother's consent is required for the acknowledgement of paternity in an unmarried couple.[15] Against this background, the argument ultimately reveals an interest in avoiding the disposal over a 'particular type' of parenthood, i.e. motherhood as genetic and gestational unity.

Recourse to natural parenthood is also made in order to criticise extensions of biological reproduction to more than just two people. However, it can be asked whether "being begotten by two parents" (Weilert 2013, p. 56) is really an indispensable component of responsible parenthood and or rather a contingent, though deeply internalised idea whose change causes intuitive repugnance. Criticism is also directed at surrogacy's contractual framing which is seen to imply a rather volitional understanding of parenthood. Specifically, the creation of children on the basis of a contractual agreement between unrelated actors rather than in the context of genuine personal relationships fuels doubts about parental responsibility. This is especially reflected in descriptions of surrogacy as "ordering" (Schleissing 2014, p. 82) or "commissioning" children (Thomale 2015, p. 10). However, the involvement of a surrogate mother and sometimes also gamete donors does not imply that children can be created by a mere act of volition. Similarly to natural conception, it may need several attempts until a pregnancy manifests and also miscarriages are not unusual (Mitra and Schicktanz 2016). Therefore, the birth of a child remains beyond a plannable result in surrogacy. The rejection of parenthood as voluntary agreement might also be too harsh in light of already existing parental constellations that involve, at least partly, intentional commitments (e.g. social fatherhood, co-parenting arrangements). This is not to derive normative conclusions from empirical facts, but rather to stress the need for moral consistency regarding the evaluation of parental responsibility. Although suspicion regarding a division of motherhood continues until today, it is not unlikely that the recent acceptance of embryo donation—a practice which also splits the genetic and gestational reproductive contribution between women[16]—as well as newly emerging debates on the prohibition of egg donation (Dücker and Hörnle, Chapter 11) might finally lead to the normalisation of split motherhood and acceptance of the resulting familial constellations in Germany. This requires acknowledging that the stability and reliability of familial relationships, especially parenthood, is not exclusively a matter of biology but also of established social practices of family life.

The Welfare of the Surrogate Mother

The perusal of arguments with regard to surrogate mother's welfare in the German debate reveals three major objections: (A) Potential infringements of surrogate's autonomy, (B) Exploitation, and (C) Unethical objectification and instrumentalization of the surrogate which is also seen as violation of her dignity. In the following, I will sketch out each of these concerns.

Infringements of Autonomy

An internationally controversial issue about surrogacy refers to infringements of the surrogate mother's autonomy. In the German debate, this concern is particularly raised with regard to transnational arrangements often involving extreme socio-economic disparities. In particular, it is doubted that women can make truly autonomous decisions because their disadvantageous social situation would practically force them into surrogacy. In fact, studies on surrogacy in India have shown that women who are ready to take on a surrogate pregnancy for a foreign couple live in impoverished conditions.[17] While this objection is based on an understanding of autonomy that requires a reasonable range of choices, the assessment of this issue might differ if autonomy is defined as freedom from coercion. From this latter perspective, a woman's decision for surrogacy would only qualify as non-autonomous if she is forced into surrogacy by others (Duden 2015). Another threat to surrogate mothers' autonomy can be seen to result from an insufficient appreciation of the consequences of surrogacy, for example due to illiteracy. Measures that ensure understanding, however, could theoretically provide a remedy for this problem. In contrast to accounts that are concerned with the procedural prerequisites of autonomy, substantive accounts (Mackenzie and Stoljar 2000) question the possibility of autonomous decision-making in the state of pregnancy, specifically with regard to the child's relinquishment, as such. Substantive accounts of autonomy, however, run the risk of paternalism. Since limitations to women's autonomy are known from other reproductive contexts (Zeiler 2004) without having led to constraints on their right of decision-making, we should be cautious to not make higher demands on autonomous decision-making in the context of surrogacy.

The Challenge of Exploitation

The objection that surrogacy exploits women is a core issue in the international debate which is also reflected in the German ethics discourse. For example, the avoidance of an ethically not justifiable exploitation of women as surrogate mothers has been stressed as major goal in the government's draft for the ESchG.[18] The exploitation of surrogate mothers is still invoked as objection to surrogacy (Schicktanz, Chapter 6). While internationally exploitation is mostly discussed with regard to economic inequalities and false incentives (Saravanan 2013), in the German ethics discourse concerns about "exploitation of underprivileged women by privileged women" tend to be raised alongside with objections of instrumentalization, for example by pointing to the use of surrogates' bodies for external purposes (Graumann 2003, p. 255). However, exploitation and instrumentalization are no identical concepts. Specifically, the objection of exploitation refers to unjust treatment, for example by taking advantage of a woman's deprived situation or by not granting her an adequate share in the context of a commercial arrangement. In contrast, the objection of instrumentalization involves a more fundamental, deontological critique, namely that it is wrong to treat human beings and their bodily functions as (marketable) objects at all. As the issue of exploitation is perhaps solvable by adequate regulation (Ramskold and Posner 2013)[19] it can be asked whether claims about women's instrumentalization represent a weightier argument to justify Germany's restrictive stance on surrogacy. Insofar claims about surrogates' instrumentalization are often equated with violations of their dignity the cogency of this argument will be examined more closely in the following section.

Instrumentalization of the Surrogate Mother

In the following, three different interpretations of instrumentalization can be distinguished. First, the blame of instrumentalization is raised with regard to the high intimacy and persistence of the surrogate's contribution (Schleissing 2014). Pregnancy cannot be interrupted just for some time; for nine months, there is no resting. This distinguishes it from almost every other paid work. It needs to be asked, however, whether nonstop surrogacy for nine months, though for a comparably high payment, is per se more instrumentalizing in nature than any other risky bodily work that is carried out for years, without regular working hours and adequate protection measures. This is not to justify inhuman working conditions. Rather, if surrogacy is criticised and even banned because of its supposed instrumentalizing character, there would be reason to require this for all kinds of bodily capacities exploiting activities.

Second, when speaking of the surrogate mother's instrumentalization, it is suggested that she is used in a morally doubtful way by the IPs who unilaterally degrade her into a mere means for the realisation of their wishes (Dietrich 1989). This disregards, however, that also the IPs are used by the surrogate mother to some extent. In other words, while the IPs use the surrogates' reproductive capacities, the IPs are used as a source of finance by the surrogate mother. For the reason that each side is giving something and receives something in return that she would not have been able to attend without the other, a reciprocal relationship is created which can be said to bring both parties on eye-level.

This mutuality in surrogacy, however, may not suffice to dissipate concerns about surrogates' instrumentalization. In particular, it may be objected that surrogate mother and intending parents will never meet on an equal footing due to the fact that pregnancy and childbirth are incommensurable activities (Schliesser 2016). From this perspective the surrogate will always be at disadvantage. However, even though human beings do not have a price, it seems important to acknowledge the considerable physical and psychological burden that is involved in bringing a child into existence. Therefore it has been pointed out that irrespective of the contract mother's motives and expectations, and even of whether she herself is satisfied with the outcome, given the enormity of her gift it would be wrong for the intending parents not to reciprocate by giving something substantial in return (van Zyl and Walker 2013, p. 375). Even though perceptions of surrogacy as an incommensurable practice suggest that financial reciprocation is insufficient to acknowledge the surrogates' contribution, payments may still be important to avoid severe asymmetries between both parties.

Claims about instrumentalization are thirdly invoked with regard to the surrogate mother's status during pregnancy. Depersonalising descriptions of surrogates as "incubator" (van den Daele 2002, p. 59) or "womb for rent" (Tieu 2009, p. 174) exemplify this concern not only in the German but also international discourse. It must be noted, however, that the perceptual reduction of a person to certain bodily functions or features is not unique to surrogacy but also affects athletes and models. However, neither sports nor fashion normally incur the accusation of instrumentalization. Therefore, the "incubator", respectively "womb for rent" argument seems to imply a weightier, though rarely specified concern. In particular, the critique suggests that surrogate mothers are instrumentalized because—in moral terms—they are entitled to more than just providing their bodily capacities. This argument is closely linked to the claim of incommensurability. In fact, if it is true that surrogates deserve "something substantial in return" (van Zyl and Walker 2013, p. 375) and money is an at least insufficient means to achieve this, the problem may only be solvable at relational level by granting the surrogate an adequate role and social identity within the familial arrangement (Beier 2015). In line with this understanding, it has been stressed that "[t]he idea that surrogacy involves treating one person as a means to the ends of others gains force to the extent that efforts are made to keep the role of the birth mother strictly circumscribed, and her relation to the child nullified after that child's birth" (Lane 2003, p. 136). Although it is not clear yet, how women will integrate the experience of surrogacy into their biography in long term, there is empirical evidence that surrogates perceive an ongoing relationship with the intending couple, and especially the intended mother (Teman 2008). From a moral perspective it appears thus as highly problematic if such relationships are systematically thwarted by the conditions of surrogacy arrangements that these particularly impose on surrogate mothers.

Conclusions

The analysis of the German ethics discourse has shown that surrogacy is ethically condemned on the basis of two major clusters of argument that focus on the welfare of children and surrogates in the broadest sense. At the same time, a major difficulty for analysing these objections results from the fact that they are rarely spelled out in detail. Particularly, claims about children's and surrogates' instrumentalization, objectification and dignity violation tend to occur as catchwords rather than full-fledged arguments. Although the identified objections to surrogacy are not significantly different from those being raised in the international debate, there remain some peculiarities regarding the normative framing of arguments in the German discourse.

First, it is striking that criticism on surrogacy often builds on juxtapositions of natural reproduction (being born) with human interference in this process (producing children). Deviations from biological parenthood and particularly the fragmentation of genetic and gestational motherhood are perceived as threat to the notion of responsible parenthood which is ultimately expected to have negative effects on children's welfare. However, from the appraisal of objections in terms of children's welfare, the concern about children's development of an own identity remained as the most critical point. Given the increasing recognition of children's right to get to know their biological origin in the context of ART in general, the identity development of a child seems most likely to be compromised if she is prevented from receiving information about the circumstances of her creation as well as the surrogate's identity.

Second, commercialisation in surrogacy appears as the most critical point about surrogacy in the German debate. Against the background of a Kantian inspired understanding of human autonomy and dignity, childbearing in exchange for money is especially rejected in terms of children's and surrogates instrumentalization. While there is no sufficient argumentative basis for suggesting that surrogacy instrumentalizes children, it turned out that instrumentalization poses a more serious threat to surrogates. However, on the basis of my prior analysis, I argued that this threat is less the result of payments being involved in childbearing, but rather of the neglect of surrogate mother's complex relationships with the newly emerging family. While this understanding of instrumentalization may also apply to non-commercial arrangements, it is interesting to note, however, that the German discourse remains undecided whether so-called altruistic arrangements are to be condemned with similar rigour as commercial ones (van den Daele 2002; Schliesser 2016).[20]

For the reminder, I do not want to take up the question whether impediments to children's right to get to know their biological origin and the threat of surrogate mother's instrumentalization provide sufficient justification for maintaining the legal ban on surrogacy in Germany, but rather outline some implications that ensue from these arguments for the practical handling of surrogacy in Germany and beyond.

First, and most obviously, it seems important to allow children to get to know the surrogate mother and to be in touch with her, even if the relationship between surrogate mother and intending parents breaks down after birth (Bleisch 2013). Specifically, the legal acknowledgment of IPs who return with a child from transnational surrogacy to their home country, as it is currently happening in Germany, may not occur at the price of the child's right to get to know its biological origin.

Second, it would be important to design surrogacy arrangements in such a way that they allow all parties involved to negotiate and develop their mutual relationships over time, rather than leaving this to exclusively to the IPs' discretion (Beier 2015). However, given that transparent information about surrogacy practices in other countries is hardly available and professionals risk the accusation of assistance to improper application of reproductive practices it is likely that German IPs are insufficiently prepared to handle these complex relational implications, particularly in the context of transnational surrogacy.

Notes

1. This estimate has been made by lawyer Thomas Oberhäuser in a radio broadcast.
2. This has even been the case before the release of the ESchG because the legal situation for surrogacy was not clear at this time either (Hass et al. 1988).
3. All following translations and paraphrases of German quotes in this chapter are the author's.
4. The survey is available at http://www.ethikrat.org/dateien/pdf/jt-22-05-2014-auswertung-vorabbefragung.pdf.
5. A graphical presentation of arguments for and against surrogacy has been developed by the German Ethics Council, http://www.ethikrat.org/dateien/pdf/jt-22-05-2014-leihmutterschaft.pdf.
6. Beyond the academic debate, this can also be observed in media reports as well as in the current government's coalition agreement. For a critical discussion of this argument, see Bleisch (2013).
7. Bundestags-Drucksache (BT-Drs.) 11/5460, p. 15.
8. For a critical discussion of this issue see Duden (2015), pp. 177ff.
9. This can also be observed for the German debate on PGD.
10. See also Hanna (2010) for a critical discussion.
11. Jürgen Habermas most prominently used this argument in his criticism of prenatal genetic interventions. In contrast to the use of this argument in the context of third-party assisted reproduction, however, his focus is on the moral relevance of freedom from alien genetic determination as prerequisite for the child's identity as autonomous person (Habermas 2003).

12. BT-Drs. 11/5460, p. 7.
13. According to a recently adopted law, a central donor sperm database will be established at the German Institute of Medical Documentation.
14. BT-Drs. 11/1856, p. 9.
15. While in married couples paternity is automatically assigned to the woman's spouse, paternity in unmarried couples requires the woman's consent that must be officially acknowledged by a civil registry office.
16. According to the predominant legal position, the donation of supernumerary embryos from IVF-treatment does not violate the ESchG, provided that these were not created with the intent of being donated (German Ethics Council 2016, pp. 41f.).
17. Even in India, however, surrogates do not belong to the poorest of the poor, especially because their health must be sufficient for carrying a pregnancy (Mitra, Chapter 5).
18. BT-Drs. 11/1856, p. 8.
19. Moreover, provided that exploitation is most likely to occur in transnational settings it can be asked whether this concern could be mitigated within (domestic) contexts that involve more equal living conditions.
20. If altruistic surrogacy is criticised at all in the German discourse, objections are mostly raised in terms of split motherhood and its assumed disturbing effects on traditional understandings of parenthood and family relationships (Schleissing 2014; BT-Drs.11/5460, p. 15) rather than in terms of instrumentalization. This is in contrast to the international debate where claims of instrumentalization also extend to altruistic surrogacy (Tieu 2009).

Bibliography

Arbeitskreis Abstammungsrecht. (2017). *Abschlussbericht. Empfehlungen für eine Reform des Abstammungsrechts*, ed. Bundesministerium für Justiz und für Verbraucherschutz (BMJV). Köln: Bundesanzeiger Verlag.

Arneson, R. J. (1992). Commodification and commercial surrogacy. *Philosophy & Public Affairs, 21*(2), 132–164.

Beier, K. (2015). Surrogate motherhood: A trust-based approach. *Journal of Medicine and Philosophy, 40*(6), 631–652.

Bernhard, A. (2014). *Kinder machen: Neue Reproduktionstechnologien und die Ordnung der Familie*. Frankfurt am Main: S. Fischer.

Birnbacher, D. (2006). *Natürlichkeit*. Berlin, New York: Walter de Gruyter.

Bleisch, B. (2013). Leihmutterschaft als persönliche Beziehung. *Jahrbuch für Wissenschaft und Ethik, 17*(1), 5–28.

Caleshue, C., Shiloh, S., Price, C., Sapp, J., & Biesecker, B. (2010). Invasive prenatal testing decisions in pregnancy after infertility. *Prenatal Diagnosis, 30,* 575–581.

Coester-Waltjen, D. (1982). Rechtliche Probleme der für andere übernommenen Mutterschaft. *Neue Juristische Wochenschrift, 35,* 2528–2534.

Diehl, A. (2014). *Leihmutterschaft und Reproduktionstourismus.* Frankfurt am Main: Wolfgang Metzner Verlag.

Dietrich, S. (1989). *Mutterschaft für Dritte.* Frankfurt am Main: Peter Lang.

Duden, K. (2015). *Leihmutterschaft im Internationalen Privat- und Verfahrensrecht.* Tübingen: Mohr Siebeck.

Felberbaum, R. E. (2009). Medizinische und ethische Aspekte der Leihmutterschaft. *Gynäkologe, 42,* 625–626.

Freeman, T., Bourne, K., Jadva, V., & Smith, V. (2014). Making connections. Contact between sperm donor relations. In T. Freeman, S. Graham, F. Ebtehaj, & M. Richards (Eds.), *Relatedness in assisted reproduction* (pp. 270–295). Cambridge: Cambridge University Press.

German Ethics Council. (2016). *Embryospende, Embryoadoption und elterliche Verantwortung. Stellungnahme* [online]. Available at http://www.ethikrat.org/dateien/pdf/stellungnahme-embryospende-embryoadoption-und-elterliche-verantwortung.pdf. Accessed 25 Jan 2017.

Golombok, S., Blake, L., Casey, P., Roman, G., & Jadva, V. (2013). Children born through reproductive donation: A longitudinal study of child adjustment. *Journal of Child Psychology and Psychiatry, 54,* 653–660.

Graumann, S. (2003). Fortpflanzungsmedizin aus ethischer Sicht: alte und neue Fragen. In M. Düwell & K. Steigleder (Eds.), *Bioethik. Eine Einführung* (pp. 246–257). Frankfurt am Main: Suhrkamp.

Günther, H.-L., Taupitz, J., & Kaiser, P. (2014). *Embryonenschutzgesetz. Juristischer Kommentar mit medizinisch-naturwissenschaftlichen Grundlagen.* Stuttgart: Kohlhammer.

Habermas, J. (2003). *The future of human nature.* Cambridge: Polity Press.

Hanna, J. K. M. (2010). Revisiting child-based objections to commercial surrogacy. *Bioethics, 27*(7), 341–347.

Hass, G., Grossmann, W., & Kallert, H. (1988). *Leihmutterschaft. Psychische und psycho-soziale Folgen.* Frankfurt am Main: Nelson Verlag.

Herrmann, B. (2006). Body Shopping? Der Körper zwischen Unverfügbarkeit und Vermarktung. In S. Ehm & S. Schicktanz (Eds.), *Körper als Maß?*

Biomedizinische Eingriffe und ihre Auswirkungen auf Körper- und Identitätsverhältnisse (pp. 207–224). Stuttgart: S. Hirzel Verlag.

Konigorski, M. (2013). *Der gemietete Bauch. Ethische Überlegungen zur Leihmutterschaft*. Beitrag am 9.5.2013 im Deutschlandfunk [online]. Available at http://www.deutschlandfunk.de/der-gemietete-bauch.886.de.html?dram:article_id=246049. Accessed 25 Jan 2017.

Lane, M. (2003). Ethical issues in surrogacy arrangements. In R. Cook, S. D. Sclater, & F. Kaganas (Eds.), *Surrogate motherhood: International perspectives* (pp. 121–139). Oxford and Portland: Hart Publishing.

Mackenzie, C., & Stoljar, N. (2000). *Relational autonomy. Feminist perspectives on autonomy, agency and the social self*. Oxford: Oxford University Press.

Meyer-Spendler, R. (2015). *Ein schmaler Grat. Erfahrungen mit Leihmutterschaft in den USA*. Mörfelden: FamART.

Mitra, S., & Schicktanz, S. (2016). Failed surrogate conceptions: Social and ethical aspects of preconception disruptions during commercial surrogacy in India. *Philosophy, Ethics, and Humanities in Medicine* [online]. Available at https://www.ncbi.nlm.nih.gov/pmc/articles/PMC5075174/pdf/13010_2016_Article_40.pdf. Accessed 10 Dec 2017.

Ramskold, L. A. H., & Posner, M. P. (2013). Commercial surrogacy: How provisions of monetary remuneration and powers of international law can prevent exploitation of gestational surrogates. *Journal of Medical Ethics, 39*(6), 397–402.

Saravanan, S. (2013). An ethnomethodological approach to examine exploitation in the context of capacity, trust and experience of commercial surrogacy in India. *Philosophy, Ethics, and Humanities in Medicine, 8*(10), 1–12.

Schleissing, S. (Ed.). (2014). *Ethik und Recht in der Fortpflanzungsmedizin. Herausforderungen – Diskussionen – Perspektiven. TTN Ethik interdisziplinär*. München: Nomos.

Schliesser, C. (2016). Körperlichkeit und Kommerzialisierung: Zur theologisch-ethischen Problematik der Leihmutterschaft. *Zeitschrift für medizinische Ethik, 62,* 107–120.

Shenfield, F., de Mouzon, J., Pennings, G., Ferraretti, A. P., Nyboe Andersen, A., de Wert, G., et al. (2010). Cross border reproductive care in six European countries. *Human Reproduction, 25*(6), 1361–1368.

Teman, E. (2008). The social construction of surrogacy research: An anthropological critique of the psychological scholarship on surrogate motherhood. *Social Science and Medicine, 67,* 1104–1112.

Teman, E. (2009). Embodying surrogate motherhood: Pregnancy as a dyadic body-project. *Body & Society, 15*(3), 47–69.
Thomale, C. (2015). *Mietmutterschaft*. Tübingen: Mohr Siebeck.
Thorn, P., & Wischmann, T. (2010). Leitlinien des BKiD "Psychosoziale Beratung für Frauen und Männer, die eine Kinderwunschbehandlung im Ausland beabsichtigen". *Journal für Reproduktionsmedizin und Endokrinologie, 7*(5), 394–402.
Tieu, M. M. (2009). Altruistic surrogacy: The necessary objectification of surrogate mothers. *Journal of Medical Ethics, 35,* 171–175.
Van den Daele, W. (2002). *Mensch nach Maß. Ethische Probleme der Genmanipulation und Gentherapie*. München: C.H. Beck.
van Zyl, L. (2002). Intentional parenthood: Responsibilities in surrogate motherhood. *Health Care Analysis, 10,* 165–175.
van Zyl, L., & Walker, R. (2013). Beyond altruistic and commercial contract motherhood: The professional model. *Bioethics, 27*(7), 373–381.
Weilert, A. K. (2013). Fortpflanzungsautonomie als Anspruch. *Zeitschrift für Evangelische Ethik, 57,* 48–61.
Wiesemann, C. (2006). *Von der Verantwortung, ein Kind zu bekommen*. München: C.H. Beck.
Zeiler, K. (2004). Reproductive autonomous choice: A cherished illusion? Reproductive autonomy examined in the context of preimplantation genetic diagnosis. *Medicine, Health Care and Philosophy, 7*(2), 175–183.

13

Parents on the Move: German Intended Parents' Experiences with Transnational Surrogacy

Anika König

Introduction

Shortly after Sandra woke up following the emergency hysterectomy, a doctor visited her in her hospital room and told her that even though they had just removed her uterus she could still have children—there still was the option of surrogacy. Three and a half years later, after one failed surrogacy, a surrogate gave birth to her and her husband's twins in California—a baby boy and a baby girl. Today, the twins are healthy and happy toddlers, living with their parents and older sister on the outskirts of a large German town (Interview with Sandra,[1] 8 October 2013).

At first sight, this would appear to be a success story. But taking a look behind the scenes, it becomes obvious that this is also a story of suffering, extreme psychological stress, and anxiety—an experience

A. König (✉)
Institut für Medizingeschichte und Wissenschaftsforschung, University of Lübeck, Lübeck, Germany
e-mail: koenig@imgwf.uni-luebeck.de

S. Mitra et al. (eds.), *Cross-Cultural Comparisons on Surrogacy and Egg Donation*,
https://doi.org/10.1007/978-3-319-78670-4_13

that many German intended parents (IPs) have in common. Germany has one of the most restrictive laws worldwide concerning the use of assisted reproductive technologies (ART), including a strict prohibition of surrogacy. As my research shows, for German IPs the experience of their surrogacy journey is strongly affected by the fact that it is illegal. Moreover, even though the use of legal ART such as in vitro fertilisation (IVF) or intracytoplasmic sperm injection (ICSI) has become increasingly common in Germany—as is the case in most industrialised countries—ART is still stigmatised. This is even more so when third parties are involved, and particularly when the procedure is illegal, as is the case with surrogacy.

This chapter is based on continuing ethnographic research I have carried out in Germany and Switzerland since 2013 and in California in 2014. Twelve IPs participated in my project. I collected data through email exchange, face-to-face and phone interviews, and written accounts some IPs sent me (such as photocopies of a personal journal or a summary of their story). While I interviewed some of them only once, I have worked with others for several years and still continue to do so. In addition, I conducted participant observation in early 2014 when accompanying a German family to California where they were making arrangements for their surrogacy.

I also interviewed two surrogates in California (one of whom also owns a surrogacy agency), and a prospective surrogate in Germany—a US citizen, who at the time of the interview planned to be a surrogate for her gay college friend. Moreover, also in California, five agency owners and staff agreed to be interviewed. In addition, I conducted expert interviews with two well-known fertility doctors in Californian clinics, as well as two lawyers in California, who specialise in surrogacy, and one lawyer in Germany, who has plead several surrogacy cases in Germany.

Drawing on this research, this chapter focuses on IPs' experiences with surrogacy abroad, especially in the US. It deals with their experiences in a national context that is informed by legal, moral, and political frameworks that are highly restrictive, and their strategies to circumvent this. Subsequently, I briefly discuss the issue of fragmented parenthood and the well-being of the child.

Legal, Moral, and Political Frameworks

The German laws regulating ART, including surrogacy, are among the most restrictive worldwide. In Germany, both the performance of the medical procedures necessary for surrogacy, as well as brokering services such as finding and matching surrogates and IPs, are strictly prohibited by the Embryo Protection Act (*Embryonenschutzgesetz*, ESchG) and the Adoption Placement Act (*Adoptionsvermittlungsgesetz*, AdVermiG) (Dücker and Hörnle, Chapter 11). Brokering also includes helping Germans to find a clinic or agency in other countries. Because of these laws, medical doctors in Germany do not perform any medical procedures that lead to surrogacy as they could risk the loss of their licence to practice medicine. In addition, IPs cannot seek official advice or make use of surrogacy agency services at home. The sources of information that are available to them are thus limited to the internet and other IPs, a situation that leads to an over-abundance of information, misinformation, and rumours, which IPs find difficult to filter and verify. Importantly, however, neither IPs nor surrogates are punished if they commission a surrogacy or agree to become pregnant for others (Dücker and Hörnle, Chapter 11). This opens up a legal loophole for commissioning surrogacy abroad.

In addition to the legal restrictions, the German public discourse, and related to this the moral regime, is highly critical of ART that include third persons. Interestingly, this mainly applies to egg donation and particularly surrogacy, though almost completely ignores sperm donation (which is legal in Germany).[2]

Criticism of surrogacy in Germany can be roughly divided into four main topics: fear of the creation of so-called 'designer babies', the exploitation of women, the 'unnaturalness' of surrogacy which is said to disrupt the bond between mother and child, and so-called 'baby selling' (Beier, Chapter 12). The first of these, the notion of 'designer babies', is certainly not limited to the German discourse, but it does play a particularly important role in the country. This is likely connected to local ideas concerning (human) life and interference with its creation. For example, one crucial notion that can be traced back to Christian morality is that of 'playing God'—an undesirable activity, since in Christian doctrine humans are seen to be subordinate to God and should not interfere with His creation. This also, and especially, applies to the beginning of life—the embryo—and therefore to the use of certain ART.

Moreover, in Germany there is widespread scepticism towards new medical technologies in general, which can, among other things, be linked to the German experience of the Holocaust and the horrific medical experiments performed by Nazi doctors on children and adults (Raz and Schicktanz 2016, p. 7). Consequently, the medical manipulation, 'improvement', and selection of human gametes and embryos are regarded with great suspicion. While surrogacy itself does not necessarily include such selection and testing processes, my research has shown that many clinics in California offer a wide range of preimplantation genetic diagnosis (PGD) options that most IPs I spoke with made use of in order to increase their chances of a successful pregnancy and a healthy child (Mitra, Chapter 5).

The second thematic thread of criticism focuses on the exploitation of women in commercial surrogacy arrangements—both egg donors and surrogates. The common assumption in this context is that women are compelled to sell their bodies and reproductive capacities due to financial distress. This issue divides feminists in particular into two camps—one that argues in favour of women's rights to make their own choices regarding their bodies (e.g. Diehl 2014), even if this means that they use it as a commodity, and the other, in Germany perhaps most prominently represented by well-known long-time feminist Alice Schwarzer, that sees reproductive services as commodifying and exploiting women (Schwarzer 2015). This latter camp argues that even if women claim to have voluntarily chosen to be surrogates or sex workers, the true reason behind such a choice is financial or psychological distress. In Germany, not only feminist discourse,[3] but the public discourse in general[4] is dominated by this latter view of commercial surrogacy, which emphasises its exploitative character.

The third thread of criticism of surrogacy concerns ideas of the 'naturalness' of pregnancy, motherhood, and an assumed inevitable bond between the pregnant woman and the foetus she carries.

This view also provided the background to the draft for amendments to the AdVermiG in 1989, where it was claimed that there is a very special relationship between the foetus[5] and the mother that prohibits carrying a child as a service (Bernard 2014).

Finally, in German public discourse, commercial surrogacy is often presented as a form of 'baby selling'. From this perspective, children are commodified and traded as a material good. Newspaper articles such as *Käufliches Elternglück* (Parental happiness for sale) (Bubrowski 2013) or *Babys übers Internet bestellen* (Order babies on the internet) (Jovanovic 2010) are exemplary of this view. Proponents of this perspective argue that commercial surrogacy poses a great threat to the idea that children are 'precious yet priceless' (Berend 2015), that they cannot and should not be bought, and that the involvement of money violates their dignity.

These arguments against (particularly commercial) surrogacy, in combination with the fact that surrogacy is not legal according to German law, not only shape the German moral regime as it becomes visible in the media discourse, but also political approaches to surrogacy. Members of all parties—ranging across the entire political spectrum, from Christian Democrat Member of Parliament Hubert Hüppe (2014) to the queer working group of the Left Party (2012)—have announced their rejection of surrogacy and of related ART such as egg donation or PGD. Similarly, the annual conference of the German Ethics Council in 2014 which focused on ART concluded with a panel discussion between members of three large political parties—again, the Christian Democrat Hubert Hüppe, as well as Harald Terpe of the Green Party, and Kathrin Vogler of the Left Party—all of whom emphasised their disapproval of surrogacy. Finally, the political rejection of surrogacy was also included in the coalition agreement of the three parties that made up the German government between 2013 and 2017 (Social Democrats, Christian Democratic Union, and Christian Social Union): 'We reject surrogacy since it is incompatible with human dignity' (CDU, CSU, and SPD 2013, p. 99).[6]

These three frameworks—the legal, the (public media-based) moral, and the political—within which surrogacy is situated are strongly intertwined and influence one another. As long as surrogacy continues to be illegal in Germany, it is unlikely that the moral framework and public discourses about it will change, or that politicians will advocate for surrogacy.

Doing Surrogacy: Circumventing Local Laws and Dealing with Moral Regimes

On an individual level, the above described discourses strongly influence the experiences of German IPs when they decide to commission a surrogacy regardless of the many constraints they face. The first, legal restrictions, can only be circumvented by travelling to countries that permit surrogacy and commissioning it there. This process has been widely referred to as 'reproductive tourism' (Bergmann 2011; Deomampo 2013; Pennings 2004)—a term that has in the last few years been increasingly criticised for its negative connotations and insinuation of pleasure (Bassan and Michaelsen 2013; Inhorn and Gürtin 2011; Shenfield et al. 2010; Whittaker and Speier 2010). Indeed, in my interviews IPs described travelling and the geographical distance between the different people involved in the process of surrogacy as a great burden. Consequently, I have chosen to use the term 'reproductive travel', as it does not imply the enjoyment that is connoted by 'tourism' (König 2017).

But where do people go in order to commission a surrogacy? Usually, German (and for that matter, most other) IPs choose one of the places that I have elsewhere called 'reproductive hubs' (König 2017; Werner-Felmayer, Chapter 2). These are geographic locations where a permissive legal setting, medical expertise, surrogacy agencies, and related services such as law firms merge in one place. In addition, women who are willing to act as surrogates live in the area or can easily travel there. The reproductive hubs that most German IPs use are California, Ukraine, and, until the recent ban on surrogacy, India.[7]

Even though at first sight the same services are offered at these different reproductive hubs, taking a closer look reveals considerable differences. Accordingly, one cannot even speak of surrogacy as one specified practice, as in fact surrogacy means many different things, is done under very diverse conditions, and the actors involved participate in it for a variety of reasons. In particular, it is the geographical location—the reproductive hub—that IPs choose that determines the ways in which surrogacy is performed and organised. But even within a reproductive hub, practices may differ greatly depending on the clinic and the agency, in terms of costs, the medical and psychological screening of surrogates, the care given to surrogates, the kind of contact that is possible between IPs and the surrogate (open vs. closed and anonymous programmes), etc.

In general, there are low-cost destinations such as Eastern European and Asian countries, where a surrogacy costs around 30,000 EUR, and high-cost destinations, mainly some states in the US, where the same process can easily cost five times as much (Werner-Felmayer, Chapter 2). Support networks and the quality of medical services are also likely to be better in places such as California as compared to Ukraine (although this is not always and necessarily the case). Finally, most low-cost destinations predominantly offer so-called 'closed' surrogacy programmes, where IPs and surrogates remain anonymous and do not meet one another. In contrast, 'open' programmes require both parties to meet (in some cases in person, in others only online, usually via Skype) before the final decision is made if they want to do the surrogacy together. In these open programmes, both surrogates and IPs can refuse to do the surrogacy with this particular person or couple. Moreover, continuous contact during, and often also after, the pregnancy is encouraged or even included in the contract between surrogate and IPs. Many US agencies prefer the use of open programmes.

Another aspect that differentiates reproductive hubs are the legal regulations in terms of awarding parental status and citizenship. Depending on how this is done, it is easier (US) or harder (most low-cost destinations) to obtain the documents necessary for the child's passport or visa, which is needed to bring it into the parents' home country. The US award citizenship based on *ius soli*, that is, every person born on US territory automatically becomes a citizen. As US citizens do not need a visa to travel to Germany, a child born through surrogacy in the US can enter Germany without any difficulty with its US passport. The situation is different in countries that offer low-cost surrogacy services, such as Ukraine or, as Werner-Felmayer (Chapter 2) and Sándor (Chapter 3) show, until recently also India. These countries regard the children born through surrogacy as citizens of their commissioning parents' country (Trimmings and Beaumont 2011). The German Civil Code (*Bürgerliches Gesetzbuch*, BGB), in contrast, rules that the woman who gave birth to the child is the mother, and thus passes on her nationality to the child, and, if she is married, then her husband is the father (Dücker and Hörnle, Chapter 11). These conflicting laws—with India and Ukraine claiming the child to be German, while Germany claims the same child to be Indian or Ukrainian—result in a refusal by either country to award citizenship and the child being effectively parent- and stateless (Trimmings and Beaumont 2011, p. 631). Without a passport, this child cannot cross national borders. One intended father (Interview with Peter, 13 June 2013), whose first daughter was born through surrogacy in India in 2010, told me that at that time, there had been an entire 'scene' of parents (of mostly European origin) who, due to legal issues, were stuck in India. In most of these cases, one partner was forced to return to Europe to earn money while the other remained in India and looked after the stateless child, hoping to finally receive a passport for it. Accordingly, for German IPs, commissioning a surrogacy in the US is much more attractive than most other locations when it comes to issues of passport acquisition and bringing the child home to Germany (Dücker and Hörnle, Chapter 11).

But in addition to the legal restrictions that IPs struggle with in Germany, some of my interviewees expressed serious moral concerns regarding the unlawfulness of the whole surrogacy process and the question of lying to the authorities. For example, an intended mother, who sent me photocopies of her diary in which she had noted her feelings and experiences concerning the surrogacy they were planning to commission, wrote:

> We were advised to go to another EU country or lie to the authorities. I can't do the latter. I cannot go abroad and stand in the German embassy and tell them lies![8] My husband had an affair, or I gave birth to the child there because I went on that trip heavily pregnant?! I cannot do that. And then the lawyer suggested we should send another couple, who also just

> had a baby, to pick up our baby in Ukraine by using their newborn child's passport. Cheating on the border officials! No, this is out of the question for me. I don't want this and I also can't deal with it psychologically. (Diary entry Monika, 18 October 2013)

Another intended mother confirmed that IPs are well aware that they are doing something that is illegal in Germany, and that in her encounters with them, she experienced that this can lead to considerable psychological distress (Interview with Martina, 16 January 2015). While the intended mother in the quote above was morally concerned about lying and acting against the law—a problem that several other IPs I talked to felt similarly about—others were more worried about the possible legal consequences of the unlawfulness of their actions. They feared that their child could be taken away from them if the authorities found out about the circumstances of its birth.

Moreover, IPs seek ways to circumvent the law, not only regarding procreation and bringing the child to Germany, but also with respect to its official registration in Germany.[9] One common strategy among my (heterosexual) research participants and as confirmed by agency personnel I spoke with in California is to find an unmarried surrogate[10] who claims to have had an affair with the intended father and officially identifies him as the father. If his sperm has been used, this can easily be proven with a DNA test. The surrogate subsequently waives her maternal rights and agrees that the child will be adopted by the father's wife (i.e. the intended mother).

Even though this is one of the easier ways of gaining parental rights for a child born through surrogacy, intended mothers I spoke with were very unhappy with this strategy. In contrast to the father, who appears as the 'real' father on the birth certificate, she becomes the 'stepmother'—an image that across many cultures has very negative connotations:

> Mothers are loving, caring, and nurturing of their children, but stepmothers are by nature cruel, vicious and jealous creatures. This view of loving mothers and cruel stepmothers has surfaced in different countries and cultures throughout history, handed down from generation to generation. (Penor Ceglian and Gardner 2000, pp. 111–112)

Most of the intended mothers I interviewed thus not only suffered from the inability to carry a child, but in the case of stepchild adoption, they also felt misrepresented in the role of the stepmother. First, all but one of the women I spoke with emphasised the fact that they are the 'real' mother, not the 'stepmother', and second, they perceived the negative image of the stepmother as an affront when in fact they have been longing for this child for a long time, have put immense efforts (psychological, physical, financial) into making this wish come true, and finally, want to be nothing else but a loving and 'good mother'.

While some observers, including myself at the beginning of my research on surrogacy, are surprised that an administrative procedure and a piece of paper (i.e. the adoption document) are met with so much aversion and experienced as so painful, it should not be underestimated that the procedure of officially designating the parents, and the documents related to it, carry immense symbolic value. This is even more so as the intended father does appear on the birth certificate as the 'real father', but together with the surrogate as the 'mother'—a combination that carries with it the subtext of adultery (Ragoné 1994, p. 110). The strategy of circumventing German laws by opting for a stepchild adoption is thus (at least for heterosexual couples) an unpopular—although often necessary and frequently employed—way of obtaining parental rights for the intended mother, as it threatens her role as a mother and a wife and thus challenges the (heteronormative) family setup.

Alternative ways of officially gaining parental rights and German citizenship for the child are complete openness concerning the surrogacy, or its opposite, absolute secrecy. The first is an approach chosen by relatively few IPs, most of whom also engage in activism that advocates for a more open approach to and the legalisation of surrogacy. These parents take literally the ESchG and the AdVermiG, which state that neither IPs, nor the surrogate, will be penalised (Sect. 14b III AdVermiG; Sect. 1 III no. 1 and 2 ESchG) (Dücker and Hörnle, Chapter 11). They also rely on the fact that until today, not one child born through surrogacy has been taken away from its parents—so far, all courts have made decisions based on the best interests of the child, which in these cases has meant leaving it with its family.

Other parents—including some of the participants in my research—have been luckier than the above couple. Martina, the mother of one child born through surrogacy in the US, told me that she had been completely open in her encounters with the German authorities and that she had still been able to register her child without a problem. In contrast to other IPs, however, she did not feel uncomfortable with being a stepmother and had decided to adopt her son. But rather than claiming the child to be the result of an affair between her husband and the surrogate, she chose not to disguise the child's true origin. Based on her own experience, she felt that other IPs' fears are unjustified (Interview with Martina, 15 August 2014). On the other hand, there have been several reports from Italy in the last few years of children who have been forcibly removed from their families (AFP 2014; Dernbach 2014). Stories such as these provoke fears among IPs and are likely to lead to more secrecy rather than openness.

Indeed, although numbers are—for obvious reasons—not available, it is very probable that the majority of IPs keep their surrogacy a secret and choose other paths to bring their child home and register it. In fact, some opt not to register the child at all. IPs who decide to keep their surrogacy a secret are, of course, also not willing to talk to a researcher about their experiences. I was, however, able to collect some information about this group via others who participated in my research. They summarised for me the discussions that took place in a closed internet forum on surrogacy, which also includes members who keep their surrogacy a secret. Another source of information about these secretive IPs was agency personnel, who were willing to share their knowledge of and experiences with such parents with me.

There are certainly many strategies to present a child as one's own, but what, according to several agency owners I spoke to in California, seems to be fairly common is to either stage a pregnancy by wearing a fake belly or to take a sabbatical or move house in order to avoid others finding out about the child's origin (also Meyer-Spendler 2015). One agency owner told me that she puts these parents in touch with a company that produces fake bellies that come in different sizes in order to plausibly stage a pregnancy. While this is a way to avoid moral judgement by friends, family, and neighbours, this nevertheless does not solve the administrative issues that need to be addressed after the child is born. One way of dealing with this latter problem is to declare having given birth during a holiday or sabbatical, and subsequently presenting to the registrar the birth certificate that names both IPs as the parents.

In the surrogacy community, there are rumours that registrars are advised to be particularly careful when checking documents certifying that a child's birth in a country that is a well-known reproductive hub offering surrogacy (especially Ukraine and until recently also India), and to possibly deny the child's entry into the German register of births. Yet, they are unlikely to become suspicious if a couple or a family has lived abroad for a longer period of time. Even if the latter is the case, however, in order to have a child added to the German birth registry, parents are required to fill in a form (*Antrag auf Beurkundung einer Auslandsgeburt im Geburtenregister*, Application for the registration of a birth abroad in the German birth register) that asks for details concerning the birth. Recently, a new question has been added to the form: 'Was the child born through surrogacy?' This places IPs in a difficult position, because ticking the box is likely to result in problems with the registration process, while not ticking it means that the parents give a false statement. Either way, the addition of this question to the form has added one more challenge to the many administrative hurdles that IPs struggle with in their encounters with the authorities. On the other hand, it also can be read as a sign that surrogacy is regularly taking place and that the German authorities have become aware of this (Dücker and Hörnle, Chapter 11).

Another group of parents avoids registration at all. This mainly applies to those who commissioned their surrogacy in the US and who, due to the child's American passport, can bring it into Germany without a problem. But the older the child gets, the more difficult it becomes if it does not have the appropriate documents—ranging from registration for kindergarten and school to the acquisition of travel documents and so on. Several of my informants, all of whom had officially registered their children in some way, claimed to find the strategy of not registering the child highly problematic and naïve, since problems are certain to arise at a later point. They maintained that running away from this merely means a postponement of the problem, often for the worse. On the other hand, they also acknowledged that people who have fought so hard for their child, often for several years, are likely to be overwhelmed and relieved when they finally have the child in their arms, to the extent that all other problems are belittled or simply ignored.

While the above described strategies are employed to overcome legal restrictions, some of them are also used in order to avoid moral judgement and social stigma, such as staging a pregnancy or going on a sabbatical and subsequently presenting the child as having been 'normally conceived'. Another tactic to avoid social stigma is to move house and cut ties with friends and/or family. Indeed, according to the participants of the internet forum and the agency personnel that I spoke with, some IPs do not even tell their closest friends and family about the origins of their child.

Other IPs are torn between their wish to talk openly to their child about its origins and the fear that it will talk to others about it. One couple I interviewed in California exemplified this concern. They had been living in the US for some time when we met in 2014 and planned to stay there for the next few years, though they did not rule out a return to Germany. They were keen to tell their child that it had been born through surrogacy, but still decided against it because they were worried about the effects it would have on the child and on themselves if the German authorities were to find out (Interview with Christian and Tina, 8 February 2014).

Another intended mother, who was generally open about the surrogacy that she and her husband had commissioned in California, nevertheless struggled with the issue of disclosure, especially in the early stages of the process. She feared that once the information was out in the open she would no longer be able to control it. This concerned their general surroundings, but also their professional lives—both she and her husband worked in high-ranking positions in large companies. However, when they needed to make several trips to the US in order to meet the potential surrogate and have the IVF done in a Californian clinic, they decided to disclose the information, particularly to her husband's superiors and close colleagues. To their surprise, they were very understanding and supportive. She had not yet informed her own employer, but because she was still on parental leave for her firstborn daughter she did not see the necessity of doing this in the near future. In their private surroundings where they openly talked about the surrogacy, the couple received a lot of support and encouragement. Nevertheless, there were also other reactions. Her husband's sister decided to cut all ties with them when she learned about their surrogacy plans. The siblings apparently already had a problematic relationship before this event, but it was nevertheless a painful experience for these IPs. Religious friends of theirs also struggled with their decision to commission a surrogacy, but were finally able to accept it (Interview with Sandra, 28 October 2013).

It must also be noted that the IPs that participated in my research are themselves critical of some of the ways in which surrogacy is performed and in our conversations regularly reflected on these issues. This applies less to the criticism that they produced designer babies, but certainly relates to the other topics mentioned above. For example, all IPs I spoke to were very concerned about the surrogate's well-being and about the issue of exploitation. They told me that they regularly sent her little gifts or vouchers for massages, birthday presents, paid for days off, etc.[11] The IPs also often presented themselves as being strongly dependent on the surrogate, who was helping them to fulfil their wish for a child, rather than the other way round. Regarding the idea of the mother-child bond and motherly love, all of them assured me that their surrogate had been very clear about not wanting another (or this) child, but feeling more like a nanny who looks after the child for a while but then does not feel hurt when she gives it away. In order to strengthen their own bond with the unborn child, some of the IPs made recordings of their voices, singing lullabies for instance, which they asked the surrogate to play to the foetus so that it would already hear their voice when still in the womb. Some intended mothers even stimulate lactation with hormones so that they can breastfeed the child themselves (although none of my research participants did this). One intended mother also told me that during the day she left baby clothes in her bed so that they would absorb her smell. She planned to have the twins dressed in these clothes right after birth so that they would get used to her and her husband's smell while they were still in hospital. Finally, all IPs I spoke to strongly rejected the idea of the commodification of the child. They argued that they paid the surrogate for her expenditure of energy and her services, but that they did not pay for the child. The fact that they would still pay the surrogate if the pregnancy did not last to term was used to support this argument.[12]

Finally, the fact that German politicians from all parties, from the far left to the far right, seem to agree on a rejection of surrogacy does not leave much space for a political discussion of the topic. Moreover, activism hardly exists. Some IPs have made cautious attempts to lobby for the legalisation of surrogacy, but the few online petitions that have been started in the last few years, with the exception of one, gathered only a handful of signatures.[13]

While the arguments employed by proponents of a legalisation of surrogacy—a better protection of the surrogate and her rights, quality control of agency services and medical care, the possibility of open contact between surrogates, IPs, and children, and the possibility of the child to know about its origin—are certainly worth discussing, there are currently only few spaces where such debates are actually taking place. One such instance was the annual conference of the German Ethics Council in May 2014 in Berlin, which dealt with ART in Germany. One of the three panels explicitly dealt with egg donation and surrogacy. In April 2017, a symposium at Heidelberg University discussed surrogacy from legal, historical, psychological, and medical perspectives. However, these events are among the rare occasions in Germany when surrogacy is openly discussed from different perspectives.

In addition to above mentioned reasons for commissioning surrogacy in the US, several research participants mentioned that they felt more comfortable with a country whose language they speak and whose culture they feel they understand. They felt that countries such as India or Ukraine are so different from their own that they would find it very difficult to connect with the country and the people, including the surrogate (also Gunnarsson Payne 2015). Nevertheless, the choice of location is often less dependent on the parents' wishes, and much more on the financial resources that are available to them. Accordingly, not all IPs I interviewed were able to afford a surrogacy in the place that met their wishes.

Fragmented Parenthood and the Well-Being of the Child

Another issue that is the subject of heated debate in Germany is the perspective of the child (e.g. Diel 2014). In this context, the debate focuses on questions of parenthood, descent, and the well-being of the child.[14] From an anthropological perspective, in different places in the world parenthood is viewed, practiced, and experienced in very diverse ways (cf. Lévi-Strauss 2013). What marks the current debate in Germany is a fairly static and conservative concept of parenthood, which is strongly based on the heteronormative notion of the nuclear family and on the idea that the well-being of the child is dependent on the existence of 'real parents', both of whom are known to the child—a knowledge that is said to shape the child's identity. From this point of view, the involvement of third and fourth persons in procreation, such as gamete donors or surrogates, is highly likely to lead to identity disorders among children born through such a configuration. This view is, for example, advanced by the German Ethics Committee, which argues that what it calls "split parenthood" may cause serious difficulties for a child's identity development (Deutscher Ethikrat 2016, p. 134). Psychological research suggests, however, that this view is not correct. As Golombok et al. conclude, '[i]n spite of the concerns that have been raised regarding the increased risk of psychological problems among children born through a surrogacy arrangement, the children in the present investigation did not differ from the naturally conceived children with respect to socio-emotional or cognitive development' (2006, p. 220). Despite this evidence, the Association of Donor Children (n.e.V. *Spenderkinder*)—a group that is organised by and represents the interest of children born in Germany through donor insemination—criticises both gamete donation and surrogacy. However, upon closer examination it becomes clear that what this organisation is in fact criticising is not the use of donor gametes per se (even though they do note that this way of building a family may pose particular and serious challenges for the many actors involved), but that donor children often do not have access to information concerning the donor and are thus unable to find out about or meet him or her. This is also the reason for such organisations (in addition to n.e.V. *Spenderkinder* this is the DI-Netz e.V., i.e. Donor Insemination Network) to reject anonymous sperm and egg donation or surrogacy (especially closed programmes).

Similarly, the parents I spoke to, who talk openly about their surrogacy, saw disclosure as an important element of good parenting and an essential contribution to their child's well-being. Some had consulted a psychologist when preparing their surrogacy, who had encouraged them to talk openly to their child or children about the surrogacy. This approach is in line with the current state of research (e.g. Readings et al. 2011; Jadva et al. 2012) that emphasises the advantages of early disclosure of a child's origins for its psychological well-being.

One way of achieving disclosure is to craft a photo album for the child or children in which the surrogacy journey is retold and illustrated. Several IPs, who participated in my research, showed me the albums they had made for their children. They, for example, included pictures of the clinic where the embryo transfer was done, of the surrogate during her pregnancy, and images of the surrogate together with the IPs and sometimes also with the child.

For some IPs, disclosure and a healthy environment for the child and the rest of the family also include continued contact with the surrogate. While many IPs slowly reduce contact with their surrogate over the years, others stay in touch with her on a regular basis. This ranges from emails or postcards for birthdays and Christmas to weekly Skype calls and yearly visits. For this latter group, the surrogate and her family may become something akin to an extended family. For example, one family that participated in my research talks on Skype with the surrogate once a week and visits her regularly in the US. In addition, the surrogate and her family have visited the IPs and their children in Germany twice already. Another family, whose twins were born two and a half years ago, does not Skype with the surrogate much, but they email about once a month, often sending her photos of the children. Furthermore, they send her and her husband gifts for their birthdays and Christmas, and are planning to go to the US to meet her in the next few years. The surrogate, too, regularly sends gifts for the children—both for the twins she carried and for their older sister, who loves the Disney jumpers and other glittery gifts she receives from the US.

How children themselves think about their origins has not yet been researched in Germany. Within the framework of my project, I have also only been in contact with parents whose children are still very small and thus cannot yet be questioned. Accordingly, I cannot make any claims concerning the well-being of these children. Nevertheless, psychological studies of families created through surrogacy in the UK—a legal and social setting that differs significantly from Germany—have shown that children can indeed feel positive about the fact that they were born through surrogacy and also about their surrogate mother (Jadva et al. 2012). Accordingly, while in German public discourse the idea prevails that the child will almost inevitably feel neglected and commodified, this ten-year-study in the UK shows that this does not necessarily have to be the case (Lamba and Jadva, Chapter 9).

Conclusion

In this chapter, I have argued that in Germany, surrogacy is strongly shaped by the prevailing legal, moral, and political setting. Rather than disappearing, however, surrogacy is relocated to other countries whose laws are more permissive. German IPs' experiences are strongly influenced by the friction this brings with it, and many of them feel torn between their wish for a child and the actions that have to be taken in order to fulfil this wish. German cultural notions of motherhood, life, and ethics further complicate the matter and due to the stigmatisation and rejection of surrogacy in both public and political discourse an unbiased discussion and assessment of this ART, with all its facets and diverse manifestations, is currently almost impossible (Schicktanz, Chapter 6).

Despite the many legal and moral hurdles, IPs in Germany nevertheless still find ways to circumvent the restrictive laws and deal with the moral condemnation of surrogacy. Their strategies can be roughly divided into three overlapping approaches: complete openness, moderate openness, and secrecy. These strategies may also differ between friends and family on the one hand and the authorities on the other, and may involve different levels of disclosure.

Finally, in this chapter I have discussed the issue of fragmented parenthood and the well-being of the child. Many commentators assume that a child that has more than two 'parents' will suffer from this situation and develop psychological problems, though previous research has refuted this claim—both regarding children's ability to cope with the fact that they have more than two parents and the fact that they have been carried by a woman other than their social mother.

Accordingly, I suggest that the current legal, moral, and political way of dealing with surrogacy in Germany should be reconsidered. Other countries, such as the UK, have shown that such a task can be fulfilled.

Notes

1. In order to protect my informants' identity, all names and identifying details were changed.
2. Critics' main argument is that sperm donation is an easy and safe procedure that does not place the donor at risk. They maintain that, in contrast, egg donation and surrogacy are medically unnecessary procedures that endanger the donor's or surrogate's health. In line with Almeling (2011), however, I suggest that the roots of the different views concerning sperm and egg donation go much deeper than this simple explanation and also involve notions of motherhood and fatherhood, and ideas about gender and sexuality.
3. Also see the public discussion entitled *Das Recht auf das eigene Kind*, organised by the Gunda Werner Institut in Berlin, November 2015.
4. See, for example, the presentation by the German Ethics Council's member Sigrid Graumann at the Ethics Council's annual conference, Reproductive Medicine in Germany in Berlin, 2014.
5. Interestingly, the German term used here is *ungeborenes Leben* (unborn life). This indicates that the foetus is already regarded as a living being, a view that is not shared by all cultures.
6. "Die Leihmutterschaft lehnen wir ab, da sie mit der Würde des Menschen unvereinbar ist."
7. There are reports of IPs going to South Africa or Latin America, but I did not meet anyone who had chosen these locations.

8. The German term she used was *und erzähle denen eine Story vom Pferd* (to tell them a horse's story).
9. While writing about these strategies would have been ethically highly problematic a few years ago, today these strategies are so well-known and have been written about so frequently that in this chapter I am not disclosing confidential information.
10. The reason why the surrogate must be unmarried is that, according to German law, the man who is married to the woman who has given birth to the child (i.e. the 'mother') is automatically regarded to be the father.
11. This is something that is, of course, only possible in open programmes, where IPs and surrogate can communicate with one another.
12. However, a surrogate is not paid the full sum if she miscarries. Usually, surrogates are paid in several instalments. The contract specifies how often these instalments are paid, and how much she receives in case of miscarriage at different stages of the pregnancy.
13. One petition was signed by 30 supporters (available at https://www.change.org/p/deutscher-bundestag-erlaubnis-zur-leihmutterschaft-in-deutschland) and another by 13 (available at https://www.openpetition.de/petition/online/gegen-das-leihmutter-gesetz-in-deutschland). Only one petition was signed by a larger number of people, namely 160 (available at https://www.change.org/p/leihmutterschaft-endlich-legal-und-staatsb%C3%BCrgerschaft-f%C3%BCr-deutsche-kinder-die-im-ausland-via-leihmutter-geboren-werden).
14. See, for example, the opinion survey conducted online by the German Ethics Council between March and April 2014, which was supplemented with comments made by participants during the Ethics Council's annual conference in May 2014. The graph presenting the results of the survey can be accessed online (Deutscher Ethikrat 2014).

Bibliography

AFP. (2014). Italienisches Paar darf Kind von Leihmutter nicht behalten. *Zeit Online* [online]. Available at http://www.zeit.de/news/2014-11/12/italien-italienisches-paar-darf-kind-von-leihmutter-nicht-behalten-12151007. Accessed 12 Nov 2014.

Almeling, R. (2011). *Sex cells: The medical market for eggs and sperm*. Berkeley: University of California Press.

Bassan, S., & Michaelsen, M. A. (2013). Honeymoon, medical treatment or big business? An analysis of the meanings of the term "reproductive tourism" in German and Israeli public media discourses. *Philosophy, Ethics, and Humanities in Medicine, 8*(9), 1–8.

Berend, Z. (2015). Misconceptions about altruism and choice in US surrogacy. *openDemocracy* [online]. Available at https://www.opendemocracy.net/beyondslavery/zsuzsa-berend/misconceptions-about-altruism-and-choice-in-us-surrogacy. Accessed 2 Oct 2017.

Bergmann, S. (2011). Fertility tourism: Circumventive routes that enable access to reproductive technologies and substances. *Signs, 36*(2), 280–289.

Bernard, A. (2014). *Kinder machen. Neue Reproduktionstechnologien und die Ordnung der Familie*. Frankfurt am Main: S. Fischer.

Bubrowski, H. (2013, May 30). Käufliches Elternglück. *Frankfurter Allgemeine Zeitung*.

Calla, C. (2017). Kinderkriegenlassen ist okay. *Zeit Online* [online]. Available at http://www.zeit.de/kultur/2017-02/leihmutterschaft-tabu-koerper-schwangerschaft-reproduktionsmedizin?page=7#comments. Accessed 24 Sept 2017.

CDU, CSU, & SPD. (2013). *Deutschlands Zukunft gestalten. Koalitionsvertrag zwischen CDU, CSU und SPD. 18. Legislaturperiode*. Berlin.

Deomampo, D. (2013). Gendered geographies of reproductive tourism. *Gender & Society, 27*(4), 514–537.

Dernbach, A. (2014). Italienisches Gericht: Paar darf Kind von Leihmutter in Ukraine nicht behalten. *Der Tagesspiegel* [online]. Available at http://www.tagesspiegel.de/weltspiegel/italienisches-gericht-paar-darf-kind-von-leihmutter-in-ukraine-nicht-behalten/10973256.html. Accessed 13 Nov 2014.

Deutscher Ethikrat. (2014). *Sollte die Leihmutterschaft in Deutschland zulässig sein?* [pdf]. Available at http://www.ethikrat.org/dateien/pdf/jt-22-05-2014-leihmutterschaft.pdf. Accessed 30 Nov 2017.

Deutscher Ethikrat. (2016). *Embryospende, Embryoadoption und elterliche Verantwortung. Stellungnahme*. Berlin.

DIE LINKE.queer Hamburg. (2012). *Fragen zum Thema "Homosexualität"* [online]. Available at http://www.die-linke-queer-hamburg.de/politik/detail/artikel/fragen-zum-thema-homosexualitaet.html. Accessed 17 Feb 2017.

Diehl, S. (2014). *Schlachtfeld Frauenkörper. Periodical Schlachtfeld Frauenkörper* [online]. Available at http://jungle-world.com/artikel/2014/35/50491.html. Accessed 17 Feb 2017.

Diel, A. (2014). *Leihmutterschaft und Reproduktionstourismus. Schriften zum deutschen und ausländischen Familien- und Erbrecht*. Frankfurt am Main: Wolfgang Metzner Verlag.
Golombok, S., MacCallum, F., Murray, C., Lycett, E., & Jadva, V. (2006). Surrogacy families: Parental functioning, parent-child relationships and children's psychological development at age 2. *Journal of Child Psychology and Psychiatry, 47*(2), 213–222.
Graumann, S. (2014). *Eizellespende – Autonomie und Ausbeutung von Frauen* [online]. Available at http://www.ethikrat.org/dateien/pdf/jt-22-05-2014-graumann.pdf. Accessed 23 Nov 2017.
Gunnarsson Payne, J. (2015). Reproduction in transition: Cross-border egg donation, biodesirability and new reproductive subjectivities on the European fertility market. *Gender, Place & Culture, 22*(1), 107–122.
Hüppe, H. (2014). *Bezahlte Leihmutterschaft ist Menschenhandel* [online]. Available at http://www.huberthueppe.de/cdu-politiker-hubert-hueppe-sieht-den-fall-einer-thailaendischen-leihmutter-die-fuer-ein-australisches-paar-gegen-geld-kinder-austrug-als-warnsignal-und-fordert-gesetzliche-regelungen/. Accessed 2 Nov 2016.
Inhorn, M. C., & Gürtin, Z. B. (2011). Cross-border reproductive care: A future research agenda. *Reproductive BioMedicine Online, 23,* 665–676.
Jadva, V., Blake, L., Casey, P., & Golombok, S. (2012). Surrogacy families 10 years on: Relationship with the surrogate, decisions over disclosure and children's understanding of their surrogacy origins. *Human Reproduction, 27*(10), 3008–3014.
Jovanovic, L. (2010). Israeli erfüllt Kinderwünsche: Babys übers Internet bestellen. *Rheinische Post Online* [online]. Available at http://www.rp-online.de/panorama/ausland/babys-uebers-internet-bestellen-aid-1.2003929%23. Accessed 22 Oct 2017.
König, A. (2017). Embryos on the move: Transnational networks of surrogacy. *Allegra Lab: Anthropology, Law, Art, World* [online]. Available at http://allegralaboratory.net/embryos-on-the-move-transnational-networks-of-surrogacy-medtech/#_ftnref1%23. Accessed 24 Jan 2017.
Lévi-Strauss, C. (2013). *Anthropology confronts the problems of the modern world*. Cambridge and London: The Belknap Press of Harvard University Press.
Meyer-Spendler, R. (2015). *Ein schmaler Grat. Erfahrungen mit Leihmutterschaft in den USA*. Mörfelden: FamART.

Pennings, G. (2004). Legal harmonization and reproductive tourism in Europe. *Human Reproduction, 19*(12), 2689–2694.

Penor Ceglian, C., & Gardner, S. (2000). Attachment style and the "wicked stepmother" spiral. *Journal of Divorce & Remarriage, 34*(1–2), 111–129.

Ragoné, H. (1994). *Surrogate motherhood: Conception in the heart*. Boulder: Westview Press.

Raz, A. E., & Schicktanz, S. (2016). *Comparative empirical bioethics: Dilemmas of genetic testing and euthanasia in Israel and Germany*. Berlin: Springer.

Readings, J., Blake, L., Casey, P., Jadva, V., & Golombok, S. (2011). Secrecy, openness and everything in between: Decisions of parents of children conceived by donor insemination, egg Donation and surrogacy. *Reproductive BioMedicine Online, 22*(5), 485–495.

Schwarzer, A. (2015). Editorial: Leihmutter? Geht gar nicht! *Emma, 4,* 6–7.

Shenfield, F., de Mouzon, J., Pennigs, G., Ferraretti, A. P., Nyboe Andersen, A., de Wert, G., et al. (2010). Cross border reproductive care in six european countries. *Human Reproduction, 25*(6), 1361–1368.

Trimmings, K., & Beaumont, P. (2011). International surrogacy arrangements: An urgent need for legal regulation at the international level. *Journal of Private International Law, 7*(3), 627–647.

Whittaker, A., & Speier, A. (2010). 'Cycling overseas': Care, commodification, and stratification in cross-border reproductive travel. *Medical Anthropology, 29*(4), 363–383.

14

Conceiving Before Conception: Gay Couples Searching for an Egg Donor on Their Journey to Parenthood

Julia Teschlade

Introduction

'Well, thinking about my child, I guess, the question of his genetic descent is more important than giving birth to him', Frank considered my comment during our interview. I had remarked that it had taken him and his partner Anton a while to find an egg donor they liked, in order to continue their journey to parenthood through gestational surrogacy with this woman. Their son Sören was born via surrogacy only four months ago in the US. Looking back, Frank already had his future child in mind when he and his partner were choosing an egg donor. 'I have the feeling', he explained, 'if the surrogate mother is a decent woman, the child will be okay. It can be this woman or that woman, it doesn't matter at all as long as she doesn't drink and is healthy. But with the genetic descent of course, there is an entire range.' He stopped his explication, leaving it to my imagination to figure out

J. Teschlade (✉)
Joint Interdisciplinary Doctoral Program "Human Rights under Pressure: Ethics, Law and Politics", Freie Universität Berlin, Berlin, Germany
e-mail: julia.teschlade@fu-berlin.de

S. Mitra et al. (eds.), *Cross-Cultural Comparisons on Surrogacy and Egg Donation*,
https://doi.org/10.1007/978-3-319-78670-4_14

what he refers to as 'an entire range'. It is exactly this vagueness that strikes me as odd. It illustrates the 'fuzzy ideas' about the significance of the donor's genes for the future child and the future family, while the gestational role seems to be less influential.[1]

Generally, same-sex male couples have several opportunities to become parents like co-parenting, foster care or adoption. The decision to realise their desire for a child through gestational surrogacy is often preceded by thoughtful deliberation processes about these alternative parenting options (Berkowitz and Marsiglio 2007; Mallon 2004). In order to become 'biogenetic' parents, it is a necessity for gay couples to receive (donated) oocytes to create an embryo with their own biogenetic material. They always require the help of a 'facilitating other' (Mitchell and Green 2007, p. 82) and usually, like in Frank and Anton's case, two women are involved in the process: the surrogate and the egg donor.[2] Surrogacy allows gay men to procreate as a couple and to build a two-parent family: They create biological relatedness to at least one partner, while having the opportunity to choose among a large number of egg donors in order to select one that, e.g. resembles the non-biological parent or shares similar cultural and/or vocational characteristics with the couple (Mitchell and Green 2007, p. 87). Hence, the couples are confronted with the unusual situation of making reproductive choices about an as of yet unknown third person to supplement their genetic material. Yet, the factors that guide their decision-making process are subject to debate. Who are they looking for? Which characteristics are important to them? And why?

There is a considerable body of research on lesbian couples and their decision-making process to find a sperm donor (Nordqvist 2012, 2014a; Mamo 2007; Chabot and Ames 2004), on sperm and egg donation accessed by heterosexual couples (Wong 2017; Becker et al. 2005; Becker 2000) as well as on how gay men decide on surrogacy (Murphy 2013; Greenfeld and Seli 2011). However, it has not yet been researched how gay couples arrive at their preferences for an egg donor. While Mitchell and Green (2007) as well as Greenfeld and Seli (2011) mention gay men's preferences for choosing an egg donor (e.g. similar physical, cultural and/or vocational characteristics), these studies do not provide an analysis of how these decisions are embedded within the wider sociocultural context of a heteronormative society that may regard these reproductive practices and families as illegitimate.

Building upon this body of research, I explore these questions on donor preferences within the German context, focusing on couple interviews (Wimbauer and Motakef 2017; Heaphy and Einarsdottir 2012) with same-sex male couples, who commission(ed) a surrogate and an egg donor in the US. Between 2015 and 2017, I interviewed 15 couples and single men from Israel and Germany. Eight couples live in Germany. In Israel, I interviewed five couples with both partners and two men without their partner. I talked to many of the couples at least twice, depending on the status of the journey, whether they had just started the surrogacy process or if they already had children. For this chapter, I concentrate on interviews with five couples. The couples are cisgendered male, economically affluent and have received a higher education. All couples are either in a Lebenspartnerschaft (civil union) or were legally married abroad. Two couples already had children when I interviewed them for the first time and three were still in the process of becoming parents. All of them are parents by now. Four of the couples chose an US-American egg donor and one couple had a close friend who donated her eggs. To protect the interviewees' privacy, the names of all interviewees, their age, professions and nationalities (if other than German) as well as places were anonymised. The interviews were conducted in German, audio recorded and verbatim transcribed by the author. The quotes used in this chapter were translated from German to English and have been edited for readability.

While egg donation as well as surrogacy are illegal in Germany (Dücker and Hörnle, Chapter 11), the field of reproductive technologies is highly professionalised in the US. Fertility clinics and egg banks offer a wide range of opportunities, connecting couples to a diverse pool of egg donors. Public debates as well as some feminist writings on assisted reproductive technologies (ART) criticise the commercial and market-driven aspect of ART as 'DNA shopping' (Stacey 2006) for a genetically engineered child (Berkowitz 2013). Indeed, the couples do have a great variety of options, which could be read as eugenic preselection, however, I show how the presumptions about biogenetic material are socially conditioned. I argue that legal, political and sociocultural parameters, like national legislation, financial resources and social expectations, influence the couples' considerations and decision-making processes.

All in all, the process of finding an egg donor is ambivalent and shaped by social assumptions about who is seen as eligible to parent a child, what constitutes a family and what is considered responsible parenting. In order to avoid external societal judgements about their non-traditional family, gay couples engage with and reproduce preconceived societal norms about normative family ideals and adjust their reproductive choices accordingly: This process can be summarised as 'conceiving before conception'. They consider 'the egg' as immutable genetic material. Their decision for a specific egg donor, which is based on their presumptions about genes, can be regarded as a first marker for good and reasonable parenting, thus making 'the egg' a symbol for inscribed culture and social notions about the family. They envision their child and their future family in a way that adheres to heteronormative expectations of kinship and parenting even before the oocyte is fertilised. Similarly, in her influential study on opposite-sex couples and surrogacy in the US, Ragoné (1994) showed that the intended parents (IPs) also refer to a traditional script of family and emphasise the importance of biogenetic relatedness to at least one partner. While Ragoné researched the decision-making processes in the context of surrogacy, I focus on the process of finding an egg donor. I do so in order to show how the heteronormative construction of kinship is already inscribed culturally in the supposedly 'natural' foundation of the egg donor's biogenetic material. In terms of a response to traditional family norms, the decision-making of gay couples is not necessarily specific for same-sex couples, because opposite-sex couples also respond to societal family norms. However, a same-sex couple has to prove their 'eligibility' as parents due to homophobic laws and social perceptions that render same-sex couples invisible as parents outside the 'desired' family norm.

In the following, I briefly introduce the legal and political context of same-sex parenting and their limited access to ART in Germany. Then, I draw from my own research data focusing on gay couple's decision-making in the search for an egg donor. I argue that the intended fathers have three main points of reference to guide their decision. First of all, the egg donor should be open to future contact with the child. Second, the donor needs a clean bill of health to minimise the risk of an inherited medical disposition. Third, the woman should have similarities with the couples' physical appearances, their sociocultural background, and educational attainment. The analysis focuses on how the couples ascribe sociocultural content to the oocyte, which in turn reflects on the legal situation in Germany, the importance of biogenetic connections as well as on presumed distinguishing factors for kinship, like family resemblance.

Legal and Political Context in Germany for Same-Sex Male Couples

In this section, I explore social and cultural norms towards same-sex parenting and medically assisted reproduction in Germany through the lens of the legal framework. Laws on sexuality and family reflect on dominant contemporary values (Dücker and Hörnle, Chapter 11, for an overview of the German Family Law). They systematically regulate social practices and impose what is seen as socially acceptable. While ARTs offer opportunities to procreate outside of heterosexual intercourse, many countries do not grant same-sex couples equal access to reproductive and parenting rights. Legislation on reproductive technologies touches upon the question of 'what constitutes a family and who has the legal right and social legitimacy to have children' (Eggert and Engeli 2015, p. 323). Limiting the access to ARTs for same-sex couples privileges the heteronormative structure of society and enables states to steer conservative orientations through legislation (Inhorn and Birenbaum-Carmeli 2008). Furthermore, it places gays and lesbians outside the framework of 'desirable parenting options'. Roseneil et al. (2013, p. 903) find that states only encourage 'good citizens to have children', which means being 'procreative within the context of the co-residential heterosexual, gender normative couple'. This logic constructs same-sex couples and non-heteronormative relationships as the 'deviant other' in relation to the norms of kinship and family formations. Riggs and Due (2013, p. 956) refer to this as 'reproductive vulnerability', defining vulnerabilities arising from 'being located outside of the norm'.

While Germany did not grant same-sex couples equal access to reproductive and parenting rights, the German Bundestag finally voted in favour of a law granting same-sex couples the equal right to marry in June 2017.

This law includes the right for couples to adopt a child.[3] At the same time, gay and lesbian couples as well as single women have limited access to reproductive technologies. Although there is no federal law on who counts as an IVF patient, the professional code by the German Medical Association limits IVF and medically ART to married heterosexual couples (Rothmayer and Ramjoué 2004),[4] and a violation of the code can be pursued in court. Furthermore, egg donation and surrogacy have been declared illegal by the Embryo Protection Act of 1990 (*Embryonenschutzgesetz*, ESchG) and the Adoption Placement Act of 1977 (*Adoptionsvermittlungsgesetz*, AdVermiG) prohibits commercial action from supporting surrogacy (Dücker and Hörnle, Chapter 11). Rothmayer and Ramjoué (2004) illustrate how the drafting of ESchG is embedded within the specific context of the German history and the experiences of World War II.

Hence, the legislation on family and ART is an indicator for dominant contemporary values and structure the reproductive choices of same-sex couples. As such, their decision-making is not just a matter of private and intimate choice, but rather a result of the lack of legal options in Germany.[5] Pursuing this path is not easy since access is limited to those with socio-economic capacities. The couples are required to familiarise themselves with the field of the cross-border fertility business and the legal situation in the different countries that offer these services. In the next section, I give a short insight into the possibilities offered by the fertility market in the US and I analyse how sociocultural conditions influence the couples' decisions.

'Conceiving Before Conception': Options for Conception and Making Choices

Having a child is often embedded within the narrative of a romantic relationship and 'true love' between the partners. In the situation of gestational surrogacy, same-sex male couples also want to start a family with the person they love, but have to expand the conception of their child beyond the two-parent family: Their child will be genetically connected to an egg donor, as well as to the couple itself.[6]

In practice, it means that the couples have to find an as of yet unknown person to supplement their genetic material in order for them to become biological parents. Furthermore, they have to work with a surrogate that carries their child to term and gives birth to it.

Finding an egg donor is usually preceded by a process where the couples familiarise themselves with other parenting options for same-sex couples like co-parenting or adoption. The story of Camilo and Hauke exemplifies how couples eventually decide on surrogacy. When I met them for our first interview, they had been matched with a surrogate only the week before. They had already had a chance to get to know her via Skype. 'After talking to her we could more easily envision that it would work out', Camilo elucidated with a smile and referred to their initial concerns about surrogacy. Camilo and Hauke are both in their late thirties. After spending the first years of their relationship abroad, they moved back to Germany once Hauke graduated from university and started looking into gay parenting options back in 2009. Gathering information at the local gay community centre about their possibilities to become parents, they consciously decided on adoption. 'We didn't have a lot of information on what surrogacy actually is', Hauke recalled. Back then, surrogacy was not an option for them, because they found it odd that the woman gives away her child after carrying it to term. However, after several years of being on the waiting list for adoption, without any indication that they would ever become adoptive parents, they started to reconsider their possibilities. 'We had never really understood that there is a difference between the egg donor and the surrogate', he explained. Thus, surrogacy only became a serious option once they understood that the woman gestating and giving birth is not biogenetically related to the child. It seems that working with the help of two women, a surrogate and an egg donor, eases moral and ethical concerns IPs often have when they start looking into ART. It feels more legitimate to commission a surrogate, because she does not give away 'her' child, which—in case she was genetically related to it—would evoke difficult images nobody wants to be affiliated with.

Once the decision on surrogacy has been made, the couples are usually matched with a surrogate by the surrogacy agency they work with. In order to find an egg donor, however, they need to familiarise themselves with the field of ART from a medical as well as economic perspective. While the agency supports them in the process of finding an egg donor, they have to educate themselves about medical technicalities as well as the costs and benefits. There are three prevalent possibilities to find an egg: Either the surrogacy agency has a data bank of egg donors, or the fertility clinic has an in-house donor bank. A third option is an egg donor agency. Working with a fertility clinic has the financial and medical advantage that the egg donors are medically and psychologically pre-screened, whereas donor agencies have a much greater variety of different donors. It is necessary to point out that the possibility to access these reproductive services abroad depends on the couple's economic situation. In the US, the cost for surrogacy can amount to up to USD 130,000 and more. As a consequence, the couple's socio-economic background also defines their route to parenthood.

For example, Camilo explained how he and Hauke pursued finding an egg donor by looking at different donor profiles in the data bank. He elaborated that finding the donor took them time: 'This is a little bit more complicated, because here, well, here the genes are important. What is genetic and what is not. We had to agree on what is important to us and this took a little bit longer.' He emphasised the relevance of genes, which suggests a different selection logic for the egg donor than for the surrogate. It resonates with what Melhuus and Howell (2009, p. 158) found for adoption and gamete donation. They suggest that even these 'unnatural' ways of procreation undergo discursive processes of naturalisation. The growing relevance of biogenetic material for constructing kinship reflects on the 'precedence to biological bonds and […] an increasing public occupation with biogenetics'. While the oocyte serves the technical purpose of creating an embryo, Camilo's statement implies the popular notion that the gene is constructed as 'the "hard-wire" in the human constitution' (Nelkin 2006, p. 1). However, by disclosing a discussion about 'what is genetic and what is not', as Camilo phrased it, he reveals that whatever they anticipated to be the determining relevance of genes, is socially undermined. Hauke and Camilo's subjective preferences ('what is important to us') do in fact define what is considered important in the genetic make-up of a donor.

Consequently, the couples have to familiarise themselves with the information provided about the donor (e.g. age, height, educational attainment, family history, etc.) and also with how to treat, validate and hierarchise this information. Only then can they align their personal preferences with the preferences they agreed on as a couple to make an informed decision on which donor to choose. Although the agencies usually channel certain criteria they consider important, like beauty, education and health, complemented by a picture of the donor (Almeling 2011), the couples are often overwhelmed by the amount of information they have to evaluate. Some of them feel uncomfortable deciding on a donor.

However, in order to show that they are responsible future parents they have to agree on their donor preferences carefully. As a consequence, some of the couples pay great attention to the personal statements of the donors. They want to learn about her personal traits and her motivation—'why she wants to do it'. In her research on the US market of egg and sperm donation, Almeling (2011) shows that 'women's sense of altruism' (p. 36) and 'donating for the right reason' (p. 129) is a selection criteria for the agencies and clinics. The couples also refer to indicators like her intrinsic interest in donating her eggs and her personal story, her level of education about the technical procedures and medical risks, and her awareness of future consequences like having a biogenetic connection with the recipients' child.

Furthermore, the dimension of connecting on an emotional and personal level is important to the couples when reading the profiles. Ingo and his partner Dieter, for example, started their route to parenthood back in 2008. Reconstructing how they found the egg donor, Ingo explained why they decided not to predetermine many criteria. 'If it works on a personal level [...] we thought "Let fate decide a little bit" [sic!]. We made a certain preselection [...] I think we were pretty open, without being perfunctory in the sense that you say "I don't care"'. Emphasising that they were not just 'perfunctory' implies, on the one hand, a need to justify their actions. It seems almost illegitimate to leave a decision to fate in a process that is as technical as ART, because it might be interpreted as irresponsible. On the other hand, by highlighting the emotional connection with the donor, the personal element becomes relevant and weakens the importance of biogenetics.

Hauke and Camilo added to this by referring to news articles and addressing an anticipated public opinion on 'genetic engineering'.[7] Hauke explained that the decision-making process is not led by the idea of producing a 'mega kid' or 'Superman'. Camilo backed him and added 'Exactly, it is not in the artificial sense that you say "I want a child with blue eyes"'. Once they have started the process of choosing an egg donor, the couples have rather realistic preferences. I identified three main principles of their process of decision-making, which I term as 'conceiving before conception'. I discuss each of these deciding factors below.

Getting to Know the Donor and Future Contact: Family Genealogy

Choosing a donor they have the chance to get to know personally and who is open to future contact with their child(ren) is important for most of the couples I talked to. A central motivation for the parents to stay in touch with the egg donor is their moral responsibility towards the anticipated future interests of the child. This reflects the presumed importance of knowing about biogenetic connections and origins in order to build a stable personal identity (Melhuus and Howell 2009), which is also regulated by law. In Germany, the right to access information about genetic relatives is part of the personality rights (Art. 2,1 and Art. 1 of the German Basic Law) and internationally it is included in the UN Convention on the Rights of the Child (Art. 7 and 8). Today, the presumption that the 'truth' about genetic origins is undeniably good for a child (Blyth and Frith 2009; Smart 2010) and prevents children from psychological problems later in life, dominates the public and scientific discourse. However, Nordqvist (2014b, p. 332) remarks critically, that the 'right to know' has become a 'powerful moral sentiment building around the need for donor-conceived children to know their genetic origin, underpinned by the idea that they might otherwise suffer harm.'

The couples I interviewed implicitly referred to these debates, showing compassion and sensitivity for their child's life trajectory, acknowledging the idea that 'a person "needs to know" his or her genetic background in order to know "who they are"' (Nordqvist 2014b, p. 323). Gert, one of the intended father I talked to at the very beginning of his journey to parenthood, explained that he and his partner were only looking into profiles of donors that they could get to know personally. They wish to present a transparent and coherent 'how-I-came-to-be' story to their children once they are old enough. 'If we are going to be asked later: "Who is my mum?"', Gert stated and clicked his tongue, 'we want to be able to provide a name and a picture and, if possible, a personal contact.' In doing so, the child has every possibility to create a genealogical lineage and can form any kind of kinship ties that might become important later in their life (Nordqvist 2014a).

Karim and Paul, a couple I interviewed when they were five months into the pregnancy, added to this mind set. It turned out that their story is somewhat different, because they received help from a close friend who donated her oocytes. Having a donor they knew personally was their precondition to decide on gestational surrogacy in the first place. They emphasise that it was particularly important for them that their children will not have a 'blank space' in their genealogy, as Paul framed it. He reasoned that

> it is good if the children can really get to know their mother and aren't in the situation 'Where is the other half of my descent?', right? And this would leave a blank space that I don't want anybody to have to put up with. Because of that, it was always clear that if we need an egg donation then from somebody we know or at least from somebody who we can get to know.

Paul clarified that their consideration also included an as of yet undefined environment the children might encounter, like kindergarten or school, where they would have to deal with traditional presumptions about kinship and family. He explained that they want to prepare their kids for (intrusive) questions they might be asked in kindergarten: 'We always wanted that the kids, once they are in kindergarten, have an answer to the question: Who is your mother?'

As a result of how important the personal connection to the donor is to them, they refrained from increasing their chances to have more viable embryos, which was suggested to them because their close friend had a low egg count. This led to a frequent questioning of their decision to receive help from her by the medical staff of the fertility clinic. Karim explained their decision by making references to the idea that characteristics can be transmitted genetically: 'When our children are born, we want to be able to tell them: "You got your bad behaviour from your mum" and we didn't want to have to puzzle over what and how and why.' Although it is not really clear if he is serious or if his remark is a facetious comment on popular beliefs, he addressed the crucial idea that human behaviour is part of the genetic make-up of a person (Finkler 2000). While his remark ignores the complex intertwining of genes and environment as well as 'the distance between the molecular level of genetic systems and actual behaviour' (Nelkin 2006, p. 2), it also reflects on the pressure they are under in their decision-making. They have to align their decisions with the contemporary sociocultural belief system. Similarly, for lesbian families negotiating family connections, Nordqvist (2012) found that genes are not disconnected from narratives of kinship and family making: The lesbian couples in Nordqvist's study and the gay couples in my study seek to link their families to the hegemonic heterosexual construction of the traditional family. Although the two-dads-family has a different structure, the narrative that every person not only has at least one dad, but a biogenetic connection to a woman—even if she is not the mother—is constructed as a determining factor on the child's life and identity.

The Relevance of a Healthy Donor

Although health is a distinguishing factor in choosing an egg donor, it is often only mentioned as a side note and treated as a factor that is taken for granted. While health is constructed as the 'absence of diseases' in the family history of the egg donor, it is problematised in the family history of the IPs, who—depending on the facilities of the fertility clinic they work with—undergo a thorough genetic screening as well as counselling. The reason for this seems obvious: The couples want their baby to be born healthy, reducing the possibilities of inherited health issues.[8] Novas and Rose (2000, p. 492) refer to the practices of genetic counselling as 'technologies of genetic selfhood' that create new 'genetic responsibilities' (ibid., p. 502). Medical professionals confront the intended fathers with a range of choices to decide on. In the following, I will sketch how the perception of health is part of a medicalisation discourse of the family (Finkler 2001).

During my interview with Hauke and Camilo, it turned out that health only became a relevant factor once they had consulted with the medical professionals at the fertility clinic.[9] Camillo recalled: 'We spoke to doctors to get advice on what is genetically relevant and what is not. There were things we didn't know. In my family, there has already been a number of cancer cases and there is the chance of doubling that risk for the child.' Having a family history of cancer, Camilo is 'at genetic risk' (Novas and Rose 2000, p. 494). Thus, Hauke and Camilo have to decrease the risk of cancer for their child when choosing the egg donor. This knowledge about possible genetic inheritance 'carries a moral load' (Finkler 2001, p. 248) for the IPs and is also the first opportunity to take care of their unborn child. In this sense, it is a matter of 'risk hedging' at a very early stage: Even before the egg is fertilised and the embryo is created, they want to find a healthy egg donor that at least does not carry the same medical dispositions as the genetic parent. Hence, the egg donor also compensates genetic and/or health-related problems of the biogenetic parent. The possibilities offered by medical professionals carry a lot of weight and cause insecurities about making the right and informed decision. Choosing a healthy donor is motivated by the wish to act in the best interest of the child. Under these circumstances, genes are constructed as 'an "ethical substance" that one works upon in relation to the self (genetic identity, reproduction, health) and in relation to others (siblings, kin, marriage, children)' (Novas and Rose 2000, p. 502). Nevertheless, medical testing is always also a question of financing. It might cause a dilemma for couples, if they cannot afford diagnostics like Comprehensive Chromosome Screening (CCS) or Preimplantation Genetic Diagnosis (PGD), because this might give the impression that they act irresponsibly and are not invested in reducing the risk for their future child.

She's Got the Look: Genotype, Phenotype and Resemblance Discourses

The resemblance between parents and children is an important feature of the 'cultural ideology of the family' (Becker 2000, p. 196) and is seen as the 'outward, bodily expression of biological relationships' (Becker et al. 2005, p. 1301). Matching the donor to the characteristics of not only the non-genetic parent, but to the family as a coherent entity, is a common pattern in conception practices. According to Nordqvist (2012), family resemblance serves as a strategy to normalise the non-heterosexual family, trying to prevent that their legitimacy as a family is questioned—a common stigma families created with the help of donated gametes face.

Furthermore, studies on donor-conceived children suggest (Nordqvist 2012) that phenotypical resemblance between the child and the parents is an important factor. Gert and Bastian, for example, who were still in the process of finding an egg donor, discussed their donor preferences as a matter of being a 'normal family'. Bastian stated that future contact to the child was the only preference they marked in the data bank. 'And then we got a list with African, Asian, Latin-American women and we selected those women we wanted to know more about according to our priorities', he explains. Gert added that they are looking for a 'Caucasian' donor. Bastian replied to Gert's interjection: 'No, our idea of family, my idea of a family, is a normal family.' Gert added forcefully:

> Yes, Bastian, but why do we have this idea of family? Because I think the child is already going to have a hard time with two fathers, which is already complicated. It [the child] doesn't need to be Black or Asian or… We don't have to unnecessarily overcomplicate it. And that is why we said, 'Okay Caucasian and everything else is a matter of taste'. We are looking for someone who is more of a dark type because Bastian is Italian.[10]

Gert questioned Bastian's idea of a 'normal family' by addressing the dilemma that they will not be 'normal' (read as 'traditional, opposite-sex family') and that they can only try to create a family that is 'close to normal'.

This implies two distinct but interlinked presumptions about them as a 'two-dads-family'. First, he suggests that someone could contest their self-understanding of being a normal family. Thus, by adhering to phenotypical resemblance they reproduce according to the 'assumed natural order of things and support a hierarchy of legitimacy, in which a clear physical resemblance to family members confers greater legitimacy' (Becker et al. 2005, p. 1301). Second, Gert anticipates homophobic consequences for a child that is raised by two fathers. He wants to protect the child from discrimination on the grounds of race. However, 'whiteness' is also considered a marker that defines the ideological (family) norm (Quiroga 2007, p. 144). Choosing a white donor reproduces racially defined privileges associated with 'whiteness' in Germany. Quiroga (2007) convincingly shows how ARTs privilege biogenetic relationships and 'white' kinship patterns. In the context of donor conception, a similar phenotype substitutes for the lack of a shared genotype between parents and children.

Moreover, Gert and Bastian cannot truly pass as 'normal' in terms of a heteronormative conception of family. Surrogacy is illegal in Germany and on top of that they are a gay couple from Germany circumventing existing laws and norms in order to create a family. In contrast to opposite-sex couples, they cannot disguise the fact that a third and a fourth party were involved in creating their family (Mitra, Chapter 5; König, Chapter 13). These circumstances place them in a difficult situation. They may be or feel pressured to explain and vindicate their decisions to society. While bending the norms, they are torn between resisting existing policies and, at the same time, choosing the path of least resistance when it comes to phenotypical traits of the donor. If the child or children resemble the fathers, they create a coherent family picture, which is a strategy of blending in and passing as a normal family. It serves the purpose of concealing deviation from normative family models.

Nevertheless, phenotypical resemblance is not the only substitute for not being able to merge the genotype of two partners and respective parents. Camilo creates a more encompassing narrative in terms of replacing the non-biological father Hauke. Camilo explained that since he will be the genetic father of their first child 'Hauke's part, to some degree, needs to be replaced. It is not only the looks but also the entire genetic package that you get, that replaces Hauke a little bit, so to speak'. Thus, looking similar is not sufficient, but the egg donor has to resemble 'Haukes's entire package' in order to get closer to non-assisted reproduction.[11] As a consequence, educational attainment is also an important distinguishing factor in defining similarities. Education serves as a proxy for intelligence, 'which lay people see both as heritable and a marker of economic success and therefore resource acquisition' (Furnham et al. 2014, p. 227). In the field of anonymous gamete donation, there is a long tradition that agencies match the parents according to similarities as well as the level of education (Bergmann 2014). Frank, for example, rates education by the grades the donor achieved and degrees she earned. He described his intuitive positive feeling towards education: 'It really gets my attention', he revealed. Similar mechanisms apply to the choice of a life partner, where socio-economic and sociocultural resemblances are relevant (Wirth 2000). Couples who have the financial means to opt for surrogacy usually pursue a successful career. In order to sustain their prestigious achievements, they transform socially achieved determinants into biological markers and see education as part of the genetic make-up of the egg donor. The parents do so in order not to risk the future career of their child-to-be.

Conclusion

Gay couples who create their family with the help of two 'facilitating others'—a surrogate and an egg donor—are confronted with having to make reproductive choices, like choosing an egg donor. While the process of defining preferences for a particular woman donating her oocytes has been critically labelled 'DNA shopping' in order to 'eugenically reproduce white infants' (Stacey 2006, p. 39), I argued that the decision-making process of the couples is much more complicated and ambivalent. I analysed the process within the legal, political and sociocultural context of the German society in order to understand how the couples arrive at their preferences. Gay parenting is still perceived as a deviant and illegitimate desire, because gay fathers challenge traditional assumptions about gender, sexuality and reproduction. Taking these circumstances into account, I explored the social and cultural norms towards same-sex parenting and medically assisted reproduction in Germany through the lens of the legal framework.

Drawing on interview data with German same-sex male couples who worked with a surrogate and an egg donor in the US, I showed how the couples ascribe sociocultural norms to the oocyte. This reflects the legal situation in Germany, the importance of biogenetic connections and presumptions about resemblance between parents and children. Preferences are usually directed along three indicators: First, they often choose a donor they can get to know personally, because they want to be sure that their children can meet the woman with whom they share a biogenetic connection. Second, health is a distinguishing factor in order not to burden their children with genes of 'poor health', like a greater risk of cancer. Furthermore, this emphasis on health is a first step in taking care of the child and making responsible and informed decisions. Third, biological relatedness is the privileged family norm. Therefore, couples try to create families with at least one biogenetic father and they have a strong preference for egg donors that share phenotypical and vocational similarities with them as a couple. This reflects the mainstream idea that families need to look alike to pass as a family.

The process of negotiating and discussing these preferences can be summarised as 'conceiving before conception'. The couples engage with preconceived societal norms about traditional family concepts and create the first mental image of their child and their future family accordingly. The process of conception starts before the technical and medical process of creating the embryo is accomplished, because they imagine what their family would look like and what kind of parents they want to be. The idea, or the 'seed' for their future child and family, is fertilised—to stick with the technical terminology of reproduction—in their head and intertwined with sociocultural presumptions about family resemblance, the paradigm of healthy children and a consistent history of family genealogy without 'blank spaces'. Thus, seemingly biological choices are deeply socially embedded. The gay fathers in this study anticipate external judgment and anti-gay prejudices and try to prevent intrusive questioning about their family, e.g. by other parents or kindergarten staff. Their biogenetic choices are made based on the notion of what they hope for in order to grant them greater recognition as a family and a better, non-discriminatory treatment of their children. Although the couples might be sceptical towards these common perceptions and they do not necessarily share the same opinion about which features are inheritable and which are ascribed socially, they anticipate these beliefs and align their choice with the societal conditions they live in. Nevertheless, they simultaneously destabilise these preconceptions, because a third and fourth party is a given necessity to create their family, which they cannot conceal. Consequently, they also constantly undermine these heteronormative preconceptions.

Notes

1. For a discussion on the 'primacy of genetics and devaluation of the gestational role' see e.g. Dillaway (2008, p. 310). Teman's (2010) study on surrogacy in Israel shows that couples build a stronger relationship to the surrogate, while the role of the egg donor turns out not to be as relevant as they expected (see also Teschlade 2016).
2. I decided to follow the language commonly used by the women themselves, who donate their eggs, despite the implications the term 'egg donor' involves. 'Donation' suggests a gift, while egg donation usually is a commercial transaction. Richards, Pennings and Appleby refer to donation as 'the process by which sperm, eggs and embryos are provided for users regardless of whether this is an altruistic gift or commercial sale' (2012, p. 3).
3. Step-child adoption was allowed in 2005 and second-parent adoption in 2013.
4. It remains an open question if this will be changed once the law on marriage for same-sex couples becomes effective on 1 October 2017. According to a private email exchange with a representative of the German Medical Association, they are working on new guidelines for ART. Furthermore, they called on the legislator to draft 'clear and consistent regulations on assisted reproductive medicine' (Brösicke et al. 2017, p. 213).

5. In their study on cross border reproductive care, Shenfield et al. (2010) found that legal reasons were a predominant factor for many patients to seek medically assisted reproduction abroad.
6. On lesbian couples see Nordqvist (2014a, pp. 272f.).
7. A recent example is the *Kinderwunschmesse* in Berlin in February 2017, a fair where amongst others national and international fertility clinics and adoption agencies presented their services. The news generally referred to the event as being 'contested', questioning the supply (especially US American and Ukrainian fertility clinics) as well as the demand side (IPs) (Schaaf 2017; Schwarz 2017).
8. A major aspect of this is the CSS. After the follicles are fertilised in vitro, a few cells from each embryo are tested and all the chromosomes are counted. All embryos with 46 chromosomes are selected for the embryo transfer. CSS is supposed to increase the 'success' of the pregnancy and reduce the risk of miscarriage. This testing also suggests a norm to parent an able-bodied child and to rule out those embryos that show chromosomal differences from the norm. For further critical reading in the field of disability studies see e.g. Waldschmidt (1995).
9. Furthermore, fertility clinics advertise healthy donors. Egg donors are usually medically screened and genetically tested before they go on the data base.
10. Gert refers to the archaic concept 'Caucasian' (that goes back to a theory of race invented by Johann Friedrich Blumenbach) to mark himself and Bastian as 'Central European', which he mentions couple of sequences later in the same interview. Furthermore, it is important to note that the direct translation would be 'black type', which does not translate very well from German to English. With 'black' or 'dark type' he refers to dark (facial) hair. In a follow-up interview he explains that it was important to them to have a rather 'dark type than a blond one', because Bastian and he are both 'dark'.
11. With regard to a gendered perspective on the egg donor, infertile opposite-sex couples usually look for a donor of the same sex of one of the partners, whereas same sex couples look for similarities in the gametes of a donor of opposite sex. This might explain why the couples I talked to preferred characteristics that are usually more common for sperm donors than egg donors (Hertz et al. 2015).

Bibliography

Almeling, R. (2011). *Sex cells. The medical market for eggs and sperm*. Berkeley: University of California Press.

Becker, G. (2000). *The elusive embryo. How men and women approach new reproductive technologies*. Berkeley: University of California Press.

Becker, G., Butler, A., & Nachtigall, R. D. (2005). Resemblance talk: A challenge for parents whose children were conceived with donor gametes in the US. *Social Science and Medicine, 61*(6), 1300–1309.

Bergmann, S. (2014). *Ausweichrouten der Reproduktion. Biomedizinische Mobilität und die Praxis der Eizellspende*. Wiesbaden: Springer.

Berkowitz, D. (2013). Gay men and surrogacy. In A. E. Goldberg & K. R. Allen (Eds.), *LGBT-parent families. Innovations in research and implications for practice* (pp. 71–85). New York: Springer.

Berkowitz, D., & Marsiglio, W. (2007). Gay men: Negotiating procreative, father, and family identities. *Journal of Marriage and Family, 69*(2), 366–381.

Blyth, E., & Frith, L. (2009). Donor-conceived people's access to genetic and biographical history: An analysis of provisions on different jurisdictions permitting disclosure of donor identity. *International Journal of Law, Policy and the Family, 23*(2), 174–191.

Brösicke, K., Knaack, J., Köppen, J., Regel, A., Rudolphi, M., & Schnicke-Sasse, P. (2017). *120. Deutscher Ärztetag. Beschlussprotokoll, Freiburg 23. bis 26. Mai 2017* [pdf]. Berlin: Bundesärztekammer. Available at http://www.bundesaerztekammer.de/fileadmin/user_upload/downloads/pdf-Ordner/120.DAET/Beschlussprotokoll_120_DAET.pdf. Accessed 31 Aug 2017.

Chabot, J. M., & Ames, B. D. (2004). "It wasn't 'let's get pregnant and go do it'": Decision making in lesbian couples planning motherhood via donor insemination. *Family Relations, 53*(4), 348–356.

Dillaway, H. E. (2008). Mothers for others: A race, class, and gender analysis of surrogacy. *International Journal of Sociology of the Family, 34*(2), 301–326.

Eggert, N., & Engeli, I. (2015). Rainbow families and the state: How policies shape reproductive choices. In D. Paternotte & M. Tremblay (Eds.), *The Ashgate research companion to lesbian and gay activism* (pp. 323–338). Surrey: Ashgate.

Finkler, K. (2000). *Experiencing the new genetics. Family and kinship on the medical frontier*. Philadelphia: University of Pennsylvania Press.

Finkler, K. (2001). The kin in the gene. The medicalization of family and kinship in American society. *Current Anthropology, 42*(2), 235–263.
Furnham, A., Salem, N., & Lester, D. (2014). Selecting egg and sperm donors: The role of age, social class, ethnicity, height and personality. *Psychology, 5*(3), 220–229.
Greenfeld, D. A., & Seli, E. (2011). Gay men choosing parenthood through assisted reproduction: Medical and psychosocial considerations. *Fertility and Sterility, 95*(1), 225–229.
Heaphy, B., & Einarsdottir, A. (2012). Scripting civil partnerships: Interviewing couples together and apart. *Qualitative Research, 13*(1), 53–70.
Hertz, R., Nelson, M. K., & Kramer, W. (2015, December). Gendering gametes: The unequal contributions of sperm and egg donors. *Social Science & Medicine, 147,* 10–19.
Inhorn, M. C., & Birenbaum-Carmeli, D. (2008). Assisted reproductive technologies and culture change. *Annual Review of Anthropology, 37,* 177–196.
Kramer, A. M. (2011). Kinship, affinity and connectedness: Exploring the role of genealogy in personal lives. *Sociology, 45*(3), 379–395.
Mamo, L. (2007). *Queering reproduction. Achieving pregnancy in the age of technoscience*. Durham: Duke University Press.
Mallon, G. P. (2004). *Gay men choosing parenthood.* New York: Columbia University Press.
Melhuus, M., & Howell, S. (2009). Adoption and assisted conception: One universe of unnatural procreation. An examination of Norwegian legislation. In J. Edwards & C. Salazar (Eds.), *European kinship in the age of biotechnology* (pp. 144–161). Oxford: Berghahn.
Mitchell, V., & Green, R. J. (2007). Different storks for different folks. *Journal of GLBT Family Studies, 3*(2–3), 81–104.
Murphy, D. A. (2013). The desire for parenthood: Gay men choosing to become parents through surrogacy. *Journal of Family Issues, 34*(8), 1104–1124.
Nelkin, D. (2006). *The DNA mystique: The gene as a cultural icon.* New York: W. H. Freeman.
Nordqvist, P. (2012). 'I don't want us to stand out more than we already do': Lesbian couples negotiating family connections in donor conception. *Sexualities, 15*(5/6), 644–661.
Nordqvist, P. (2014a). Bringing kinship into being: Connectedness, donor conception and lesbian parenthood. *Sociology, 48*(2), 268–283.

Nordqvist, P. (2014b). The drive for openness in donor conception: Disclosure and the trouble with real life. *International Journal of Law, Policy and the Family, 28*, 321–338.

Novas, C., & Rose, N. (2000). Genetic risk and the birth of the somatic individual. *Economy and Society, 29*(4), 485–513.

Quiroga, S. S. (2007). Blood is thicker than water: Policing donor insemination and the reproduction of whiteness. *Hypatia, 22*(2), 143–161.

Ragoné, H. (1994). *Surrogate motherhood. Conception in the heart*. Boulder: Westview Press.

Richards, M., Pennings, G., & Appleby, J. B. (2012). Introduction. In M. Richards, G. Pennings, & J. B. Appleby (Eds.), *Reproductive donation. Practice, policy and bioethics* (pp. 1–12). Cambridge: Cambridge University Press.

Riggs, D. W., & Due, C. (2013). Representations of reproductive citizenship and vulnerability in media reports of offshore surrogacy. *Citizenship Studies, 17*(8), 956–969.

Roseneil, S., Crowhurst, I., Santos, A. C., & Stoilova, M. (2013). Reproduction and citizenship/reproducing citizens: Editorial introduction. *Citizenship Studies, 17*(8), 901–911.

Rothmayr, C., & Ramjoué, C. (2004). Germany: ART policies as embryo protection. In I. A. Bleiklie, M. L. Goggin, & C. Rothmayr (Eds.), *Comparative biomedical policy: Governing assisted reproductive technologies* (pp. 174–190). London: Routledge.

Schaaf, J. (2017). Man macht ja alles. *FAZ* [online]. Available at http://www.faz.net/aktuell/gesellschaft/menschen/erste-kinderwunsch-messe-findet-in-berlin-statt-14884805.html. Accessed 1 July 2017.

Schwarz, P. (2017). Mach mir ein Kind. *Die Zeit* [online]. Available at http://www.zeit.de/2017/09/kinderwunsch-reproduktionsmedizin-berlin-messe. Accessed 1 July 2017.

Shenfield, F., de Mouzon, J., Pennings, G., Ferraretti, A. P., Andersen, A. N., de Wert, G., et al. (2010). Cross border reproductive care in six European countries. *Human Reproduction, 25*(6), 1361–1368.

Smart, C. (2010). Law and the regulation of family secrets. *International Journal of Law, Policy and the Family, 24*(3), 397–413.

Stacey, J. (2006). Gay parenthood and the decline of paternity as we knew it. *Sexualities, 9*(1), 27–55.

Teman, E. (2010). *Birthing a mother. The surrogate body and the pregnant self.* Berkeley: University of California Press.

Teschlade, J. (2016, July 22). *(Un-)intentional pioneers: Same-sex parents oscillating between heteronormative family narratives and queering kinship*. Paper presented at the EASA panel 'Kinning from the edges: LGBTQ doing and undoing families'.

Waldschmidt, A. (1995). "Lieber lebendig als normal!": Positionen der Behindertenbewegung zu Humangenetik und Pränataldiagnostik. In E. Schindele (Ed.), *Schwangerschaft. Zwischen guter Hoffnung und medizinischem Risiko* (pp. 333–362). Hamburg: Rasch & Röhring.

Wimbauer, C., & Motakef, M. (2017). *Das Paarinterview. Methodologie – Methode – Methodenpraxis*. Wiesbaden: VS Springer.

Wirth, H. (2000). *Bildung, Klassenlage und Partnerwahl. Eine empirische Analyse zum Wandel der bildungs- und klassenspezifischen Heiratsbeziehungen*. Wiesbaden: Springer.

Wong, K. A. (2017). Donor conception and 'passing', or: Why Australian parents of donor-conceived children want donors who look like them. *Journal of Bioethical Inquiry, 14*(1), 77–86.

Part IV

State Supported System: The Case of Israel

15

In the Throes of Revolution: Birthing Pangs of Medical Reproduction in Israel and Beyond

Carmel Shalev

Introduction

My first encounter with surrogate motherhood was in the early 1980s. I was then a graduate student at the law faculty of the Hebrew University in Jerusalem, and the topic I had chosen for my master's thesis was abortion. The question that interested me was related to the judgments of supreme courts in three different countries—the US, UK and Israel—which had all dismissed legal actions taken by men seeking to prevent women from terminating a pregnancy. I grew up and came of age in a period that lauded equal opportunity and imagined formal equality for women and men. Thus, the question that presented itself was: Why is it so clear that men do not have standing in the question of terminating a pregnancy? Should not a norm of sexual equality apply? (Shalev 1984).

C. Shalev (✉)
Retired, University of Haifa, Haifa, Israel
e-mail: cshalev@012.net.il

S. Mitra et al. (eds.), *Cross-Cultural Comparisons on Surrogacy and Egg Donation*,
https://doi.org/10.1007/978-3-319-78670-4_15

In searching for an answer to this question, I came across substantive differences between men and women—some biological and some social. I reached the conclusion that decisions about abortion must be made by women in cases of disagreement, because she is the one who carries the pregnancy in her body, and she is also the one who bears the main social responsibility for raising children after birth. I discovered patriarchal prejudices about the division of labour between the sexes (the word 'gender' was not yet in use) and double standards—men were cast in the role of breadwinners and warriors, and women as mothers and either wives or whores. I also discovered writing, some of which was science fiction like Aldous Huxley's Brave New World, which foresaw revolutionary developments through assisted reproductive technologies (ART).

For several decades, since the middle of the twentieth century, doctors had been performing artificial insemination as treatment for infertility in married men. They used the sperm of men who were outsiders to the marriage and called them 'donors' despite the fact that they were usually paid for giving their sperm. In the early years of the practice, these third-party men were often medical students; in exceptional cases, the doctors used their own sperm. The 'donors' identity was shrouded in a norm of anonymity, to protect the marital relation against insinuations of adultery even though there had been no sexual infidelity, and to maintain a pretence that the children were 'legitimate' (Shalev 1989).

Then came the 1960s and 'the pill', together with the civil rights movement and anti-war protest, flower children and rock festivals. These were times when human rights and their core principles of dignity, liberty and equality were extended to women, and a new movement of feminism was born on the basis of women's right to reproductive and sexual freedom (Shalev 2000, 2015b).

Towards the end of the 1960s, there was a worldwide trend to decriminalise abortion. To this day, women's freedom of choice as to whether or not to be mothers is essential and key to gender equality. This is a fundamental right, not just because of her right to bodily integrity, but also because of her right as a human being to the negative freedom of protection from extraneous interference in her private decisions and to the positive freedom of personal autonomy to imagine, plan and design the course of her life (Berlin 1969) as a political, economic and social being, and as a sovereign moral agent (Shalev 1989).

At the same time, in the literature of the late 1970s there was mention of a new practice—surrogate motherhood. This seemed to be the mirror image of artificial insemination with donor sperm, except that the problem to be solved here was female rather than male infertility. If the wife in a marriage could not bear children, a third-party woman could be artificially inseminated with the husband's sperm and carry a child for the couple. Even before the development of in vitro fertilisation (IVF), surrogate motherhood was envisaged as a possibility of nonsexual reproduction by one woman for another by a simple procedure of artificial insemination.

But whereas donor insemination seemed to have been accepted, there appeared to be strong objection to surrogacy. So again the question arose: Why should they be treated differently? It appeared that surrogacy posed a greater challenge than donor insemination to traditional patterns of reproductive relations within the patriarchal 'natural' family. Both donor insemination and surrogacy were perceived as a deviation from and violation of the sacred bond of marriage. In donor insemination, a norm of anonymity could protect the integrity of the marriage and the offspring from the stigma of illegitimacy. It was relatively easy to make the sperm provider disappear, pretend that the husband was the 'natural' father, and create a legal fiction that dissociated legal from biological paternity. In surrogacy, on the other hand, the presence of the carrying mother throughout the process of conception, pregnancy and birth could hardly be concealed.

The role of the surrogate mother posed a radical challenge to the basic norms of reproduction within and without marriage—adultery and bastardy. The emergence of new families was one result from the proliferation of a range of intrusive nonsexual reproductive technologies, which might involve third-party collaborations, with surrogates and sperm and egg cell providers.

Thus, ART raised new questions about the legal definition of parenthood and the 'legitimacy' of children. From time immemorial legal parenthood had derived for men from marriage, and for women from birth, as a matter of status. Marriage served to resolve the uncertainty of male parenthood (there being no material connection between the act of copulation and the resultant pregnancy): the husband was presumed to be the father of the children of his wife in marriage. It was also extremely difficult to establish paternity in extramarital relations and the 'illegitimate' offspring were not entitled to go by the father's name or to inherit him. But in donor insemination and in surrogacy, parenthood was to be ascribed according to agreement. To what extent could relations of conception, pregnancy and birth be governed by a legally binding contract? What were the implications of moving from status to contract in reproduction? (Maine 1861).

Furthermore, there were moral concerns around the payment of money. In adoption, a woman could not be held to a promise to give her child to strangers, and she certainly could not sell the child. Since the middle of the nineteenth century, common law had granted women rights of parenthood (guardianship and custody) and children had ceased to be regarded as the property of their fathers. Hence, the sale of children was against public policy and morality, and thus could not be the subject of a legally binding contract. Was there any difference between a surrogate and a woman who conceives and gives birth to a child on her own and sells her child to strangers? (Shalev 1989).

In both adoption and surrogacy, there is a woman who for payment gives an infant to others and concedes her status as a mother to them. But there is an important distinction between the two cases: the surrogate mother conceives from the very start on the basis of an agreement with others and the common intention that they will be responsible for raising the child. As opposed to adoption, conception is deliberate within reproductive relations that are intentional and might be viewed as a contract for personal services. The application of a contractual model to reproductive relations could mark a paradigm shift from the discriminatory sexual reproductive regime of patriarchal marriage to an egalitarian regime of personal reproductive agency and autonomy (Shalev 1989).

Women as Reproductive Agents

The idea to pay women for their reproductive labour within a birthing contract was radical. To this day, women go mostly unpaid for their 'informal' domestic work. Surrogacy presented a possibility of women as rational, economic and moral agents in relation to their reproductive capacity. The fear of 'commercialisation' and 'commodification' in ART arose only in relation to women. There was no concern about paying men to provide sperm, a volitional and self-controlled act as opposed to the invasiveness of IVF treatments that women undergo. Nor was there any objection to paying medical professionals for their services. Birthing was perceived as a natural process in which 'woman' played the passive role of a vessel for an embryo that seemingly developed of itself, in contrast to the active creativity that characterises the work of 'man' when harnessing the powers of nature to realise his aspirations and ambitions (Shalev 1996a, b).

Recognition of the economic value of reproductive labour had far-reaching implications, and the possibility of women exercising autonomy as reproductive agents was a profound challenge to long-standing social conventions and gender stereotypes. This generated fears that women would not be able to go through with their undertaking and separate from the newborn children, and the paternalistic suggestion that they should not be bound to keep their reproductive promises in order to protect them against unconsidered or irrational decisions. There was no evidence base for such unease, and we know today from experience in Israel and other countries around the world that surrogates do not withdraw from their agreements (Teman, Chapter 4). They respect their word, they are responsible, and they take pride in their labour, even when they are treated poorly.

Why could not the promise of a woman in regard to reproductive labour be a matter of binding agreement? Why should she not be subject to the principle of honouring one's obligations? Reproductive labour, as all other forms of human activity, takes place in the context of social relations within which one acts as a conscious, rational, moral and political being. The unwillingness to acknowledge the binding effect of the surrogate mother's agreement implied that there was something in reproductive labour that impaired a woman's autonomy. As if she were imprisoned in the subjectivity of the womb (DeBeauvoir 1949). The paternalistic protection of women reinforced the prejudice that women as such are frivolous and fickle, and implied a rejection of their human capacity because of their biology (Shalev 1989).

Twenty years of legally regulated surrogacy in Israel teach us that women can indeed act as autonomous and sovereign moral agents in relation to their reproductive capacity. There has not been one instance of conflict around relinquishing the child, and there are moving stories of intimate bonds and respectful relationships that were formed between the women who carried the pregnancies and the parents who are raising the children as their own (Teman 2010) even though in other cases they can be wrought with tension. Yet concerns about commercialisation and commodification have returned in new guise in the face of a growing worldwide industry of 'reproductive tourism' since the start of the twenty-first century (Mitra, Chapter 5; Nahman, Chapter 16). This global market is capitalising on the increasing and diversified uses of medical fertilisation, including repro-genetics, and is fuelled by a technological imperative to consume ever more sophisticated techniques.

To a certain extent, we envisaged such developments, even though we could not imagine the tangible details. On the one hand, private regulation of social reproductive relations by means of binding agreements presumed the free will of the parties as autonomous beings. On the other hand, there was concern that a woman might agree to use her body in the service of another from a bargaining position that was socially vulnerable, within the constraints of limited economic opportunity. The idea of the surrogate mother's autonomy was appealing in terms of personal empowerment. But it was doubtful whether in the long term it could overcome the cultural determinants that disadvantage and discriminate women as such (Shalev 1989).

Israel's Surrogacy Law

The will of any individual woman to choose to undertake reproductive intervention—whether for her own sake or that of others—is constructed by powerful social forces, including deep-set patriarchal residues of gender discrimination, medicalisation of reproduction and market forces. In Israel—renown as a superpower of ART with the highest per capita rates of usage worldwide—it is also influenced by a religious and nationalist pronatalist culture of 'be fruitful and multiply' (Shalev and Gooldin 2006; Shalev 2008, 2013; Rimon-Zarfaty, Chapter 17).

Israel's 1995 law on surrogacy was the first of its kind worldwide, and the conditions it laid down for lawful surrogacy agreements were mandated in part by religious norms of *halakha* (Jewish law) such as *mamzerut* (bastardy), a stigmatic and discriminatory status that adheres only to the extramarital child of a married woman, and not to that of an adulterous married man or of an unmarried woman. *Mamzerut* carries severe impediments to marriage with other Jews—a *mamzer* and his or her offspring may marry only other *mamzerim* or converts to Judaism, for 'ten' generations. This amounts effectively to social excommunication. The stigma was to act as a deterrent and punishment for women's adulterous conduct, which could undermine the legal presumption that the children she bore were her husband's offspring. The effect was to brand children for the 'sins' of their parents (Shalev 1998).

Because the child of an unmarried woman is not tainted by the impediments of *mamzerut*, Israel's law provided that only unmarried women would be allowed to act as a surrogate mother (Rimon-Zarfaty, Chapter 17). This meant that unmarried women would service married women, perpetuating the patriarchal dichotomy of Madonna/whore, and exploiting the ongoing socio-economic vulnerability of single and divorced mothers in contemporary times, and the intrinsic imbalance in her relation as an individual to a married couple.

It is important to note that there was some *halakhic* debate as to whether nonsexual reproduction also amounted to adultery. Some considered that *mamzerut* occurred only if conception was the result of forbidden sexual intercourse, which is not the case in surrogacy. But Israel's law incorporated the restrictive and cautious rabbinical view that feared for *mamzerut* even in the case of 'genetic adulteration', so as to be on the safe side and preclude any doubt as to the child's capacity for marriage (Shalev 2013). Thus, the law allowed rabbinical registrars of marriage access to an official registry of surrogacy children, since marriage and divorce in Israel are governed by religious authority. Notably, the children themselves do not have a right under the law to access information about the circumstances of their birth.

In addition, the law reflects a bias that favours the primacy of male over female genetic parenthood. On the one hand, it provided that the sperm must be the husband's, and on the other that the egg cell must not be the surrogate's. In other words, it forbids a maternal genetic relation between her and the embryo. Biological fatherhood is genetic, whereas biological motherhood is both genetic and physiological. The latter, pregnancy, is similar to social motherhood in that it calls for the care and nurture of a foetus as one would care and nurture a child. Thus, the importance attached to genetic parenthood reflects a gender bias. Although surrogacy could be easily performed through the use of artificial insemination with the husband's sperm, because the surrogate was not allowed to be the genetic mother, egg cells would have to be obtained from another woman and fertilised in vitro, and the surrogate would undergo a procedure of embryo implantation. In other words, women were subjected by law to hi-tech medical interventions rather than less invasive and less dangerous treatments.

Interestingly enough, the rule that the egg cell may not be provided by the surrogate was not because of any *halakhic* concern. One might imagine it to be justified had the law also determined that surrogacy could only take place if the embryo was conceived from the egg cell of the infertile married woman. However, this is not the case. In fact, the rule was the result of pressure from a lobby of childless women who wanted to be able to have children with a surrogate but also feared that if she were the genetic mother, she might become emotionally attached to the foetus and refuse to surrender the child after birth (Shalev 1998). It assumed that the lack of a genetic relation would lessen the emotional experience of carrying the pregnancy.

As already mentioned, there was no scientific base for such fear, and surrogate mothers usually honour their agreements, so that it might be no more than another expression of the gender bias about the importance of genetic parenthood. True, in the first known case of surrogacy in the US, the *Baby M* case, the parties fell out, the surrogate ran away with the child, and she was then ordered by a court of law to relinquish the child to the custody of her father—on grounds of the child's best interests (despite the fact that the court considered the agreement to be contrary to public policy and thus unenforceable) (Shalev 1989). Since then, there have been numerous cases in the US in which it was actually the intended parents (IPs) who refused to accept responsibility for the children after birth, as opposed to extremely rare cases of surrogates rescinding their agreement and claiming the children as their own. Mostly, the women who act as surrogate mothers, all around the world, insist that the children are not theirs. Perhaps the absence of a genetic connection makes it easier to say so. Nonetheless, regardless of genetics, surrogacy is a highly complex practice in terms of the emotional and social human relations it entails.

Bio-Social Revolution

Surrogacy is one practice of many that mark a bio-social revolution in the wake of innovations in repro-genetic medicine. The technological developments were accompanied not only by radical change in social mores of sexual and reproductive relations, but also by the evolution of a discourse of reproductive rights. Reproductive liberty, the right to privacy and autonomy in reproductive relations, is a negative right and political by its nature, since it creates a zone of privacy in which individuals are protected from undue interference by society and the state in essentially personal affairs. In Israel, however, the Supreme Court created a positive right to parenthood, which entails the collaboration of others. The judicial pronouncements are dramatic. The right to parenthood is regarded as a 'basic existential need' and as 'the embodiment of man's selfhood'. Thus, freedom of choice whether to be or not be a parent transformed into a positive right to become a parent by any and all possible medical means. Under the spell of a compelling medical technology and the influence of a consumerist culture, the use of medical fertilisation in Israel became a social imperative such that nonuse requires power of will to withstand pressure, almost as an act of social dissidence (Shalev 2015a, p. 162).

Reproductive consumerism in Israel reflects cultural messages of mandatory motherhood, a cult of genetic parenthood and over-consumption of technology, to a degree of dependence and addiction. Surrogacy provides an illustration of this general trend. At first, it was designed as a response to a medical need in extreme and unusual circumstances, for childless women unable to carry their own pregnancies due to uterine pathology. Today, it is perceived as a consumer choice to satisfy personal preferences. Repro-genetic medicine in general and surrogacy in particular are no longer mere means to treat childless people, but rather an accepted tool for creating multiple-child families. Women are active participants in this arena. They demand a right to parenthood for themselves and for their partners; wish to realise themselves as responsible mothers, doing everything possible to have a 'perfect' baby; and submit to invasive medical procedures, again and again, despite the risks (Shalev 2015a, p. 157).

The technological revolution of ART also brought about repro-genetic medicine, together with a paradigm shift from family planning to planning children. Repro-genetic medicine allows the classification of embryos as desirable or undesirable, good or bad, either in the womb or in vitro. A liberal discourse of human rights, gender equality and reproductive privacy evolved into a neoliberal market discourse of medical consumerism, which offers a 'right' to choose the features of 'my' embryo. In a global industry fuelled by greed for profit, satisfaction of consumer desire masquerades as a messenger of enlightened scientific progress in the service of self-gratification and 'me' medicine (Dickenson 2013). The statement 'I would like' or 'I want' translates into 'I need' and 'I am entitled'. The critical distance between 'I want' and 'I ought' collapses in a supermarket of personalised lifestyle preferences (Sacks 2002).

Under patriarchy, woman's reproductive capacity was a husband's estate. Today, it is largely controlled by a global industry. The twenty-first century has seen the growth of a private market in medicine, and globalisation has produced a thriving business in medical tourism, including repro-genetics. Shady medical practices, such as trafficking in human organs, circumvent national laws in shifting business networks. Reproductive tourism in particular involves the cross-border movement of women, gametes and embryos in various third-party reproductive collaborations. Recently, there are signs of a new practice for selecting the sex of embryos through pre-implantation genetic diagnosis. For the future, we might foresee engineering of embryos by techniques of cloning, mitochondrial replacement and genetic 'editing', which would afford unprecedented control over the manipulation and design of children's characteristics according to idiosyncratic preference (Werner-Felmayer and Shalev 2015).

Meanwhile, the revolution has produced a new form of human life—hundreds of thousands of human embryos—fertilised eggs frozen in time and stored in medical laboratories around the globe. ART might have held a promise of liberating women from the tyranny of their biology (Firestone 1972), but did not do so in fact. Instead, women have been captivated as eager consumers and subjected to medical interventions. All the applications of repro-genetic technology are mediated necessarily through female bodies, which remain the *sine qua non* of human reproduction.

Costs of Medicalisation

Repro-genetic medicine enjoys uncritical support from Israel's parliament and government, medical and rabbinical establishments, courts of law, and the general public, while ethicists provide justifications of individual liberty and freedom of science. However, this has been at significant cost in terms of the health of women and infants.

At first glance, it might seem that choices for women have expanded. The doctors offer a bounty of sophisticated and acronymised technical options: IVF, micro-manipulation of sperm for men with low sperm count (ICSI), pre-implantation genetic diagnosis of fertilised eggs (PGD), third-party reproduction (surrogacy and gamete 'donations'), 'social freezing' of egg cells or 'fertility preservation' for pre-menopausal single women, and posthumous use of sperm. In the beginning, IVF was developed for the treatment of female infertility due to blockages in the fallopian tubes. Today, on the most part, it is performed on healthy women. More often than not, healthy women undergo medical intervention be it for the sake of their infertile male partners, or to diagnose their embryos, or to hedge against future decline in fertility, or in the case of third-party reproductive arrangements—for the benefit of strangers.

In other words, women have become the object of intrusive medical procedures which entail physical discomfort and risk of harm, and necessarily involve invasion of privacy. Moreover, they face complex challenges to their emotional well-being and self-esteem, even when they are the beneficiaries of treatment. They experience stress and anxiety in their relations with partners, children, family relatives, employers and communities, and feelings of shame, anger, depression, inadequacy, grief, loss and despair due to the failure of treatment cycles characterised by low success rates (Shalev 2012).

When women help others to have a child, there are additional sensitivities around the relations with the commissioning parents. Furthermore, they are submitted to even greater medical control and subjected to double standards of medical care compared with women who are treated for their own fertility needs, especially but not only in inter-country settings (Shalev et al. 2016). In Israel, for example, a typical surrogacy contract provides that the surrogate may not smoke or have sexual intercourse, and is required to get a certificate of approval from a doctor if she wants to travel abroad during the pregnancy. She will be expected to have daily hormonal injections for three months after verification of pregnancy. Decisions about medical treatment are taken usually by the intended parents and medical professionals. In most cases, for example, the surrogates do not have a choice as to the number of embryos that will be implanted, which is the reason for the relatively high rates of multiple births, despite the associated health risks for both women and foetuses (Lipkin and Samama 2011).

Likewise, egg cell procurement in third-party arrangements is tinged with a double standard of medical care compared with women who undergo the procedure for their own need. There is a great demand for egg cells, coming mostly from women over the age of forty, and the global market for this commodity is far larger than that for surrogacy (Shalev and Werner-Felmayer 2012). Also when gay men wish to have families through surrogacy, egg cell procurement is an essential element, because of the bias against a genetic connection between the surrogate and child and because of eugenic preferences. Thus, an Israeli surrogacy clinic operating in Nepal at the time of the earthquake in the spring of 2015 recruited the surrogates from an impoverished state in India, while the egg cell providers came from South Africa, because western couples prefer Caucasian offspring (Shalev et al. 2017).

This throws light on another troubling aspect of third-party reproduction—the adroitness of the technology and the market to fragmentise and depersonalise the delicate and intimate process of mothering children into the world (Shalev 2012). It is all too easy to deface and erase the personae of collaborating women, and to reduce them to bodily functions or organs, as if 'the lady vanishes' (Dickenson 2007). The norm of anonymity, adopted from sperm procurement, is easily applied to egg cell providers in the global market of inter-country medical reproduction. And, while domestic settings of surrogacy make it hard to conceal the identity of the pregnant woman, that is not the case in inter-country settings where it is not unlikely that she will meet the commissioning parents face-to-face only once, and where no one has any legal responsibility to maintain medical records and preserve identifying information relevant to the child's birth, such as the identity of the gestational-birth mother (Shalev et al. 2016).

Israel and the Global Market

Israel is considered to have devised an effective regulatory model for domestic surrogacy, yet between 2012 and 2015 more children were born from transnational surrogacy than the overall number of in-country surrogacy children.

In inter-country settings, it is important to make sure the gestational mother is present, not only for the best interests of the child but also for her own human dignity. She has a right to be acknowledged as having birthed the children and to choose whether and how to have ongoing contact with them. She might not want to be responsible for the child as a mother; she might not want to raise the child herself; she might not want to be recorded as a birth mother in the population registry. And yet she might want the child to know who she is, and she might want to know the child or be in touch with the family from time to time. If she has no choice at all in the matter, she is degraded from a person in her own right to a means in the service of others. Notwithstanding technological developments, the essential nature of reproduction remains the same: it is relational. The practice of rendering surrogates invisible after birth dehumanises them as mere instruments in the hands of others.

People consume transnational services because of legal restrictions in their own countries and also because of global gaps in the costs. If people cannot get the commodity they want in their country of residence, they will buy it abroad. If local law restricts access to medical reproduction—such as the law in Israel, for example, that restricts surrogacy to straight couples—there are alternative services across the globe (Mitra, Chapter 5). While the US remains the preferred country of destination for those who can afford it because of the relative autonomy of the surrogates, the operations of transnational surrogacy agencies servicing Israelis spread in the last decade to India, Georgia, Thailand, Nepal and Mexico. India became a destination for surrogacy because the US was too expensive. When India disallowed surrogacy for single-sex couples, business moved to Thailand, and when it was banned there, operations moved to Nepal. Now, Mexico appears to be the current venue, while egg cells for Israelis come from women living in South Africa or the Ukraine (Shalev et al. 2017). The old patriarchal order has crumbled. Women now more than ever before enjoy equality and freedom, and exercise choice as sexual and reproductive agents. But antiquated patterns of controlling women have been replaced by free market practices that are lawless and potentially exploitative. Across the globe, a dynamic and mobile network of medical entrepreneurs offers routes that sidetrack the law.

Israel is considered to be relatively liberal and permissive in regard to surrogacy, but even it sets limits on what is legally allowed and thus it does not answer all the domestic needs. In addition, differential global pricing creates economic incentives to consume transnational services. So women and men, singles and couples, gay and straight travel abroad in their quest to become parents—to jurisdictions where the practices are under-regulated and to countries where labour is relatively cheap. While Israel's domestic law permits commercial surrogacy for its residents, it excludes tourists from abroad. At the same time, it exports unlawful or 'illegitimate' reproduction to foreign lands. In this way, we launder our reproductive delinquencies through a market model of outsourcing. But, as opposed to the regime of patriarchal reproduction, we do not brand the children; rather, we embrace them with love. While the children are considered ‘kosher’ as it were (legitimate), the process that brings them into the world is fraught with practices that are harmful to women due to double standards of medical treatment and unequal relations of power, which are magnified in inter-country settings by geographical distance, cultural alienation and structural injustices (Shalev et al. 2016).

Against this background, in recent years public debate about surrogacy in Israel has centred on whether to expand the criteria of eligibility for intended parents [IPs]. After a gay couple petitioned the high court of justice to challenge the constitutionality of restricting surrogacy to straight married couples under the 1995 law, a public commission was appointed to review the matter. The entire commission seemed to agree that surrogacy should be allowed for gay parents, but was divided about the terms and conditions (Mor-Yosef Report 2012). The controversy touched upon the distinction between commercial and noncommercial surrogacy.

The majority position was that gay surrogacy should be regulated through an altruistic track, as opposed to the law’s commercial model which allows payment to the surrogate mother for her services in the amount agreed upon by the parties. The minority view was that no distinction should be made between IPs, and all surrogacy in Israel should be performed solely on an altruistic basis. However, the difference between commercial and altruistic surrogacy is not clear cut. There seems to be a naïve view that payment and altruism are mutually exclusive, which is not the case (Schicktanz, Chapter 6; Rimon-Zarfaty, Chapter 17, for a problematisation of the distinction between commercialism and altruism). On the one hand, to judge from Israel’s organ transplant law, an altruistic model does not preclude payment of ‘compensation’ for out-of-pocket expenses. On the other hand, to judge from statements of surrogates in multiple studies, they have strong altruistic motives (helping others) together with a sense of entitlement to be paid for their work. It might be helpful to reframe the question: Should the payment be for personal services and left to contract and the market, or should it be ‘compensation for her time and effort’ and the amount regulated by the state?

Be that as it may, the question which intrigued me from the very start—why only women may not be paid for reproductive labour—seems to arise once again. From my perspective, paying the women is not the problem. The root causes of objectification and exploitation are the social forces that construct the relationship between the commissioning parents and the women who help them have children. Israel is no different from other countries in the pronounced socio-economic disparity between surrogates and IPs according to measures of age, education, occupation, place of residence and country of birth (Lipkin and Samama 2011). Nonetheless, there are remarkable stories of healthy relationships that form around the birthing of the children. Global market conditions make it nigh impossible to form such a relationship, and the problem with outsourcing third-party reproductive services might be reframed in terms of distributive justice in the allocation of scarce resources. Not many women want to provide egg cells or carry pregnancies for others. How can we divide these limited human resources so as to allow new families without discrimination, and without creating a market in which reproductive services are sold to the highest bidder?

Surrogacy as Relationship

What is it that motivates women to help others have a child? In addition to economic opportunity, most women say they are driven by altruistic sentiments. They empathise with and want to alleviate the suffering of those who cannot have children by themselves. There are numerous testimonies of women in Israel who acted as autonomous agents, supported by mothers and sisters, and bonded with the IPs to develop open and respectful relations with one another that continue as a long-standing friendship (Teman, Chapter 4). Israeli studies showed that from the surrogates' point of view the success of the process depends on the quality of the relationship with the commissioning couple, and this is no less important than the money she gets paid (Lipkin and Samama 2011).

The surrogates perceive themselves as offering a priceless gift: the realisation of motherhood for a woman who cannot carry a pregnancy and give birth by herself. For some, it is the most meaningful accomplishment of their lives. They serve their sisters with care and responsibility and regard their undertaking as no less than heroic. But the relationship with the commissioning family is critical for their sense of self-esteem. Beyond the financial reward, the surrogate gains in two respects: satisfaction from a heroic act of altruism and intimate connection.

As in any relationship, there is also pain. The women need recognition and appreciation, and if that is lacking, they might feel ill-treated and misused. Throughout the long and emotionally complex process of conception, gestation and delivery, they receive a large amount of attention from the IPs and medical caregivers. Immediately after birth, after the umbilical cord is cut, there is an abrupt distancing sometimes to the degree of total disconnection. In a way, when the baby is given to the IPs, the surrogate separates not only from the newborn infant but also from the couple. If she fails to receive the recognition she would like, feelings of duplicity, betrayal and abuse can arise. She might be hurt and offended simply because the couple do not invite her to the *brit milah*, the circumcision ceremony for the newborn child.

Then and Now

Surrogacy unleashes radical questions about the powerful social forces that control women's reproductive relations. Thirty years ago, when the practice first began, the issues were women's reproductive freedom to choose whether or not to become mothers, and their agency to enter a legally binding agreement as to their reproductive capacity. Since then, the hold of patriarchy over women's lives has weakened, and reproduction has undergone a process of medicalisation that is proliferating through an unregulated global market involving the cross-border movement of IPs, third-party collaborators, human gametes and embryos, and newborn children. The normative framing has shifted from a liberal articulation of a right to privacy and freedom from state intervention in reproductive decisions to a neoliberal claim for a consumerist and eugenic right to parenthood. Surrogacy takes place within this context and also exemplifies it. The questions around surrogacy with which

Israel is now struggling reflect the normalisation of reproductive medical intervention and are a result of globalisation.

Although domestic surrogacy in Israel is governed by an intrusive bureaucracy (Shalev 2015a), it has not created much notable case law. In the lower courts, one judgment upheld an administrative limitation on the age of the commissioning parents. The cases that have come before Israel's Supreme Court mainly challenge the statutory restriction of surrogacy to heterosexual married couples. The first such case, on behalf of a single woman, was dismissed on grounds that this was a matter of policy and not of constitutional law. The second, as already mentioned, was on behalf of a gay couple and led to the Mor-Yosef commission that reached a deadlock as to whether surrogacy should be 'commercial' and discriminatory or 'altruistic' and uniform. Meanwhile, the Supreme Court continues to address questions about the rights of gay couples to have children by means of surrogacy arrangements, expressing serious doubts about discrimination, but referring the matter again and again back to the legislature.[1]

The most striking domestic issue that has arisen in Israel's courts is whether surrogacy may be used to bring children into the world posthumously? (Hashiloni-Dolev and Triger 2016; Shalev 2002). In one unusual case, documented by a television investigative journalist, a widower had a child from embryos created with his deceased wife, with the assistance of a surrogate mother outside Israel. However, the instances that have reached the courts typically concern an argument for a 'right to grandparenthood'. These cases involve the posthumous use of sperm from a deceased son, usually portrayed as a soldier killed in action or on reserve duty (Hashiloni-Dolev and Schicktanz 2017). Here, too, surrogacy illustrates broad themes of medical reproduction: a bias for male genetic continuity, a tendency to expand the use of reproductive technology for individual gratification and in Israel particularly—a nationalist pronatalist culture.

Policy on this odd yet sensitive matter was laid down in 2003 by an Attorney General directive, which gave much credence to the presumed desire of men to father children even after their death, and thus laid the ground for sperm preservation as a trade-marked 'biological will'. It allowed posthumous fathering solely on the request of the deceased's female partner, but the emerging trend is that of the deceased's parents demanding a right to surrogacy. The first family court decision in the matter allowed the commissioning grandparents to enter an agreement with a single woman on grounds of her right to have a child with sperm from a known donor. But recently, the Supreme Court dismissed an action by grandparents where their deceased son's widow had chosen not to have children from the sperm that was procured after her husband's death in a road accident.[2]

By and large, however, it is the global market that presents the greatest challenges to Israel's law and policy on surrogacy. Israel's parliament has yet to address the moral challenges of this market, as described above. As for the Supreme Court, so far it has given one major decision setting clear limits on the acceptability of cross-border surrogacy. This was in a case that concerned a physically disabled woman, who arranged the birth of a child who was conceived with sperm and egg cells from donors, and carried and birthed by her niece with whom she had a surrogacy agreement. The court ruled that she could not be regarded as the child's parent in the absence of any genetic relation, and the child was removed from her custody by child welfare authorities.[3]

Conclusion

Despite the revolutionary innovations of repro-genetic medicine, the essence of human reproduction has not changed—it is profoundly relational. Indeed, promises must be kept, but the law of contract and rules of conduct that befit a free market of commodities appear to be inadequate when it comes to the delicate and intimate human collaboration that is needed to usher a child into the world. A market mentality is not the most apt paradigm for regulation of collaborations designed to birth a child. The creation of new human life is incomparable to any other feat. The new life has singular needs, meaning and dignity. The idea of personal liberty, of a self-standing individual situated alone as an autonomous creature distinct from its surroundings, is inappropriate for reproduction which cannot be accomplished singly or independently. Reproductive relations ought to be based on values of respect, trust and responsibility both between the adult collaborators and towards the newborn child (Shalev 2015a). The relations between a pregnant woman and the people who intend to raise the child she is carrying in her womb are not business relations for the purchase of a product. There are things that cannot be bought and can only be given out of love and the goodness of one's heart.

The language of human rights is of supreme importance. Women would not enjoy reproductive freedom or social equality were it not for it. Nonetheless, liberal theory recognises that harm to another marks the limit of individual liberty (Mill 1859). Rights holders are empowered individuals, and in exercising their freedom and autonomy, they wield power over others who are affected by their actions. Once our actions impinge upon others, we are no longer completely free to do as we wish. Choices that we make with regard to our reproductive collaborations involve and affect others by definition. Therefore, choice carries responsibility (Shalev 2012).

The neoliberal values that underlie the claim for a right to be a parent or a grandparent view persons as essentially separate from others, abstract them in isolation from their relationships and value independence and self-interest rather than connectedness and cooperation for mutual gain. The view is self-regarding, and hence instrumental and exploitative of the other. One acts for one's own benefit and makes use of others and profits from that use at their expense. On the other hand, the disposition of caring is other-regarding and relational. Care implies responsibility for the welfare of others who are affected by our choices. Care is altruistic and collaborative, responsive and reciprocal (Gilligan 1982; Shalev 1989).

Women have proven they are reproductive agents, who care for the welfare of the child and the IPs. Domestic surrogacy can be physically and emotionally demanding. Transnational surrogacy all too often involves violation of women's human rights and dignity (Shalev et al. 2016). IPs need to be cognisant of such conditions if they are to act as responsible caring agents. An ethic of care and responsibility requires self-restraint with a shift in consciousness: from self-interested calculation to contemplation of one's relation to others and to an understanding of our connectedness in a seamless web of nature, human being and technology.

Notes

1. Supreme Court, *Itay and Yoav Arad-Pinkas vs. The Committee for Approving Surrogacy Agreements*, app. no. HCJ 781/15, judgment of 3 August 2017.
2. Supreme Court, *Anonymous vs. Anonymous*, app. no. BAM 7141/15, judgment of 22 December 2016.
3. Supreme Court, *Anonymous vs. Ministry of Welfare and Social Services*, app. no. BAM 1118/14, judgment of 1 April 2015.

Bibliography

Berlin, I. (1969). Two concepts of liberty. In H. Hardy (Ed.), *Four essays on liberty* (pp. 118–172). Oxford: Oxford University Press.

DeBeauvoir, S. (1949). *Le deuxième sexe* [The second sex] (H. M. Parshley, Transl. from French 1953). Harmondsworth, Middlesex: Penguin Books.

Dickenson, D. (2007). *Property in the body*. Cambridge: Cambridge University Press.

Dickenson, D. (2013). *Me medicine vs. we medicine: Reclaiming biotechnology for the common good*. New York: Columbia University Press.

Firestone, S. (1972). *The dialectic of sex*. Frogmore, St. Albans: Paladin.

Gilligan, C. (1982). *In a different voice*. Cambridge: Harvard University Press.

Hashiloni-Dolev, Y., & Schicktanz, S. (2017). A cross-cultural analysis of posthumous reproduction: The significance of the gender and margins-of-life perspectives. *Reproductive Biomedicine & Society, 4*, 21–32.

Hashiloni-Dolev, Y., & Triger, Z. (2016). Between the will of the deceased and the will of the survivors: Posthumous use of sperm for reproduction, patriarchy, pronatalism and the myth of sperm continuity. *Iyunei Mishpat, 39*, 661–706 (in Hebrew).

Lipkin, N., & Samama, E. (2011). *Surrogacy in Israel: 2010 situation report and proposals for legislative change*. Isha L'Isha—Haifa Feminist Centre [online]. Available at http://isha.org.il/wp-content/uploads/2014/08/surrogacy_Eng001.pdf. Accessed 9 Oct 2017 (in Hebrew).

Maine, H. S. (1861). *Ancient law*. Reprint 2005. New York: Cosimo Classics.

Mill, J. S. (1859). *On liberty*. Reprint 1970. Cleveland and New York: Meridian Books.

Mor-Yosef Report. (2012). *Recommendations of the public commission for considering the legislative regulation of fertility and obstetric medicine in Israel.* Jerusalem: Ministry of Health (in Hebrew).

Sacks, J. (2002). *The dignity of difference* (p. 35). London, New York: Continuum.

Shalev, C. (1984). A man's right to be equal: The abortion issue. *Israel Law Review, 18,* 391.

Shalev, C. (1989). *Birth power: The case for surrogacy.* New Haven: Yale University Press.

Shalev, C. (1996a). Surrogate motherhood. The commerce in reproductive services. *Bitachon Sotziali, 46,* 87 (in Hebrew).

Shalev, C. (1996b). Surrogate mother arrangements. Legal and normative aspects. In A. Ben-Zeev & S. Almog (Eds.), *An unusual pregnancy. A multidisciplinary study on artificial insemination* (pp. 191–215). Tel Aviv: HaKibbutz HaMe'uhad (in Hebrew).

Shalev, C. (1998). Halakhaand patriarchal motherhood. An anatomy of the new Israeli surrogacy law. *Israel Law Review, 32,* 51–80.

Shalev, C. (2000). Rights to sexual and reproductive health. The ICPD and the convention on the elimination of all forms of discrimination against women. *Health and Human Rights, 4,* 39.

Shalev, C. (2002). Posthumous insemination. May he rest in peace? *Refuah U'Mishpat, 27,* 96–99 (in Hebrew).

Shalev, C. (2008). Reflections on human dignity and the Israeli cloning debate. In M. Düwell, C. Rehmann-Sutter, & D. Mieth (Eds.), *The contingent nature of life. Bioethics and the limits of human existence* (pp. 323–344). Berlin: Springer.

Shalev, C. (2012). An ethic of care and responsibility: Reflections on third-party reproduction. *Medicine Studies, 3,* 147–156.

Shalev, C. (2013). Seminal reasoning: Ultra-orthodoxy and the biopolitics of medically assisted reproduction in Israel. In U. Auga, C. von Braun, & J. Husmann (Eds.), *Fundamentalism and gender: Scripture-body-community* (pp. 220–243). Eugene, OR: Pickwick Publications.

Shalev, C. (2015a). Repro-genetic technologies in Israel. In G. Siegal (Ed.), *Blue-white bioethics* (pp. 140–164). Jerusalem: Bialik Publishing House (in Hebrew).

Shalev, C. (2015b). Reproductive ethics. In H. Have (Ed.), *Encyclopedia of global bioethics.* Dordrecht: Springer.

Shalev, C., & Gooldin, S. (2006). The uses and misuses of in-vitro fertilization in Israel: Some sociological and ethical considerations. *Nashim—Journal of Jewish Women's Studies & Gender Issues, 12,* 151–176.
Shalev, C., & Werner-Felmayer, G. (2012). Patterns of globalized reproduction: Egg cells regulation in Israel and Austria. *Israel Journal of Health Policy Research, 1,* 15.
Shalev, C., Eyal, H., & Samama, E. (2017). 'Get us out of here!' Practices of transnational surrogacy from Israel. In M. Davies (Ed.), *Global babies: Transnational surrogacy and the new politics of reproduction* (pp. 49–64). London: Zed Books.
Shalev, C., Moreno, A., Eyal, H., Leibel, M., Schutz, R., & Eldar-Geva, T. (2016). Ethics and regulation of inter-country medically assisted reproduction. A call for action. *Israel Journal of Health Policy Research, 5,* 59.
Teman, E. (2010). *Birthing a mother: The surrogate body and the pregnant self.* Berkeley, Los Angeles, and London: University of California Press.
Werner-Felmayer, G., & Shalev, C. (2015). Human germline modification: A missing link. *American Journal of Bioethics, 15*(12), 49–51.

16

Repro-Migration: Lessons from the Early Days of Cross-Border Migration Between Israel and Romania

Michal Nahman

Introduction[1]

> I've been travelling almost all over the world, really. And I am a very open person… I am a very friendly person, I am not afraid to speak any language that I don't know… I did not want to go to Bucharest; I just did not want to go. I had to go… I wanted to be… non–seen… invisible… like I've never been there. (Dorit 2002)[2,3]

One of the things anthropology is good at is interpreting things in a multi-scale manner and in this way creating new stories, and new strategies for understanding and sometimes even assisting humanity. The anthropological study of assisted reproductive technologies (ART) has not traditionally focused on migration, but focused rather on studies of gender, kinship, infertility, culture, religion and economics (Ginsburg and Rapp 1995; Franklin 1997; Strathern 1992; Nahman 2016).

M. Nahman (✉)
Department of Health and Social Sciences, University of the West of England (UWE Bristol), Bristol, UK
e-mail: Michal.Nahman@uwe.ac.uk

S. Mitra et al. (eds.), *Cross-Cultural Comparisons on Surrogacy and Egg Donation*,
https://doi.org/10.1007/978-3-319-78670-4_16

Yet, more recently scholars have been interested in the kinds of 'journeys' made by those seeking reproduction (Speier 2016; Inhorn 2015; Nahman 2016; Kroløkke 2014). I argue that these comings and goings need to be situated in histories of colonisation, racialisation and racializing assemblages (Weheliye 2014) and in the affective economies of these practices.

In this chapter, I juxtapose cross-border reproductive practices that occurred in 2002 with practices and imaginaries of border making, for the dual purposes of comparison and of deepening our knowledge of both theoretical 'domains' which I call 'repro-migration'. The notion of going to 'a country that is not mine' for egg donation and having a baby 'that is not mine' are co-present in these stories.

Having begun with a provocation of: 'Is the egg a synecdoche of the nation', I explored (Nahman 2013) the ways in which transnational practices, narratives, policies and crises of egg donation between Israel and Romania could tell us something; both about the politics of the Jewish State—as a Zionist settler colonial project—and about contemporary egg donation across borders. It was researched in 2002 at the time of the *Al Aqsa Intifada.* I was trying to interrupt the genre of writing about ART to make them more situated in global bio politics and state politics. Another aim was to disrupt Israeli anthropology of reproduction to include Palestinians, and to challenge the notion of a 'Jewish kinship' (Kahn 2000) and to pose instead, the notion of an 'Israeli kinship' (Nahman 2013). This 'Israeli kinship' was intended to undermine the idea of a Jewish 'local biology' (particularist approach) in Israel. Instead, I have argued we might want to suggest that the State in conjunction with religious ideologies, and neoliberalism dominate reckonings of relationality for both the colonial elite and colonised Palestinians. This is a way of including notions of occupation, military violence, racializing thinking, histories of migration into ideas of kinship. Repro-migration is shorthand for this assemblage.

In order to make palpable the occupation of Palestine and the resistance of Palestinians that occurred during the researching of this book I included 'extracts' from my field notes. These were strategically placed in the text to disrupt the genre of ethnographic telling and remind you there is always a teller that someone is holding the 'reproscope' for viewing these reproductive practices (Nahman 2016).

Transnational trade in human eggs has led to many bioethical and anthropological debates around the mode of these exchanges, and their definitions. The desire for discretion in the pursuit of egg donation is well documented (Bharadwaj 2008; Inhorn 2003, p. 263). But the plea of the woman quoted above, to be invisible, was also about not having to go through the egg donation at all. The Israeli women I interviewed did not want to have to go through what they often referred to euphemistically as 'this thing'. The technological availability meant that, *ein breyra*. (Trans. 'there's no choice', which in Israel is often used with reference to military and security measures.) In Israel I got the sense that one is impelled to use technology if it is available. And because of a lack of available ova 'back home' women felt compelled to travel, or buy eggs imported transnationally. Their own personal histories of migration to Israel, and the racial politics of their settlement and 'inclusion' into the collectivity are enmeshed with the desire for a child and the contemporary narratives of border defence. Border defence and egg recipient narratives are not always or necessarily linked, this connection is an ethnographic interference. These 'synecdochal connections' enable a more critical account of the wider social context in which all of this was taking place.

Becoming Repro-Migrants: Pain, Travel, Genetics and Motherhood

> My name is Sharona, I'm 46 years old, single, I want to be a mother. It's very important to me to raise a family, a home. Very simply I feel emptiness from not having a child. I feel very bad. So I decided to bring a child to this world... So that I will feel that I am a mother. And I have a right. I want to fulfil that right... So very simply the time is running out. I don't want to miss... I've about two or three years left. I'm scared. (Sharona, August 2002)[4]

The process of becoming an ova recipient involves the decision about whether one would accept ova from another woman, and strategies of negotiating various issues: the importance of genetics, the relevance of gestation of the ova, and how aspects of a kind of biological thinking come in and out of recipients' ideas about having a child through ova donation. As this chapter demonstrates, Israeli egg donation is tied intimately to the history of Jewish migration to Israel from various parts of the world including Arab states, Europe, Asia and Africa.

An ova recipient is usually a woman who has undergone many in vitro-fertilisation (IVF) cycles that have 'failed'. If these unsuccessful attempts at conceiving a child are deemed to be due to the 'quality' of her own oocytes, a physician might suggest the woman to try ova donation. Sitting with me in her kitchen in a town south of Tel Aviv, the head of the Israeli patient advocacy group CHEN[5] an ova recipient herself, characterised three 'groups' of ova recipients. The first includes women who have reached an age at which their ova are considered to be 'unviable'. These women are generally over forty years old. The second category consists of women who, for reasons which are unexplainable (since they are young and considered otherwise to be 'healthy') have not been able to conceive with their own ova. The third category exists of women who were born without ova, or who had suffered cancer previously.[6] These are the main groupings of women in Israeli ova donation programmes.[7]

Balaban 'mapped' or 'grouped' ova recipients for me, a sociological mapping that was intended to help me 'make sense' of the population. This kind of 'category'-thinking pervades Israeli society and military (Handelman 2004) and has been a bio political tool for managing populations. I perform it here with tongue firmly in cheek to enter a discussion of the politics of race and racism in Israel: The 25 women I interviewed who painstakingly decided to become ova recipients have varied ethnic, religious, class and personal backgrounds. Most of them were born in Israel. Most of them are Jewish. 20% of the interviews were with recipients who self-identified as Palestinian, Arab or Druze. It was more difficult to distinguish among ethnic 'groups' in Israel as couples are often 'mixed' and many people consider themselves to be both, Mizrakhi[8] and Ashkenazi.[9] However, of those who did self-identify as Mizrakhi there were six interviewees (24%). The remaining fifteen interviews were conducted with people who were either of 'mixed' ethnicity, Ashkenazi, or couples in which partners self-identified as coming from different ethnic backgrounds. A large proportion of people called themselves 'mixed', by which they meant they had parents who are Ashkenazi and Mizrakhi. The majority, then, of my interviews with egg recipients come from non-hegemonic subjects, people who have historically been on the margins of Israeli society. Interviewees came from different socio-economic backgrounds as well. Most of the women lived in central Israel, cities such as Tel Aviv, Herzliya, Rishon Le'tsion, Holon, Jerusalem, Netanya, and Petah Tikvah. Palestinians, Druze and those who self-identified as Mizrakhi in this study came from more northern and southern cities, Palestinian villages inside Israel, Jewish settlements in the Occupied Palestinian Territories of the West Bank, and *moshavim* (different kinds of Jewish settlements).[10] These demographics will be important to bear in mind later on in this chapter, they hint towards the questions of mixture, and difference that materialise Israeliness (Nahman 2006). Attention to the details of ethnicity means undermining the social marginalisation and categorisation of people, while at the same time enabling a critique of how gender, race, class and borders are made in Israel (Alcalay 1993; Lavie 2011a, b; Motzafi-Haller 2001; Shohat 1989). Putting Mizrakhi women in the heart of the interpretive description here is an attempt to combat the tendency to have them/us as a category of analysis but not as 'speaking subjects' (Motzafi-Haller 2001).

In similar ways to egg donation in other social contexts such as the US and parts of Europe, becoming an ova recipient in Israel involves usually painful decisions about whether to stop trying to have a baby altogether, whether it is important to gestate the pregnancy oneself, how to tell others about this decision (König, Chapter 13). They resonate in some ways to the 'I can't have a baby stories' that are part of a Western tradition of narrating motherhood or the difficulties encountered in trying to become a mother (McNeil 1993a, b).

Dorit, a Jewish woman of Romanian background in her late-forties told me that:

> ...it's a process of coming to accept that you are in treatment. When I was doing the IUI I didn't, spiritually speaking, even want it to succeed... I was sort of divided into two people, the logical person and the emotional person. The logical one decided to go for it because I didn't have any other choice. But the emotional person did not connect.

Despite accounts of the seeming out-of-control way in which Israelis undergo IVF, the women I spoke to portrayed a painstaking decision-process, as well as much social and familial pressure. Her frustration with what she framed as her own failure to succeed in the dream of heterosexual coupledom and childbearing is indicative of some of the ways in which self-blame is integral to the experience of becoming ova recipient. The affective aspects of taking an egg from another person and accepting the idea or prospect of this was often coupled with the knowledge the actual process of getting an egg would involve transnational travel. For Rinat, a 44-year-old soldier, was the fact that she 'needed' egg donation and there were none in Israel at the time "destroyed her whole world". She commented:

> And then I saw in the paper about two months ago, an article about egg donation. And it said they might make … [private import of eggs] illegal, so I contacted Ofra. And she told me of this place, which still performs ova donations. And I knew that I had other options like going to Cyprus but I already knew that there was this simple option where I didn't have to go back and forth to a country that is not mine and so that's how I got here.

Beyond the reticence to travel abroad for egg donation, which very many of my interviewees expressed, was a concern with the 'genetics' of their potential future child. In fact, at the time, Israeli law was that the genes did not decide who is a Jew, but rather being born to a Jewish mother (Kahn 2000). The egg recipients I spoke to argued repeatedly that they were like an adoptive mother to their future child, or that they worried about the donors' genetic history affecting their future child. This is one of many instances in which the interviews in 2002 anticipated already the new egg donation law of 2010, which has a more genetic notion of citizenship, religion and relatedness (Shalev, Chapter 15; Rimon-Zarfaty, Chapter 17). They often struggled with balancing the fact that they wanted a child with the fact that that child would not be related to them. Dorit expressed these ideas as such:

> At the beginning I was against it, but then I connected with it. Once I got to the end of the IVF it was the child, any child that was important to me… I went through all the procedures but the emotional side of me

> really objected to this because I would never have a genetic child of my own. And I don't know why this bothered me so much. What's so fucking important, excuse me, about my own genes? There's no logical explanation for it. But it was crucial for me to accept that I would not have a genetic child.

Dorit so movingly questioned her own need for a genetic connection with her future child. 'Natural motherhood' seems so deeply connected here to flesh and blood in the traditional Western kinship sense (Strathern 1992). And this flesh and blood would, according to Rabbis and doctors, of course be 'hers' in that she would produce it in her own body, with her own cells if we took a strictly 'biological' view on the process. Yet, her understanding is deeply genetic. Blood and flesh are imagined as genetic in Israel.

Part of the issue is that the women I interviewed are embattled within themselves about the origins, geographical and genetic, of the eggs. The wording here is crucial, they are in a battle. So more central here than genetics perhaps, is the idea of the symbolic war involving destruction and survival deeply embedded in their 'can't have a baby' narratives.

> There have been shootings and bombings daily in the settlements and at army posts along the roads. Many people have died. The government has been striking. I've become addicted to Israeli TV and this is not good! (22 Feb 2002)

Waging a War to Make a Baby: Militarising Egg Recipients' Relationships

One day I was in the clinic in Tel Aviv observing and the recovery ward was full, approximately six beds were occupied. Sharon, the nurse who emigrated from the US, and with whom I sometimes chatted in English, ushered me into observe from behind the secretaries' desk facing the people who were waiting. Orit came by at one point and commented that I could not really see anything from my vantage point, so we found a convenient place for me to sit. She and I chatted for a bit.

Later I 'hung out' in the recovery ward. A woman named Shoshana was wheeled out of surgery having just had 'embryo transfer'. The nurse, as she wheeled the patient out of surgery, told me that Shoshana was willing to converse with me. I thus followed the nurse and patient into the cubicle where the woman was placed and conducted an interview. Shoshana was a 49-year-old woman at the time of her embryo transfer. She told me that she was a grandmother and had two daughters aged 29 and 26. She was divorced from her first husband in 1994 and has remarried since then. Her current husband also had children from a previous marriage. She explained that her treatment/pregnancy history was long. In her first round of treatments she had become pregnant three times but the pregnancies did not advance. After having had an extended break from treatments she decided to try again. In this second round of treatment cycles she had had ten IVF cycles, and eight pregnancies. None of them developed to term. One pregnancy developed to twenty-four weeks but, she said that her 'cervix opened and the babies died'. It was twins, and *hakol halackh*, 'all was lost'. She commented that the only person who knew she was having egg donation was her husband. If people asked her what she was going to the clinic for she would tell them, she was going for a new IVF treatment. She chose not to tell the rest of her family, 'because they will call [her] stupid'. Her co-workers did not know because she has quite a senior position and fears how her attempts at pregnancy through egg donation will appear. She told me that she was of 'Romanian extraction'. We talked about how poor Romania is today. She then told me that when her family arrived in Israel (in the 1940s) they lived in a *ma'abara*, a temporary camp for new immigrants.

I asked her how she came to seek treatment at this clinic. She told me she used to be a patient of one of the doctors who was accused of stealing eggs.[11] So she left that clinic and came here, '*ki lo hayta breyra. Nizgar haberez*', 'because there wasn't any other option, the faucet closed'. And so she was compelled come here because she knew it was the only place that was performing egg donation at the time. She had had three of four ova donation cycles. This was her first try at this particular clinic. Tova commented that 'it was not easy to decide to do egg donation because [her] husband found it hard to accept'. When I asked her about the 'genetics of the egg', she said that she was a lot less bothered about that, because 'the goal sanctifies the means'. And 'when you want something you overturn worlds'. But, she added that not knowing the origin of the egg was, in fact, preferable. She told me that, from a Jewish perspective, it is better to have an egg from a non-Jew or an Arab because then there's no chance of the child that is born later marrying its genetic sibling. Most of the Jewish women I had interviewed prior to this (and indeed after this as well) had stated that they could never accept an egg from a Palestinian or Arab woman (see below for a discussion of ambivalence around this issue). Here, the 'means' of acquiring a child may be slightly problematic from her perspective, but the end goal makes it an acceptable. I returned to the issue of her existing children and clarified how many she and her husband have between them. She answered, 'three, but none together'. I asked why it was necessary to have one together, and she replied, 'because it's a family, it's togetherness, it's a cell, nuclear'. We then return to the topics of treatment and the donor. She told me she did not wish to know anything about the donor except her age. She waited six months before her eggs were ready the treatment itself cost her 11,000 shekels, approximately 1500 GBP. She told me that she does not know where she got the courage to do this but that, *ani osa milkhama lehavi yeled*, 'I am waging a war to make a child'.

> A 'suicide bomber' was 'neutralised' in a café on Emek Refaim. 14 people were injured in the West Bank town of Ariel in a suicide bombing. This morning I woke to the sound of loud booms of the army bombing Bethlehem. (7 March 2002)

I wondered, after interviewing Shoshana, what it meant in this particular time of the *Al Aqsa Intifada*, to say that one is waging a 'war to make a child'. The symbolic resonance was so strong, and seemed even more than the usual militarisation of Israeli IVF and pregnancy documented by other anthropologists (Ivry 1999). To make her relationship to her husband truly 'nuclear' she must wage war against the limitations of her body, her age and availability of reproductive materials.

For some egg recipients this created a militarised bond among themselves. Dorit:

> It's not like you were with someone on a trip/holiday, it's not the same. Understand? It's not like you studied with someone and you have common memories, joint experiences. There is something not good here, very difficult, that connects between you. The pain, the knowing that you will never be able to have a child of your own. This is the glue. And I'm telling you, this happens on every trip.... one of the women got pregnant and her entire family does not know it's an egg donation! I know it. I have a secret about her. I know something about her that no one else knows...We in Israel say: 'We parachuted together on the *Mitleh*'... In the Sinai war, of 1956, the (IDF) troops landed on Sinai. "*Mitleh*" is the name of the place. ...The only time in Israeli history that regular fighters had to parachute into the battlefield. ...it's like...we have something connecting us, it's something secret something very intimate, very discreet and something very, very painful. Awfully, awfully painful. And this connection is made, temporarily.

Egg donation and IVF, have been theorised by Dorit as being about a deep pain, a kind of bonding of sisters, and a military bond. The experience of being, becoming, not wanting or wanting to be an egg recipient tells another story of the relationship of some women to the Israeli state. The kinship between recipients is such a strong bond according to Dorit. The secret that they will not have a baby that is genetically theirs is a similar fear of women globally who undergo IVF with egg donation. The difference here is that it is being figured as akin to military secret.

On Friday 12 April Tahani Assad 'Ali Fatuah, a pregnant woman, aged 35 who works as a pharmacist and lives in Nablus went into early labour. She was in her 8th month of pregnancy after undergoing fertility treatments having initially tried to conceive without assistance for four years. Nablus was under curfew on this day. Red Crescent was called several times, but was unable to reach her in time. A local doctor was called into deliver the baby at home. The baby was born and then quickly died minutes after being born, having required an incubator. The hospital was two kilometres from her home. B'tselem—the Israeli Information Centre for Human Rights in the Occupied Territories—attributed cause of death to the IDF who blocked access to medical assistance (B'tselem 2002a).

'Intimi': Intimacy, the State, and the Birth of the Israeli IVF

'Egg Donation is an intimate thing, they won't speak about it'. This is what I heard repeatedly from ordinary people and doctors in Israel when I first went to study this topic in 2002. In the following, I explore Foucaultian biopolitical aspects of the state and the reproduction of people. Foucault believed that sexuality was not a taboo for Victorians in his iconic work on the History of Sexuality. Here I argue that, likewise, even though Israelis argued that egg donation is a taboo subject it isn't a taboo really. Rather they have a role in the discursive masking of the illegitimate and colonial nature of the State. Above, Dorit made the connection herself between a secret military mission in grabbing land in the Sinai desert, with getting donated eggs in Romania. Here I explore this further.

On a Friday morning in 1984 donors and recipients sat together in a waiting room at an Israeli hospital. They had answered an advertisement for ova donation that had appeared for the first time in the Israeli national newspapers. The donors and recipients sat together in the clinic's waiting room. In this, the first case of Israeli egg donation, no protocols were in place in order to facilitate anonymity for donors from recipients.[12] Today, Israeli ova donation is practiced with many levels of secrecy and is often narrated as 'intimate'. This occurs in everyday talk, clinical practice, responses in interviews and at the national legislative level. Israeli bio politics means that ova donation is institutionalised as a secret that must be secured, and everyday discursive practices narrate it as an intimate thing that cannot be spoken about. Before my arrival in Israel I was told that there would be no women for me to interview about experiences of ova donation because of the 'crisis' of an 'egg shortage' (Nahman 2013). When I did arrive and spent several months observing IVF clinical practices and interviewing doctors and nurses in the Jerusalem fertility clinic, they told me repeatedly, 'ova donation is an *intimate thing*, they won't talk to you about it'. I found this idea quite strange at the time since it was clear to me that Israelis do talk about 'intimate' things in quite public ways. Already upon my arrival, and increasingly throughout the duration of my research I watched countless television programmes about issues of reproduction, sex and the body. Family members whispered to me about so and so who 'must have had egg donation'.[13] Despite the warnings of the intimacy of egg donation (and that therefore it was not something people spoke about) people *did* talk to me about this issue.

Given the importance of motherhood and 'the family' in Israel it is no surprise that women who have experienced difficulty in conceiving children attempt to have them with the assistance of technology.[14] It is further not surprising that there might be a certain level of stigma attached to infertility.

Yet it seems that Israeli ova donation is surrounded by discourses and practices that construct it as an intimate thing, but at the same time constantly reveal it. Can it be that the discursive construction of ova donation as 'intimate' is closely connected to the discourses that construct Israel as 'pronatalist'? Is the 'intimacy' of ova donation connected to the repetitive refrain of 'Israel has the most IVF clinics per capita in the world'? Is all this talk of '*intimiut*' a way of making the nation and the border something that is both mentionable and unmentionable?

I interviewed Rina and her husband Albert in their family home. The television was blaring loud comedy programmes and friends and teenage children walked in and out of the room freely. Rina, in her mid-forties, was one of the few pregnant ova recipients I met in Israel. It was clear Rina and Albert did not keep their IVF treatments hidden from friends and family, but the egg donation was not something they shared with everyone.

> [A]bout the egg donation, we didn't tell them these details. It doesn't matter […]. It's not a secret, but it's personal; you don't run and tell people. (Rina 2002)

For the majority of ova recipients this secrecy extends to their future child as well. For instance, Shoshana, who was still trying to become pregnant and whom I interviewed in a private clinic's conference room said: 'we will not tell our child he is from a donated egg. Why should he know? This is not interesting. This is not an adopted child! It has no relevance'.

The statement that it is 'not interesting', sitting alongside what many women told me regarding their concerns that the child would not be 'their own', can be read as a synecdochal[15] moment (Nahman 2007, 2013) between the inception of the State and their own conception. Clearly, ova donation carries some relevance; I am referring here to the matrilineality of Jewishness, and the importance of Jewishness to the concept of this State (Rimon-Zarfaty, Chapter 17, for relevant discussion in the context of Israeli legislation). In Israel, the 'Public-Professional Committee on the Subject of Ova Donation' which deliberated the matter of egg donation in the year 2000 before tabling an Egg Donation Law was intent on having a secured donor offspring registry. This would be a database that contained the identities of donor offspring for the purpose of preventing future genetic sibling marriage. A company called 'InfoFORT' was hired to present the committee with the plans for a secured database. Their presentation to the committee opened with an image of a blue 'ovum' secured by a golden lock. The annual maintenance of this registry was to cost $208,734. The emphasis of the proposed database was on security in maintaining the secrecy of donor offsprings' identity.

The very fact that such a registry was proposed is not surprising as there is a desire to maintain anonymity and information on gamete donation worldwide. Yet, I want to suggest that guarding and securing the identity of the donor offspring (as 'illegitimate' children of their parents) materially-semiotically stands in for the extensive Israeli practices of securing and guarding the imagined nation-state (other, more literal examples are the security 'fence' or rather, the wall being built to separate the West Bank from Israel).

Israeli Extraction: Racialised Borders, Difference and Mixtures

Dr. Barukh was on the phone, 'yes, it's fine to give her eggs to a Jewish woman'. My interest was piqued. Dr. Ezra knocked on the door to Barukh's office and let himself in. Ezra looked at me and said, 'so, you say you want to see egg donation? Come with me' (Barukh 2002). He took me to the ward to meet Maryam. Maryam is a Palestinian citizen of Israel who comes from a trendy village just outside Jerusalem called Abu Gosh. She is 27 years old, works at home, and her husband owns a local shop. Ezra rushed into the recovery ward and told her that she had produced 26 eggs. He explained to her that the normal number is five, and he asked if she would be willing to 'donate a few in order to help another couple become pregnant'. Maryam asked whether this would reduce her own chances of becoming pregnant. Ezra replied that this would not harm her chances of getting pregnant. She agreed. She was just beginning her eighth IVF cycle.

Yudit is a 37-year-old Jewish woman of Indian family background who has been married for twelve years. She began tests for infertility after six months of marriage. There was a lot of pressure from her mother-in-law. She works at the hospital as a computer programmer and systems analyst. Having had one daughter already through ova donation, she had been waiting for another egg donation for over a year after several failed attempts. The nurses telephoned Yudit and exclaimed, 'we have eggs for you!' They began making all the arrangements, telling her to find her husband so he could come in as soon as possible to provide the sperm. Finally, the nurse said, 'oh and one more thing: the donor is Muslim, is this ok?' There was a pause. Disappointment registered on the nurse's face. Yudit had rejected the eggs; she did not want a baby from an 'Arab woman'.

I interviewed Yudit a week later (Yudit 2002)[16]. She talked about always having wanted to experience a pregnancy. But initially she had immense difficulty dealing with the fact that, 'it's my husband's sperm, and I know it's my husband's child but it's not my child. It's like, I'll carry the child, I'm sort of like a surrogate mother, it's not exactly mine'. But she took heart that at least the child would be her husband's.

Once she resigned herself to egg donation, she registered with a few hospitals. I ask her what criteria she requested in the donor. She said that she did not know what to put down, except that she knew she wanted someone Jewish. I asked 'why'. She commented:

> It's something… I don't have an explanation, but I didn't want a non-Jew… We are Sephardim. And the Sephardim have a problem with ova donation: the donor has to be single[17] or not Jewish. I had difficulty with

> this, like, with taking a donated egg from a non-Jew because I thought to myself, in Israel, it could only be from a Muslim woman, so I was very scared. I didn't want it. This is something I can't explain... I really didn't want this. Like, I wouldn't have minded if it was a tourist who came to Israel, and she was Christian, this really didn't bother me.

When I switch off the tape, Yudit tells me that she slightly regrets her decision to reject the egg from 'the Arab woman' but is sticking firm with this decision.[18]

Ethnographies of Israeli reproduction tend to parse 'Israeliness' along lines of Jewish and non-Jewish, examining the role of religion in Israelis' thinking about making babies. Some have demonstrated an awareness of racism within Israel among Jews (notably, Teman 2010; Birnbaum-Carmeli and Carmeli 2002). The tendency is to suggest that Israeli ART are unique and particular. The question remains, when history of settler colonialism, the attempts at whitening the Mizrakhim and class inequalities are attended to can anything broader be extracted?

Ethnicity and genetic thinking are highly relevant category to Israelis, and operate in similar ways to some Euro-American ideas because Israel is an outcome of European settler colonialism.

Conclusion

Like other forms of migration, such as travelling to seek work in wealthier economic zones, repro-migration can signal that the travel has not been simply done for 'fun'. Indeed, migration due to global inequalities is an important reference point for seeing who becomes a source of reproductive stuff and who travels. In my current research on migrant egg providers in Spain, Eastern European women's desirability as white egg donors comes at the expense of their own commodification. As shown in my work on Romanian egg donors (Nahman 2013) many of the women in Romania whom I interviewed expressed the desire to migrate elsewhere because life was difficult for them in Romania. Today I am observing the outcomes of that with migrant women in Western Europe being desirable donors.

Repro-migrations attends to the manifold movements across symbolic-concrete (material-discursive) bodily, national borders that occur in egg donation without simplifying who 'these women' are. There are 'local biologies' here that are about 'migration', the border and the state-nation (Lock and Kaufert 2001). That is, how to deal with the nuance, complexity of the concepts 'Israeli women' and 'experience' of egg donation, without eliding the universal in the anthropological search for 'the particular'. More specifically, what are some of the personal and financial costs of egg donation? What is allowed to pass into and what is kept out of the imagined 'Israeli body'? These are questions that engage the process of writing and genre as much as they are about gender, race and nation.

Extracting eggs from women's bodies is a complex technical-material practice. Oocyte (egg) extractions make 'the national' and 'the political' palpable. Equally, 'the national' and 'the political' make ova palpable. Here, the Israeli notions of ethnicity, in the guise of ideas of 'mixture' and 'difference', are found in egg recipients' selection and rejection of imagined traits and donors. 'Biosocial ethnographic moments' (Nahman 2007) from the clinics and preferences for traits of donor/baby/self-make national borders palpable: They are an enactment of a repro-migration, where the donor is 'over there', the baby 'in here' and the self 'right here'. Distances, both geographic and 'racial' are telescoped.

Notes

1. This chapter is a substantially shortened version of my former chapter Nahman, M. R. (2013). *Repro-Migrants. In Extractions* (pp. 84–127). Palgrave Macmillan UK.
2. All names have been anonymised.
3. Dorit self-identified as an Ashkenazi Jew of Romanian descent. Throughout this piece I identify these self-descriptions in order to indicate the diversity of the interview population. This study covered the experiences of Palestinians as well as other Others in the Israeli context. Working against the normalising of the study of Jews separately to colonised and minoritised populations in this fieldsite was a central aim of my ethnography.

4. Sharona self-identify as a Kurdish Jew.
5. Chen L'Piryon is the name of a patient advocacy group in Israel that has worked quite closely with legislators to develop the Egg Donation Law of 2010. http://www.amotatchen.org/english/homepage/homepage.htm.
6. Another group of women who often cannot conceive, but were not discussed with me in interviews are those who were born with 'ambiguous' sexual organs. In North America at least such women have often had their ovaries, or ovo-testes removed at an early age because of cultural anxieties about sexual ambiguity. One woman whom I interviewed in Israel had had this done to her as a child. Such a 'condition', which in North America and the UK is termed 'intersex' has wide reaching implications for the individual in question. For a discussion of the medical construction of sex through the category of intersex, and for an in-depth discussion of the variety of experiences of individuals who have been diagnosed with such sexual variation see Kessler 1998.
7. This is based on an interview with Ofra Balaban, founder of CHEN.
8. Mizrakhi and Sephardi are often used interchangeably to denote Jews from North African, Arab or Muslim countries such as Iraq, Iran, Morocco, Turkey, Egypt, Libya etc. There is a politics to the use of Mizrakhi rather than Sephardi. The latter, which means 'Spanish' lends a European veneer to what are largely non-European people.
9. By mixed I mean that people often were not one or the other, they could be both. Again, I want to stress that I see identity as a process, that is always incomplete, and so whether or not people identified as Palestinian, Mizrahi, and Ashkenazi I recognise the complexity of identity categories, which can never fully describe people (Hall 1991). People are not Mizrakhim—a reified category available for objectification. Rather Mizrakhiut is a place to spring into social action and criticise Euro Israeli academic theorising (Motzafi-Haller 2004).
10. It tended to be difficult to categorise people into sociological groupings, for they themselves often resisted such categorization. I am interested in how discourses of race, geography and economy play a part in the discursive practices of ova donation. I identify people by name, and I tend to include information about them if it seems relevant to the particular passage. Nonetheless, in this study I set out deliberately to include women of different religious, ethnic and class backgrounds, which I ascertained through their own self-identification. Also, I did not screen out non-Jewish respondents, for example. In analysing the

responses to my interview questions I have tried to account for these varied subject positions but also not to over-attribute people's responses to questions about their ethnicity and class. The following discussion of the ways ova recipients construct the racialized boundaries of the nation, relates to the 'choices' they made about which kinds of donor they wanted and what kind of child they imagined themselves having. That is, the majority of my analysis is not about the identities of the recipients per se. Instead, I focus on the kinds of ideologies about appropriate/desirable national subjects that were produced and reproduced through processes of 'choosing' donors and eggs.

11. For more on the egg stealing crisis in 2000 please see https://www.theguardian.com/world/2000/may/19/suzannegoldenberg.
12. This is how it was recounted to me by the social worker who was present for the first egg donations in Israel.
13. These were references to the advanced age of women who conceived after long periods of time spent—'trying'.
14. This is of course also the case in other national contexts such as the US (Becker 2000); Britain (Franklin 1997); Egypt (Inhorn 2003).
15. Synecdoche is a kind of metaphorical relationship. In the case it is about how 'parts' relate to 'wholes'. In the book I suggest that the narratives of egg donation have a relationship to narratives of the State. They they 'reveal' social anxieties and truths about the State. This is similar to Mary Douglas's thinking about peoples' anxieties about what is inside and outside or the norms of society (Nahman 2013).
16. Yudit self-identified as an Israeli of Indian heritage.
17. To avoid the possibility of 'adultery' (Kahn 2000).
18. The significance of such ambivalence is discussed further in Nahman (2013).

Bibliography

Alcalay, A. (1993). *After Jews and Arabs: Remaking Levantine culture.* Minneapolis: University of Minnesota Press.

Becker, G. (2000). *The elusive embryo: How women and men approach new reproductive technologies*. California: University of California Press.

Bharadwaj, A. (2008). Biosociality to bio-crossings: Encounters with assisted conception and embryonic stem cells in India. In S. Gibbon & C. Novas

(Eds.), *Genetics, biosociality and the social sciences: Making biologies and identities*. London: Routledge.
Birenbaum-Carmeli, D., & Carmeli, Y. (2002). Physiognomy, familism and consumerism: Preferences in donor insemination. *Social Science and Medicine, 54,* 363–376.
B'Tselem. (2002a). *IDF causes death of infant by preventing mother's evacuation to hospital* [online]. Available at http://www.btselem.org/testimonies/20020413_death_of_tahani_fa-touhs_premature_baby. Accessed 14 Sept 2017.
B'Tselem. (2002b). *IDF rubber bullet causes head injury to seven-month pregnant woman* [online]. Available at http://www.btselem.org/testimonies/20020708_injury_of_suheir_shhada. Accessed 14 Sept 2017.
Franklin, S. (1997). *Embodied progress: A cultural account of assisted conception*. London: Routledge.
Ginsburg, F., & Rapp, R. (Eds.). (1995). *Conceiving the new world order: The global politics of reproduction*. Columbia: University of California Press.
Hall, S. (1991). Ethnicity: Identity and difference. *Radical America, 23*(4), 9–20.
Handelman, D. (2004). *Nationalism and the Israeli state: Bureaucratic logic in public events*. Oxford: Berg.
Inhorn, M. (2003). *Local babies, global science: Gender, religion, and in vitro fertilisation in Egypt*. New York: Routledge.
Inhorn, M. C. (2015). *Cosmopolitan conceptions: IVF sojourns in global Dubai*. Duke University Press.
Ivry, T. (1999, June). Reproduction as martial art. In *International Institute of Sociology, Annual Conference*. Tel Aviv, Israel.
Kahn, S. M. (2000). *Reproducing Jews: A cultural account of assisted conception in Israel*. Durham: Duke University Press.
Kessler, S. J. (1998). *Lessons from the Intersexed*. Rutgers University Press.
Kroløkke, C. (2014). Eggs and euros: A feminist perspective on reproductive travel from Denmark to Spain. *International Journal of Feminist Approaches to Bioethics, 7*(2), 144–163.
Lavie, S. (2011a). Mizrahi feminism and the question of Palestine. *Journal of Middle East Women's Studies, 7*(2), 56–88.
Lavie, S. (2011b). Staying put: Crossing the Israel–Palestine border with Gloria Anzaldúa. *Anthropology and Humanism, 36*(1), 101–121.
Lock, M., & Kaufert, P. (2001). Menopause, local biologies, and cultures of aging. *American Journal of Human Biology, 13*(4), 494–504.
Mcneil, M. (1993a). Editorial: Procreation stories. *Science as Culture, 17,* 477–482.

McNeil, M. (1993b). New reproductive technologies: Dreams and broken promises. *Science as Culture, 3,* 483–506.

Motzafi-Haller, P. (2001). Scholarship, identity, and power: Mizrahi women in Israel. *Signs, 26*(3), 697–734.

Motzafi-Haller, P. (2004). Negotiating difference in Israeli scholarship: Towards a new feminist discourse. In A. Kemp et al. (Eds.), *Israelis in conflict: Hegemonies, identities and challenges*. Brighton: Sussex Academic Press.

Nahman, M. (2006). Materializing israeliness: Difference and mixture in transnational ova donation. *Science as Culture, 15*(3), 199–213.

Nahman, M. (2007). Synecdochic ricochets: Biosocialities in a Jerusalem IVF clinic. In S. Gibbon & C. Novas (Eds.), *Biosocialities, genetics and the social sciences: Making biologies and identities*. London: Routledge.

Nahman, M. (2013). *Extractions: An ethnography of reproductive tourism.* Basingstoke: Palgrave.

Nahman, M. (2016). Reproductive tourism: Through the anthropological 'reproscope'. *Annual Review of Anthropology, 45,* 417–432.

Shohat, E. (1989). *Israeli cinema: East/west and the politics of representation.* Austin: University of Texas Press.

Speier, A. (2016). *Fertility holidays: IVF tourism and the reproduction of whiteness.* New York: New York University Press.

Strathern, M. (1992). *After nature: English kinship in the late twentieth century.* Cambridge: Cambridge University Press.

Teman, E. (2010). *Birthing a mother: The surrogate body and the pregnant self.* Berkeley: University of California Press.

Weheliye, A. G. (2014). *Habeas Viscos: Racializing assemblages, biopolitics, and black feminist theories of the human.* Durham: Duke University Press.

17

Parochial Altruism: A Religion-Sensitive Analysis of the Israeli Surrogacy and Egg Donation Legislation

Nitzan Rimon-Zarfaty

Introduction

Assisted reproductive technologies (ARTs) challenge and reconstruct social understandings of bodily boundaries, family and kinship. This is prominent in the cases of surrogacy and egg donation that represent different possibilities for 'third-party' reproduction (Birenbaum-Carmeli 2003)—and thus have been raising bioethical controversies and debates.

One major focus in the bioethical debates surrounding those practices has to do with the motivations of women donating their eggs or serving as surrogates and the interests behind those actions. Within those debates, two main motivations appear as prominent, and sometimes as contradicting: financial and altruistic (Raymond 1990). Those bring to the fore the issue of consumerism and commercialisation of reproduction. While supporters of egg donation and surrogacy have raised the legitimacy of such practices as providing the surrogate or the egg donor

N. Rimon-Zarfaty (✉)
University Medical Center Göttingen, Göttingen, Germany
e-mail: Nitzan.Rimon-Zarfaty@medizin.uni-goettingen.de

S. Mitra et al. (eds.), *Cross-Cultural Comparisons on Surrogacy and Egg Donation*,
https://doi.org/10.1007/978-3-319-78670-4_17

with economic opportunities while constructing them as autonomous rational moral agents (Rosenblum 2013; Shalev 1998), others have criticised the very idea of exchanging money for such reproductive abilities (Steinbock 2004). Such critics have been raising concerns regarding the 'industrialization of reproduction' (Drabiak et al. 2007, p. 303). Relating to egg donation for example, it has been claimed that some forms of market activities should not be performed on certain human categories, arguing for the illegitimacy of selling human body parts (Steinbock 2004). Other critics raise concerns from coercion and exploitation of surrogate mothers and egg donors for the benefit of the wealth (Cattapan 2014; Rosenblum 2013; Steinbock 2004; Raymond 1990), as well as the exploitation of poor women from poor countries in the context of transnational uses of those practices (Shalev and Werner-Felmayer 2012; Nahman 2011). If so, this debate emphasises the making of reproductive abilities into a commodity (ibid.), or even the making of children into commodities (Raymond 1990; Shalev, Chapter 15).

Within this context, altruistic motivations are constructed as legitimate (Drabiak et al. 2007). In some studies conducted in the western world (the US and Europe), it was found that even when receiving financial compensations, most surrogates or donors reject money as their main motivation, while referring to altruistic motivations—perceived as more socially excepted (Almeling 2006; Drabiak et al. 2007). Such altruistic motivations have also been referred to as reflecting gender social expectations from women (ibid.), to meet and fulfil the needs of others (Raymond 1990).

If so, generally speaking, many of the bioethical discussions relate and generally contrast commercialism and altruism—condemning commercial contracts while presenting more supportive positions regarding altruistic agreements.

Those debates are also manifested in relevant legislations around the world. Previous classifications of egg donation and surrogacy laws roughly distinguished between countries that prohibit any use of such practices (e.g. Germany and Austria); countries that allow only altruistic forms of surrogacy and egg donation (e.g. the UK and Canada); and countries that allow for both altruistic and gainful uses of those practices (e.g. California) (Cattapan, 2014; Saxena et al. 2012; Shalev and Werner-Felmayer 2012; Almeling 2006).

However, a study of the distinction between commercialism and altruism—examining, blurring and understanding its demarcation lines, using a specific culturally sensitive analysis—can contribute to this debate. In this chapter, I wish to examine this classification by analysing the cases of the Israeli surrogacy and egg donation laws. For this purpose, I will focus on a dominant normative traditions in the Israeli bioethics—the Jewish orthodoxy (Boas et al. 2018; Shapira 2013), informed by the Jewish law—the *Halacha*. Based on literature review, I will present a detailed analysis of the Jewish law's influences on the Israeli laws of surrogacy and egg donation. My analysis will first focus on restrictions informed by the Jewish law and will further present a detailed overview of the Jewish religious perspective on family integrity and formation. I will then move to analysing the meanings of those influences or restrictions as representing mechanisms of social and political power relations, which brings to fore social categories of (levels of) religiosity as well as religious affiliation and nationality. If so, this chapter relates to the concept of religion in the wider sense to include not only direct religion-related influences but also religion-based social groups and power relations. Finally, I will demonstrate how this analysis (1) challenges the classification of the legal frameworks as commercial or altruistic driven and (2) holds a contribution to the more general bioethical discourse on surrogacy and egg donation by suggesting the relevance of the concept of 'parochialism' also in framing altruistic motivations.

The Jewish Religion and the Israeli Bioethical Debate: Sociopolitical Reflections

In order to understand the claim that the Jewish tradition holds a decisive role in shaping Israeli legislation in bioethical issues (e.g. surrogacy and egg donation), as well as the claim that Jewish orthodoxy holds a prominent political role in those contexts and beyond, one must gain a general insight into the Israeli bioethical debate and its nature.

Israeli society is very pluralistic. It includes different publics and social groups with very different cultural and religious backgrounds and therefore with different normative traditions (Shapira 2013). Furthermore, the very definition of Israel as a Jewish-democratic state demonstrates the extent to which it is continually seeking to negotiate or fuse Jewish orthodox religious values[1] with liberal ones (Shapira 2013; Gross and Ravitsky 2003).On the one hand, Israel represents a liberal approach in many issues, such as formal recognition of single-parents' families. On the other hand, the *halachic* tradition receives extensive political and legal power (Sperling 2010). The situation is far more complicated and conflictual since Israeli society involves not only Jewish but also people from other religions.

This negotiation between different sets of values that have essentially a lot to do with social categories of religion, religiosity and secularism is demonstrated in the Israeli bioethical debate (ibid.). Hence, this debate has been identified as representing a constant tension between liberal individualist occidental values or stances and traditional Jewish religious (Shapira 2013) or else conservative communitarian and national ones (Boas et al. 2018; Gross and Ravitsky 2003).

If so, Jewish orthodoxy, its values and the Jewish law informing them are of great relevance when it comes to analysing Israeli legislation in bioethical issues. Legislation represents an expression of the decisions made within the bioethical debate with its negotiation between different ethical perspectives (Boas et al. 2018). An analysis of the surrogacy and egg donation legislation can thus be productive in gaining insights into the influence of the Jewish tradition and the political power relations involved.

The Israeli Regulation of ART: General Reflections

The Israeli regulations of reproductive technologies have been identified as rather permissive, in ways which enable the accessibility as well as the funding of many of them (including in vitro fertilisation (IVF)) to all women in all statuses (e.g. from all familial statuses, Jewish, non-Jewish, etc.) (Shalev, Chapter 15; Rosenblum 2013). This regulatory framework is also accompanied by a high demand from the Israeli women themselves. As a result, the usage of medical technologies at the beginning of life and genomics in Israel (including selective reproductive technologies) is exceptionally high (Hashiloni-Dolev 2006), while Israel holds the highest rate of fertility clinics, fertility treatments and IVF cycles per capita worldwide (Shalev, Chapter 15; Kahn 2000).

Various scholars have presented critical reflections on reproductive technologies and their usage in Israel. Some referred to the Israeli public discourse and policy as promoting the development and the usage of biotechnology and science, constructed as 'natural' and as necessary means for the survival and the continuance of the Jewish existence in the Middle East (Prainsack and Firestine 2006). Others drew attention to the relatively under discussed risks involved in the usage of such technologies (Gooldin 2007). The Israeli public was thus described as trustful of science, perceived as 'progress' (Hashiloni-Dolev 2006).

The extensive use of reproductive technologies in Israel and the enabling regulation should also be regarded in the context of a prominent cultural script—the Israeli 'Pro-Natalism' (Teman 2016; Sperling 2010; Gross and Ravitsky 2003; Kanaaneh 2002; Kahn 2000). The Israeli Government policy, including reproductive technologies' related regulation, has been identified as aimed at encouraging reproduction (Teman 2016; Sperling 2010). According to Gooldin (2007), the Israeli regulation promotes a liberal discourse focusing on the parent's basic rights to emotional welfare and 'happiness'. Within this discourse, parenthood is constructed as an obvious and unconditional human right.

Procreation also holds a Jewish religious importance. It is considered a fulfilment of the basic commandment *Pru u'Rvu* (be fruitful and multiply), while barrenness represents a biblical motive of suffering equivalent to death (Shalev, Chapter 15; Rosenblum 2013; Schenker 2008). Similarly, in the Israeli–Jewish society, infertility is perceived as form of disability and as a source of suffering (Kahn 2000), while singlehood and childlessness are highly stigmatised, especially within more religious or conservative social groups (Rosenblum 2013).

Fertility and reproduction are also perceived in Israel in terms of national 'demographic' goals. Reflecting upon the Jewish history, the mass extermination in the holocaust and the ongoing Israeli–Palestinian conflict, this discourse reflects a perception of the strength and survival of the Jewish collective as depending upon its quantitative or numerical growth (Shalev, Chapter 15; Gooldin 2007; Kanaaneh 2002).Therefore, the Israeli pro-natalism has been identified as selective and as holding a Jewish–Zionist nationalist agenda (Teman 2016).

Against this background, this chapter will present and discuss the relatively limiting Israeli legislation of egg donation and surrogacy (Birenbaum-Carmeli 2016). My analysis suggests that due to the nature of the Israeli bioethical debate, Jewish religion with its related traditional norms plays an important role in the legislation.

The Jewish *halachic* Perspective on Reproductive Technologies

Judaism includes different voices and streams as well as different stances and *halachic* interpretations when it comes to bioethical issue. Since within the Halacha, Talmudic precedents are used to solve contemporary problems, it is very often that different rabbinical authorities (*Poskim*) reach very different rulings (Marx-Stolting 2009). Considering this complexity, previous analysis of the 'Jewish' perspective on reproductive technologies has shown that it is nevertheless possible to detect certain joint tendencies that are common to the different streams and interpretations. Those suggest that Judaism (in comparison, for example, with the Catholic doctrine) can be identified as generally supportive of the medical profession as well as the usage and development of medical technologies (Marx-Stolting 2009; Barilan 2007; Hashiloni-Dolev 2006).[2]

Focusing on reproductive technologies, aside from the abovementioned strong *halachic* emphasis on procreation through the explicit commandment to be fruitful and multiply, another important issue for understanding the *halachic* somewhat permissive tendencies is its interpretation of the moral status of the (pre)embryo. Unlike the Catholic doctrine, which gives full moral status to the embryo from the moment of fertilisation, the Jewish tradition grants the foetal entity with its moral status gradually along the stages of pregnancy.[3] Within this perception, the in vitro embryo is not considered a life. Therefore, the extermination of pre-embryos is not considered as equivalent to abortion and practices like stem cell research are *halachically* allowed (Gross and Ravitsky 2003), as the status of the pre-embryo *halachically* enables the use of medical manipulations for therapeutic reasons on such entities (ibid.).

For all those reasons, the Jewish law was referred to as a very dominant cultural script and therefore as one central explanation of the 'Israeli' acceptance of reproductive technologies. However, there are two claims to be made in this context: first is that the Jewish religious or *halachic* perspective is not merely an implicit cultural script affecting Israeli social perceptions regarding reproductive technologies or explaining such permissive tendencies. Jewish orthodoxy is also an active political actor holding a political power.

Second, many of the rabbinical authorities will support the usage of reproductive technologies, as long as it meets the Jewish law's restrictions and their traditional world views. Those include, for example, the *halachic* prohibition on incest as well as the maintenance of the principle of 'family integrity' (Teman 2016, p. 166; Kahn 2000). Indeed, Judaism allows the practice of all ART when the oocyte belongs to the wife and spermatozoon originates from the husband (Schenker 2008). Egg donation and surrogacy, however, challenge the Jewish law and traditional values. In what follows, I will first present the Israeli surrogacy and egg donation laws and will then demonstrate the extent to which religious conditions or specific restrictions affect the relevant Israeli legislation.

The Israeli Surrogacy and Egg Donation Laws: Jewish *halachic* Influences[4]

The Israeli Surrogacy Law

Israel was the first country to have regulated state-monitored gestational surrogacy arrangements under a state law devoted specifically to surrogacy and entitled the 'Embryo Carrying Agreements Law' (1996).Within this law, the surrogacy agreement is strictly monitored by a special state approval committee on a case-by-case basis (Sect. 3). The law set several restrictions for such an agreement (partial list): all parties should be adults and legal residents of the Israeli state (Sect. 2); surrogacy is available only to married or otherwise legally paired heterosexual couples (Sect. 4) and is conditioned to medically proven infertility or inability to carry a pregnancy to term. The surrogate must be unmarried (meaning single, divorced or widowed, unless the committee approves otherwise if convinced that the couple was unable to find an unmarried surrogate) and cannot be related to the prospective parents. The sperm used in the medical procedure must belong to the prospective father, and the egg must not belong to the surrogate; the egg can either belong to the prospective mother or to an egg donor. In addition, the surrogate mother must share the same religion as the prospective parents, unless all parties are non-Jewish, in which the agreement may be approved based on the opinion of relevant religion representation (Sect. 2). The law forbids the payment of a salary to the surrogate; however, a so-called financial compensation is allowed (Sect. 6). The financial arrangement is also subjected to the approval of the Ministry of Health (Rosenblum 2013) (Teman, Chapter 4; Shalev, Chapter 15).

In 2011, the 'Embryo Carrying Agreement Committee' (2011) formulated additional conditions/restrictions, among them: the surrogate must be between 22 and 38 years of age; the age of the intended parents will also be considered if it is more than the 'natural' parenthood age; the surrogate mother must not go through more than two embryo-carrying processes—even if not resulted in births; the surrogate must have at least one child of her own to which she gave birth, but must not have gone through more than three births.

The Israeli Egg Donation Law

Until the legislation of the Israeli egg donation law (2010), egg donation was only permitted in the case of IVF patients undergoing fertility treatment for their own benefit (Birenbaum-Carmeli 2016; Shalev and Werner-Felmayer 2012; Gruenbaum et al. 2011). This situation, combined with the famous 'egg affaires' scandal in 2000, led to a severe shortage of eggs. As a solution, apart from cross-borders egg donations (Birenbaum-Carmeli 2016; Nahman 2011; Nahman, Chapter 16, for an ethnography and discussion of cross-borders egg donation and repro-migration), and after a whole decade, the new law allowed for healthy women in Israel, to donate their eggs (Shalev and Werner-Felmayer 2012; Rosenblum 2013).

The law determines (partial list): that retrieval of eggs from a volunteer donor must be subjected to a special local hospital committee's approval. The recipient must be between 18 and 54 years of age and medically unable to conceive with her own eggs (Sect. 11). The donor must be between 21 and 35 years of age and may undergo up to 3 retrieval cycles with no more than two recipients for each cycle (Sect. 12). All parties should be legal residents of the Jewish state (Sect. 11, 12). The donor will be compensated by the state, and the amount is to be determined by the Ministry of Health (Sect. 43). As in the surrogacy law, the egg donor cannot be either married or related to the recipient or the genetic father. In addition, the donor must belong to the same religion as the recipient (Sect. 13). In case the donor is either married or does not belong to the same religion, the doctor must inform the recipient and her partner and receive their consent in writing for the procedure. In this case, the procedure is subjected to the approval of a special national 'exceptional committee' (Sect. 20). The law also instructs the establishment of a database. The physician in charge is instructed to deliver the database, the personal details of the donor, the recipient and the genetic father in order to clarify lack of familial relatedness (Sect. 31, 32). The personal detail of all parties including the resulting child will be registered in the newborn registration (Sect. 34) in order (among other reasons) to prevent sibling marriage (Sect. 36). The donor's identity will remain confidential and is not to be revealed neither to the recipient nor to the future child, and vice versa, the donor cannot receive the identity of the recipient or the future child (Sect. 36, 39).

Jewish *halachic* Influences

Despite the perceived differences both in the content of the two laws and in their levels of restrictiveness (Teman 2010), a close reading of the laws reveals a few clear similarities in some of the restrictions they present (Rosenblum 2013): the demand that both the surrogate and the egg donor will be unmarried; the demand that the surrogate and the egg donor will not be related to the prospective parents/the egg recipient and the genetic father; and the demand that surrogate and egg donor will belong to the same religion as the prospective parent or the recipient. In what follows I will describe, based on literature review, how those restrictions can be analysed as reflecting an adherence to the Jewish law in three main issues: the religious concerns regarding illegitimacy, incest and religious identity, respectively.

Religious Concerns Regarding Illegitimacy (*Mamzerut*/Bastardy)

Both the surrogacy law and the egg donation law instruct that the surrogate and/or egg donor should ideally be unmarried. This specification can be attributed to rabbinic concerns regarding adultery and the status of a child born to a married surrogate or egg donor (Gruenbaum et al. 2011; Schenker 2008; Kahn 2000).

The issue at stake is of *halachic* illegitimacy (*Mamzerut*/Bastardy: a status given to a child conceived as a result of illicit sexual relations—meaning an adultery between a married woman and a man who is not her husband). A so-called *Mamzer* is subjected to sever social stigmatisation, as well as practical difficulties, such as capacity of marriage (Kahn 2000; Shalev 1998). Even though IVF, egg donation and gestational surrogacy are by all means artificial procedures, which do not involve sexual relation (and in the case of gestational surrogacy does not even involve the surrogate's own egg), some rabbis might indeed rule that a married surrogate or egg donor is illegitimate (Shalev, Chapter 15). Therefore, the legislator chose to adopt a restrictive approach by dictating that the surrogate should be unmarried (ibid.; Gruenbaum et al. 2011) and by that avoid any potential religious objections to the law.

Religious Concerns Regarding Incest-Like Relationships

The second issue is the *halachic* concerns about an incest-like relationship—i.e. marriage or sexual relation between relatives (Shalev 1998). This concern has led to the restriction determining that both the surrogate (ibid.) and the egg donor cannot be related to the prospective parents/the egg recipient and the biological father.

The egg donation law also reflects a concern regarding genetic sibling marriage as it regulates the newborn registry aimed at preventing sibling marriage. Furthermore, according to some rabbinical authorities, a marriage of two children, born to the same mother (i.e. from the same womb, even if not genetically related), is considered an incest and should thus be forbidden (Teman 2016). This situation may happen if a surrogate's genetic child will want to marry the couple's genetic child—gestated by the same surrogate. For that reason, all children born by surrogates are listed in the rabbinate in order to prevent the marriage of two children carried and born from the same womb (ibid.).

Religious Concerns Regarding Religious Identity

The laws also demand that the surrogate mother and/or egg donor and the prospective parents/the egg recipient will share the same religion. Kahn (2000) defined this restriction as a concession to rabbinic concerns about the (religious) status of the resulting child. Since according to the Jewish law, the religion of the child is determined according to the religion of the mother, certain reproductive technologies such as surrogacy and egg donation have raised a profound rabbinic debate about the significance of genetic, gestation or parturition motherhood (Birenbaum-Carmeli 2016; Schenker 2008; Shalev 1998). Therefore, the religion of the surrogate mother and/or the egg donor is of *halachic* importance (Rosenblum 2013; Schenker 2008; Kahn 2000). The restriction dictating both parties must belong to the same religion can thus be analysed as a mean to ensure a clearance regarding the child's religion (Schenker 2008) while avoiding any potential rabbinic problem with determining the religion of the future child (Gruenbaum et al. 2011).

Jewish Religious Perspective on Family Integrity and Formation

In what follows I wish to suggest that the above-mentioned *halachic* concerns bring to the fore the religious importance of the nuclear family, its integrity (in the sense of legitimacy and relatedness), stability and unity (Kahn 2000). I therefore wish to move beyond the specific *halachic* concerns to analyse the Jewish religious view of the 'family' and its connection to the legislation.

The claim that the legislation restrictions result from the Jewish law has been contested. Teman (2010, 2016) related in this context to a case in which a married surrogate was *halachically* approved by a leading rabbinical authority (Israel's Chief Rabbi Shlomo Amar) in 2006, but nevertheless rejected by the state committee (who only in recent years started to allow married women to become surrogates). Nevertheless, I will make a claim for the relevance of the Jewish orthodox stance regarding the concept of 'family' in understanding the legislation. I wish to suggest that the legislation represents a case in which the legislator and the enforcement of the laws take a stricter form than that of some religious authorities to preserve a state of consensus. The laws' restrictions were meant to avoid religious oppositions by preventing potential rabbinic problems related to the status of the future children and family integrity (including incest and *Mamzerut*) and thus answer the widest range of rabbinical views (Teman 2016; Kahn 2000; Shalev 1998).

The idea of family integrity is also connected to a traditional or conservative family model. The legal restriction that determines surrogacy is only allowed for heterosexual, legally paired couples is very prominent in this respect, while emphasising the importance of heteronormativity as the practical meaning of this restriction is that surrogacy is prohibited in cases of single-parents or same-sex couples (Teman 2010, 2016).

The state's preference of traditional heteronormative family formations over other alternative family formations can be understood as resulting from two main motivations: national and religious. Relating to the national motivation, Teman (2010, 2016) claimed that the existing legal restriction on surrogacy represents gatekeeping mechanisms of the nation. Following this line of argumentation, the nuclear family may be analysed as constituting the cornerstone of the nation. In the light of the challenges posed by surrogacy on concepts of family and motherhood, the body politics' protection of the nuclear family boundaries becomes a measure for protecting the national boundaries (ibid.).

However, one may also argue that such conservative tendencies are also rooted in the religious law and tradition. In trying to explain the rejection of alternative family models using the Jewish law or tradition, it seems that in regard to same-sex couples, the picture is quite clear: the Torah explicitly forbids homosexual relations. Thus, for the orthodox Jews, homosexuality is considered a sin that leads to a severe punishment (Schenker 2008). Jewish orthodoxy also disapproves lesbian sexual relations (Schenker 2008; Esther 2005) and does not acknowledge same-sex marriage.

Traditional Judaism perceives heterosexual marriage as a commandment and as the only route to enable religious and personal fulfilment (Schenker 2008; Kahn 1989). Jewish life exists in three main spheres: the synagogue, the study hall and the home. Home and family are therefore the core of religious life, within which the emotional and religious centre is the union between a man and a woman—a husband and a wife. Heteronormative relationship thus becomes an aspect of religious experience. According to the Torah, understanding the husband–wife relationship and love is as close as understanding what love to God can be (Esther 2005). Marriage is also considered the proper framework for bringing children and following the commandment to be fruitful and multiply (Schenker 2008; Kahn 2000).

The situation in case of single-parent's families is more ambivalent. In spite of the religious importance of marriage, it has been claimed that the Jewish law does not present any clear restriction on single motherhood (Teman 2016, 2010). However, a close look at the rabbinic debate on the usage of reproductive technologies to enable single motherhood reveals a complicated picture. Since there is no clear *halachic* problem with the status of a child born to an unmarried woman, one might expect rabbis will approve the use of ARTs (e.g. artificial insemination or IVF) to an unmarried woman seeking their approval. However, there is no coherent orthodox rabbinic response to this question (Kahn 2000, pp. 57–58). Israeli media's coverage of the issue (Hirshfeld 2011) also presents a rabbinical controversy regarding single women's use of donor insemination for example. On the one hand, some rabbis allow such a practice, while still clarifying that heteronormative families are considered a preferable framework for procreations. On the other hand, others forbid it, claiming that from a religious point of view reproductive technologies are aimed at helping to create only heteronormative nuclear families. It generally seems that rabbinical authorities and *halachic* rulers prefer not to publicly discuss the issue and avoid any general rulings. Importantly, in this context, the usage of sperm donation also poses the *halachic* challenges of 'sperm waste' and incest, in case the donor is Jewish (Birenbaum-Carmeli 2016; Schenker 2008).

It is important to note, however, that the restrictions presented in the surrogacy law, limiting the use of the practice only to married or otherwise legally paired heterosexual infertile couples, are lacking in the Israeli regulation of egg donation as well as other reproductive technologies such as IVF, which are principally open for single and lesbian women (Rosenblum 2013). Teman (2010) explains this difference between the legislations, by pointing out that in the case of the surrogacy law, by appointing a state committee, the state becomes directly and explicitly accountable for each and every surrogacy agreement. Thus, every approved agreement may be viewed as a public statement about who should be encouraged to reproduce and who should not. I wish to suggest that for that very reason, in which the practice and its use also become transparent, following religion-based considerations becomes politically crucial.

Yet, the lack of similar prohibitions in the egg donation law seems from a religious sensitive perspective inappropriate and therefore requires an explanation. While the absence of the restriction on single mothers can be perceived as more acceptable—as this is a religiously ambivalent issue, the lack of prohibition when it comes to same-sex couples is more surprising. However, even though no clear prohibition exists, one may notice that in practice, the current legal situation still limits same-sex couples' potential use of the procedure. Since same-sex couples cannot use surrogacy, homosexual couples cannot in any case make use of egg donation. Furthermore, though the law itself does not set any limits on providing the practice to lesbian couples, a new suggestion for amendment of the law aimed at enabling egg donation between same-sex female partners (Egg Donation Bill 2016) was rejected in August 2016 (Prime Minister's Office 2016).

Religiosity, Religious Affiliation and Nationalism: Social and Political Reflections

Moving beyond the Jewish law's influences on the legislation and the specific concerns they represent, the above-mentioned influences also hold specific social and political meanings.

In her analysis of the Israeli surrogacy legislation, Elly Teman (2016) related to the Israeli situation of state-controlled surrogacy as a mechanism of symbolic control, prioritising the Jewish–Israeli, heteronormative, two parents' family. Taking a religion–social sensitive perspective, this observation can be perceived as bringing to the fore two types of power relations related to two main concepts: 'religiosity' (the dynamic between different categories of religiousness and secularism) and 'religious Affiliation and Nationalism' (e.g. the dynamics between Jewish and non-Jewish Israelis) (Nahman, Chapter 16, for a related discussion on *ethnic* social categories in the context of colonialism and repro-migration).

(Levels of) Religiosity

One very central observation apparent from the analysis of the laws is the clear resemblance in the restriction they pose, restrictions that are informed by the Jewish law and its main or obvious relevant concerns. Those legal restrictions serve as clear testimony to the political power held by the Jewish orthodoxy and its direct influence on the ways both egg donation and surrogacy are regulated and practiced in Israel.

The restrictions presented in the surrogacy law, enabling the use of the practice only in the case of two-parent heteronormative families, can also be explained as resulting at least to some extent from *halachic* concerns and therefore highlight political power relations.

In recent years, there has been a salient media and public debate in Israel, concerning single women and same-sex couples' access to surrogacy. In 2012, the Ministry of Health has formed a governmental committee—the Mor-Yosef committee—which recommended allowing surrogacy for single women and same-sex couples. However, the implementation of those recommendations is dependent upon the Israeli Knesset (Rosenblum 2013), which to this date has not corrected the existing law. The former Minister of Health, Yael German, who is known for her liberal stances, also initiated in 2015 a relevant amendment to the surrogacy law to enable surrogacy for same-sex and single-parents' families as well as to allow men to receive egg donation in Israel and abroad for surrogacy processes. However, unlike the existing surrogacy law, this amendment did not receive a wide acceptance of rabbis. It was in fact contested and eventually rejected in March 2016 (The Knesset 2016; Teman 2016).

Reflecting upon the relevant Jewish–Israeli sociopolitical power relations, it is the more secular–liberal policy makers that are attempting to enable single women and same-sex couples the accessibility to different reproductive medicine practices. Indeed, Israeli gay liberation movement has been identified with and gained legitimacy within the secular–liberal (often 'left') sector (Ariel 2007).

This analysis demonstrates in a quite direct manner, the extent to which bioethics is inherently political (Boas et al. 2018). The discussions over surrogacy and egg donation confront liberal and religious worldviews which in the Israeli context, at least to some extent, take the form of secular and religious political parties (ibid.).

Within this context, the legislative effort of secular–liberal policy makers can be interpreted as holding a symbolic meaning of legitimising what is being constructed as more non-traditional-'secular' family models. This is of course a problematic definitions as there are increasingly more single-parents and same-sex couples within the Israeli religious sectors (Ariel 2007) representing more accepting religious interpretations (Kahn 1989). However, I wish to suggest that within this specific sociocultural and political context, the legal framework of surrogacy (and perhaps to some extent of egg donation) may be analysed as prioritising the heteronormative 'traditional'—religiously acceptable family model over other 'secular' family forms, in ways which bring to the fore levels of religiosity.

However, discussing sociopolitical religiosity-related power relations entails only part of the picture. This discussion should also take into account existing regional power relation in the context of religious affiliation and nationalism.

Religious Affiliation and Nationalism

The restriction presented in both laws, and dictating that the surrogate/egg donor and the prospective parents/egg recipient must belong to the same religion, highlights the concept of religious affiliation.

Within the Israeli context, this concept is also strongly connected to nationalism in ways which highlight the Israeli regional politics.

Zionism has made a connection between collective Jewish genealogies and national rights. Within this context, the Jewish religious identity is constructed as a national identity that is also based, according to the Zionist discourse, on a shared biological origin (Birenbaum-Carmeli 2016). From this point of view, surrogacy and egg donation, which enables novel forms of kinning, challenge the bio-genetic Jewish identity of not only a particular future child but also of the Jewish collectivity as primordial. The attempt to maintain and guarantee biological Judaism can thus be analysed as indirectly promoting national agendas (ibid. 2016, p. 22).

Taking this argument a step further, Teman (2016, 2010), in her analysis of the Israeli surrogacy law, claimed that since Islam prohibits surrogacy, finding a Muslim surrogate or a Muslim Kadi to authorise the process becomes very difficult. Therefore, the law is formulated in a way that enables the use of the practice to help Jewish couples. Similarly, the Islam generally prohibits the use of donated eggs, while raising concerns related to infidelity (Gruenbaum et al. 2011). This situation can thus be regarded as exemplifying an important aspect of the selectivity of the Israeli pro-natalism (Teman, 2016, 2010). If so, the surrogacy and egg donation legislation involves complex mechanisms of inclusion and exclusion.

Between Commercialism and Altruism: The Relevance of Parochialism

Going back to the bioethical debate surrounding egg donation and surrogacy, which focuses on the concepts of commercialism and altruism; In what follows, I will suggest a possible contribution, resulting from the religion-sensitive perspective on the Israeli legislation.

A close read of the Israeli surrogacy and egg donation laws reveals that concerns regarding commercialism are indeed prevalent in the legislation which thus uses the term 'compensation'. This terminology suggests that payment to egg donors or surrogates is acceptable.

However, donors, for example, are not paid for their eggs as such, but are rather reimbursed for direct costs and compensated for the burdens involved in eggs retrieval (Steinbock 2004).

On the other hand, however, in the case of surrogacy, the committee does not approve purely altruistic agreements and requires the exchange of funds. Furthermore, the legislation (due to *halachic* concerns) also instructs that both the egg donors and the surrogates should not be related to the prospective parents or the egg recipient (and the genetic father). Consequently, some forms of altruistic surrogacy and egg donation become impossible in Israel. By doing so, the possibility that a parent's sister, for example, or another family member will become their surrogate or egg donor, which indeed exists in other countries, becomes impossible (Rosenblum 2013). It is thus very difficult to define the Israeli legislation within the framework of altruism and commercialism (Shalev, Chapter 15).

This difficulty becomes even clearer when taking into account the relevant political power relations with their related social categories of religious affiliation, nationalism and (levels of) religiosity. Both legislations restrict the usage of those practices to situations in which both parties involved belong to the same religion. This restriction de facto creates a type of biomedical segregation based on religious affiliation. At the same time, the existing legal framework with its concerns regarding family integrity and formation also prioritises certain family models over others. It legitimises the heteronormative 'traditional', religiously acceptable family model over other alternative 'secular' family forms, in ways which highlight religiosity-based sociopolitical power relations while raising claims for other forms of biomedical segregation.

Following this line of argumentation, while the bioethical debate identifies egg donation and surrogacy motivations as well as legal classifications in terms of altruism and/or commercialism, the Israeli legislation challenges the relevance of them both. On the one hand, this legal framework creates a limitation on the types of financial agreements that it enables—within this legal framework: No financial agreement can be purely commercial as the legal situation brings religious and national sentiments as well as religiosity-related legitimations into it. On the other hand, those limitations challenge altruism in its ideal type by restricting it to certain religion-related social categories. This situation brings to the fore other relevant, yet neglected concepts to the bioethical debate on surrogacy and egg donation—those include restricted social reciprocity and national–religious solidarity that are socially constructed by the Israeli legal framework.

In other words, aside from the commercial and/or altruistic motivations to serve as a surrogate or donate eggs, the Israeli legislation reflects different type of motivations related to specific social reciprocity and solidarity. Such motivations are based on predetermined religion-related collectivities.

In particular, this new contribution calls for a more careful perspective on altruism, while taking into account the concept of 'parochialism'. Altruism is highly related to social norms as part of cooperation and social order. However, norm compliance and enforcement (as in the case of legislation) are often being made in the context of inter-social conflicts (as in the Israeli situation), in ways which might lead to parochialism—a preference of one's own social, ethnic, racial, national, religious or language group (Bernhard et al. 2006). Patterns of stratified altruistic concerns are thus strongly connected to exclusivist social mechanisms (Persson and Savulescu 2017).

The issue of 'parochial altruism' has been extensively studied in fields like psychology, neuroscience, human biology and economy, in relation, for example, to evolutionary theories (Bernhard et al. 2006). It was also discussed in bioethical literature as arguably leading to a need to pursue research on 'moral enhancement' (Persson and Savulescu 2017). This chapter demonstrates its relevance for understanding specific surrogacy and egg donation legislation.

I do not mean to imply that on the individual level, Israeli Jewish women who donate their eggs or serve as surrogates are not motivated by non-parochial altruistic motivations and that their action should be less respected or encouraged. Nevertheless, I wish to contribute to the bioethical discussion surrounding the concept of altruism.

The analysis I presented in this chapter, an analysis of the Israeli surrogacy and egg donation legislation from a religion-sensitive perspective grounds the abstract bioethical debate concerning commercialism and altruism in a concrete context. By doing so, sociopolitical dimensions are brought into the discussion. Consequently, the analysis highlights the ways in which the enforcement of altruistic norms or behaviours may take a parochial nature (Bernhard et al. 2006) and thus in turn may shape human altruism in the wider sense.

Notes

1. Though there are different streams in Judaism (the main three are the orthodox, the conservative and the reform), in Israel, it is generally the orthodox stream and rabbis that are invested with state power and jurisdiction (Kahn 2000).
2. To support this claim, Marx-Stolting highlights, for example, the Jewish perception of man as God's partner and as a creator whose role is to keep on creating, improving and developing the world. This leads in turn to a positive view of medicine, technology and science as means which enable such a sacred mission. Another example is the *Pikuach Nefesh* commandment (the obligation to save a life in jeopardy) which leads in turn to a positive view of medicine (for further discussion: Marx-Stolting 2009).
3. Jewish tradition grants the embryo/foetus with its moral status gradually throughout the evolvement of pregnancy: starting from the 40th day after fertilisation (the moment of ensoulment) and until the moment of birth (when the greater part of the body is born) in which he gains full personhood (Barilan 2007; Steinberg 2005). According to the Jewish Halacha, an embryo implemented in the uterus during the first 40 days after fertilisation is considered 'nothing but water' (Steinberg 2005).
4. For part of the laws' analysis, see also Rosenblum (2013). For the analysis of the surrogacy law: Teman (2016, 2010).

Bibliography

Almeling, R. (2006). 'Why do you want to be a donor?': Gender and the production of altruism in egg and sperm donation. *New Genetics and Society, 25*(2), 143–157.

Ariel, Y. (2007). Gay, orthodox, and trembling: The rise of Jewish orthodox gay consciousness, 1970s–2000s. *Journal of Homosexuality, 52*(3–4), 91–109.

Barilan, M. Y. (2007). 'Tza'ar Gufa Kodem' (her pain prevails): Selective abortions in the Halacha and in the Israeli law. In J. Davies & A. Sahar (Eds.), *The right to life with no malformation: From embryo formation to human being: Scientific, ethical and legal aspects* (pp. 81–135). Tel Aviv: Dionon, Probook Publication (in Hebrew).

Bernhard, H., Fischbacher, U., & Fehr, E. (2006). Parochial altruism in humans. *Nature, 442*(7105), 912–915.

Birenbaum-Carmeli, D. (2003). Contextualizing a medical breakthrough: An overview of the case of IVF. *Health Care for Women International, 24*(7), 591–607.

Birenbaum-Carmeli, D. (2016). Thirty-five years of ART in Israel. *Reproductive Biomedicine and Society Online, 2,* 16–23.

Boas, H., Hashiloni-Dolev, Y., Davidovitch, N., Filc, D., & Lavi, S. J. (2018). Introduction: Bioethics in Israel. In H. Boas, Y. Hashiloni-Dolev, N. Davidovitch, D. Filc, & S. J. Lavi (Eds.), *Bioethics and biopolitics in Israel: Socio-legal, political and empirical analysis*. Cambridge: Cambridge University Press.

Cattapan, A. (2014). Risky business: Surrogacy, egg donation, and the politics of exploitation. *Canadian Journal of Law and Society, 29*(3), 361–379.

Drabiak, K., Wegner, C., Fredland, V., & Helft, P. R. (2007). Ethics, law, and commercial surrogacy: A call for uniformity. *The Journal of Law, Medicine & Ethics, 35*(2), 300–309.

Egg donation law. (2010). *Sefer HaChukkim 5770, SH No. 2242, p. 520* [online]. Available at http://www.knesset.gov.il/privatelaw/data/18/3/289_3_2.rtf. Accessed 7 Mar 2017 (in Hebrew).

Embryo Carrying Agreement Committee. (2011). *General information for applying the embryo carrying agreement committee* [online]. Available at http://www.health.gov.il/Services/Committee/Embryo_Carrying_Agreements/DocLib/pundInfo0210.pdf. Accessed 10 Mar 2017 (in Hebrew).

Embryo Carrying Agreements Law. (1996). *Sefer HaChukkim 5756 (hereafter "Surrogacy Agreements Law"), SH No. 1577, p. 176* [online]. Available at http://www.health.gov.il/LegislationLibrary/Poriut_05.pdf. Accessed 3 Mar 2017 (in Hebrew).

Esther, D. (2005). Orthodox lesbians: Not a contradiction in terms. In C. Frankfort-Nachmias & E. Shadmi (Eds.), *Sappho in the holy land: Lesbian existence and dilemmas in contemporary Israel* (pp. 87–102). New York: State University of New York Press.

Gooldin, S. (2007). Technologies of happiness: The management of fertility in a pro-natal welfare state. In Y. Yonaand & A. Kamp (Eds.), *Citizenship gaps: Migration, fertility and identity in Israel* (pp. 167–206). Tel-Aviv: Van-Leer/Hakibutz Hameuhad Publishing (in Hebrew).

Gross, M. L., & Ravitsky, V. (2003). Israel: Bioethics in a Jewish-democratic state. *Cambridge Quarterly of Healthcare Ethics, 12*(3), 247–255.
Gruenbaum, B. F., Pinchover, Z. S., Lunenfeld, E., & Jotkowitz, A. (2011). Ovum donation: Examining the new Israeli law. *European Journal of Obstetrics & Gynecology and Reproductive Biology, 159*(1), 40–42.
Hashiloni-Dolev, Y. (2006). Between mothers, fetuses and society: Reproductive genetics in the Israeli-Jewish context. *Nashim: A Journal of Jewish Women's Studies & Gender Issues, 12*, 129–150.
Hirshfeld, Z. (2011). Is there a solution for a single religious women who wants a child? *Ynet Judaism* [online]. Available at http://www.ynet.co.il/articles/0,7340,L-4144759,00.html. Accessed 19 Sep 2017 (in Hebrew).
Kahn, R. Y. H. (1989). Judaism and homosexuality: The traditionalist/progressive debate. *Journal of Homosexuality, 18*(3–4), 47–82.
Kahn, S. M. (2000). *Reproducing Jews: A cultural account of assisted conception in Israel*. Durham: Duke University Press Books.
Kanaaneh, R. A. (2002). *Birthing the nation: Strategies of Palestinian women in Israel*. Berkeley: University of California Press.
Marx-Stolting, L. (2009). Ethics in health care chaplaincy: A Jewish perspective. In W. Moczynski, H. Haker, & K. Bentele (Eds.), *Medical ethics in health care chaplaincy: Essays* (pp. 67–88). Munster: LIT Verlag.
Nahman, M. (2011). Reverse traffic: Intersecting inequalities in human egg donation. *Reproductive Biomedicine Online, 23*(5), 626–633.
Persson, I., & Savulescu, J. (2017). Moral hard-wiring and moral enhancement. *Bioethics, 31*(4), 286–295.
Prainsack, B., & Firestine, O. (2006). 'Science for survival': Biotechnology regulation in Israel. *Science and Public Policy, 33*(1), 33–46.
Prime Minister's Office. (2016). *Egg donation bill (amendment- egg donation between female partners), of the Israeli Kensset members Michal Biran et al. Government Decision No. 1331* [online]. Available at http://www.pmo.gov.il/Secretary/GovDecisions/2016/Pages/dec1788.aspx. Accessed 8 Mar 2017 (in Hebrew).
Raymond, J. G. (1990). Reproductive gifts and gift giving: The altruistic woman. *Hastings Center Report, 20*(6), 7–11.
Rosenblum, I. (2013). Being fruitful and multiplying: Legal, philosophical, religious, and medical perspectives on ART in Israel and internationally. *Suffolk Transnational Law Review, 36*, 627–648.
Saxena, P., Mishra, A., & Malik, S. (2012). Surrogacy: Ethical and legal issues. *Indian Journal of Community Medicine, 37*(4), 211–213.

Schenker, J. G. (2008). Assisted reproductive technology: Perspectives in halakha (Jewish religious law). *Reproductive Biomedicine Online, 17*(3), 17–24.

Shalev, C. (1998). Halakha and patriarchal motherhood: An anatomy of the new Israeli surrogacy law. *Israel Law Review, 32*(1), 51–80.

Shalev, C., & Werner-Felmayer, G. (2012). Patterns of globalized reproduction: Egg cells regulation in Israel and Austria. *Israel Journal of Health Policy Research, 1*(1), 15–26.

Shapira, A. (2013). The dying patient law: Is it an appropriate bioethical balance? *Refuah Umishpat, special issue: Ethics and Law in Health – Comparison, Rivalry and Dialogue*, 59–63, January 2013 (in Hebrew).

Sperling, D. (2010). Commanding the be fruitful and multiply directive: Reproductive ethics, law, and policy in Israel. *Cambridge Quarterly of Healthcare Ethics, 19*(3), 363–371.

Steinberg, A. (2005). Jewish perspective. In S. Blazerand & E. Z. Zimmer (Eds.), *The embryo scientific discovery and medical ethic* (pp. 21–39). Basel: Karger.

Steinbock, B. (2004). Payment for egg donation and surrogacy. *The Mount Sinai Journal of Medicine, 71*(4), 255–265.

Teman, E. (2010). The last outpost of the nuclear family: A cultural critique of the Israeli surrogacy policy. In D. Birenbaum-Carmeli & Y. Carmeli (Eds.), *Kin, gene, community: Reproductive technologies among Jewish Israelis* (pp. 107–126). New York and Oxford: Berghahn Books.

Teman, E. (2016). Surrogacy in Israel: State-controlled surrogacy as a mechanism of symbolic control. In E. Scott Sills (Ed.), *Handbook of gestational surrogacy* (pp. 165–173). Cambridge and New-York: Cambridge University Press.

The Knesset. (2016). *Rejected in the preliminary vote: Allowing surrogacy for same sex couples: The Knesset announcements* [online]. Available at http://m.knesset.gov.il/News/PressReleases/pages/press090316-o21.aspx. Accessed 6 Mar 2017 (in Hebrew).

Index

S. Mitra et al. (eds.), *Cross-Cultural Comparisons on Surrogacy and Egg Donation*,
https://doi.org/10.1007/978-3-319-78670-4

The manufacturer's authorised representative in the EU is Springer Nature Customer Service Centre GmbH, Europaplatz 3, 69115 Heidelberg, Germany. If you have any concerns regarding our products, please contact ProductSafety@springernature.com

Printed and bound by CPI Group (UK) Ltd, Croydon, CR0 4YY
25/09/2026
02222848-0003